COMPUTER HARDWARE AND SOFTWARE

AN INTERDISCIPLINARY INTRODUCTION

MARSHALL D. ABRAMS
University of Maryland, National Bureau of Standards
PHILIP G. STEIN
National Bureau of Standards, Montgomery College

ADDISON-WESLEY PUBLISHING COMPANY
Reading, Massachusetts
Menlo Park, California · London · Amsterdam · Don Mills, Ontario · Sydney

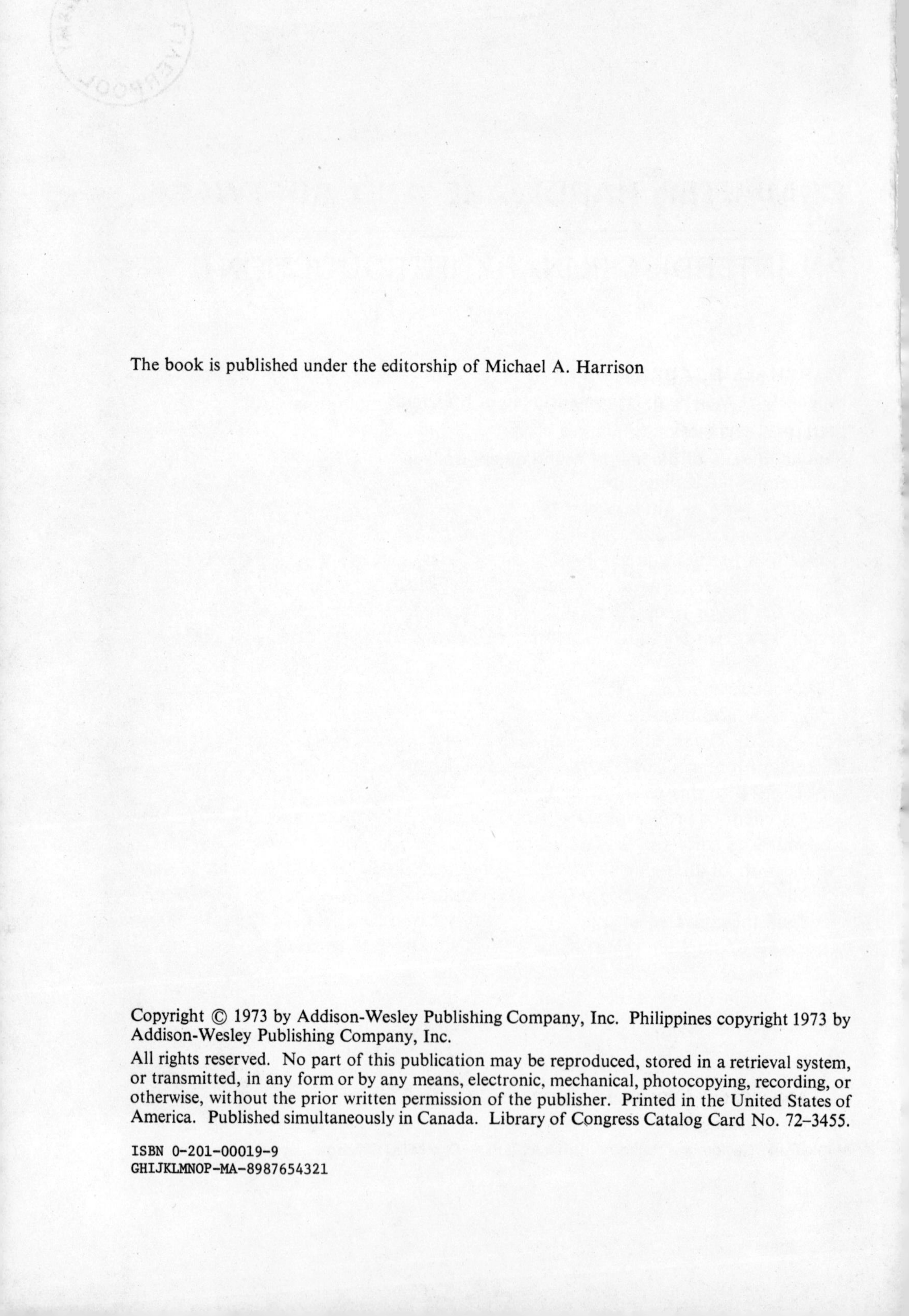

The book is published under the editorship of Michael A. Harrison

 Printed in the United States of America. Published simultaneously in Canada. Library of Congress Catalog Card No. 72-3455.

ISBN 0-201-00019-9
GHIJKLMNOP-MA-8987654321

Operation code	Description of operation
DOUBLE SUBTRACT FROM REGISTER	(REGISTER i, REGISTER i + 1) − (ADDR, ADDR + 1) → (REGISTER i, REGISTER i + 1)
DOUBLE STORE REGISTER	(REGISTER i, REGISTER i + 1) → (ADDR, ADDR + 1)
Floating-point operations	
FLOATING ADD TO REGISTER	(REGISTER) + (ADDR) → (REGISTER)
FLOATING SUBTRACT FROM REGISTER	(REGISTER) − (ADDR) → (REGISTER)
FLOATING MULTIPLY IN REGISTER	(REGISTER) · (ADDR) → (REGISTER)
FLOATING DIVIDE REGISTER	(REGISTER) ÷ (ADDR) → (REGISTER)
Logical operations	
LOGICAL AND TO REGISTER	(REGISTER) **AND** (ADDR) → (REGISTER)
LOGICAL OR TO REGISTER	(REGISTER) **OR** (ADDR) → (REGISTER)
LOGICAL EXCLUSIVE OR TO REGISTER	(REGISTER) **EXCLUSIVE OR** (ADDR) → (REGISTER)
Transfer instructions	
JUMP TO ADDR	ADDR → (PC)
JUMP TO ADDR, INDIRECT	(ADDR) → (PC)

(Continued on rear endpaper)

FOREWORD

The last two decades have seen the freshet of published computer information grow to a torrent, a flood. Popularizations, programming manuals, management literature, engineering tests have been created, distributed, used, and discarded in considerable quantity and praiseworthy quality. But the demand, as well as the need behind it, still increases.

In my view, the three reasons for this increased need are continuing expansion into new application areas, a fantastic upwelling of fresh technology, and increasing recognition of the interaction of subdisciplines—processor hardware, communications hardware, software, documentation techniques. This book addresses itself to the major interaction, the one between computer hardware and software design, and attempts to do so in terms of today's, or at least this year's, technical advances.

It follows an ancient (four computer generations—eternity, by computer standards) and honorable tradition. One of the very first (1950) hard-cover publications in the field, "High Speed Computing Devices," supervised by Tompkins and Wakelin of long-vanished but not forgotten Engineering Research Associates, addressed itself to this very interaction.

A student or professional will not often actually work *at* the interface between hardware and software. In fact, as the authors emphasize, the existence as well as the location of that interface is the subject of much discussion. But to learn efficiently or work effectively in software, familiarity with hardware matters is vital, and the converse is equally true. The hardware expert usually has had some exposure to computer usage, since the machines are now essential design tools; there are only a few software experts who continue to regard the boxes of processor equipment as entirely "black." This book facilitates work on either side of the common territory.

Gaithersburg, Maryland
February 1972

H. R. J. Grosch

PREFACE

This book is an introduction to the organization and implementation of hardware and software for digital computers. The only prerequisite is a knowledge of algorithmic methods and mastery of an algorithmic programming language. Exposure to the various topics contained herein will enable the student to select his own area of greatest interest. It should also provide him with some background in the other specializations with which he must interface. The authors feel that it is impossible to write really good software without an awareness of the hardware on which it is to be implemented. Similarly the hardware designer can make his best contribution to a design effort only when he knows how the equipment is to be used and what functions it is to perform. Good hardware design can make software easier to write and more efficient to run. The engineer needs to understand software before putting pencil to paper if this concept is to succeed. Trade-offs between hardware and software implementation of functions are discussed, and reasons for choosing each alternative are presented.

Hardware and software are so intimately interrelated that it is extremely difficult to discuss a new topic in one without lots of background in the other. Our attempt to solve this dilemma divides the material into sections. Each section encircles a depth of knowledge of both hardware and software, and they are developed in parallel.

The first section, Chapters 1 through 5, constitutes an introduction to the design and function of computer systems. The basic building blocks and organization are developed without excessive detail of either. Chapters 1 and 5 treat hardware/software interrelationships and trade-offs. Chapters 2 and 3 introduce the separate functions of each. Chapter 4 provides background necessary for both.

Section 2 constitutes another complete hardware/software circle. Chapter 6 provides an introduction to internal machine organization and detailed steps involved in program execution. The interrelationship of hardware and software is again noted. Chapter 7 presents the authors' very strong views on the teaching of assembly language programming. It is a most inefficient use of the students' time to learn a hypothetical language for a hypothetical machine! Our supposition is that there is sufficient commonality in assembly languages to permit generalization.

The objective in teaching assembly language is not to produce high levels of expertise, but rather to increase the student's awareness of the hardware and its software interactions. The language introduced in this chapter is *not intended to be used* (even as an exercise programming language). That is why mnemonic abbreviations have been avoided. We believe that assembly language is best taught by using our general language to explain the assembly language for the machine the students are going to program. Most manufacturers' assembly language manuals make good references but poor instructional material.† Chapter 7 is designed as an explanation of the manufacturer's manual in a format which allows easy correlation with specific machine commands. The commands discussed in that chapter will generally be a subset of those available. When the student is able to, he will increase his vocabulary from the manufacturer's reference material. If no machine is available—a regrettable situation—the language presented may be used as a hypothetical language; the use of mnemonics is then recommended.

Section 3 treats only hardware, this time to the full depth of discussing the actual electronics applicable to software, but the section is primarily an introduction to hardware design. Chapter 9 is concerned with optimization, both of the hardware design and designer effort. There is also discussion of the fact that changing technology affects the extent to which various methods of implementation can be used. Chapters 8 and 9 should be of interest to all readers, even if they do not care to study the electronics in the following two chapters. Chapters 10 and 11 are concerned with the methods of implementing the logic functions in hardware devices. These chapters have traditionally belonged to the electrical engineer, but computer scientists are encouraged to learn this material, too. The problems the hardware engineer faces in making pencil and paper designs work when implemented in physical components have an important influence on software. Increasingly complex software has been possible only with the hardware to support it. Chapter 12 is devoted to the storage of information in primary and secondary media. Not all of the various current technologies are explored; rather, a foundation is built for future exploration.

Section 4 returns to software, but software that serves as the man-machine interface. Chapter 13 introduces language translators as methods of accommodating human users, one step farther away from the hardware. Supervisory functions, discussed in Chapter 14, provide additional conveniences for users of the computer. The dependence of software-provided functions on the hardware is also discussed here.

Appendix A contains necessary information often taught as "modern math": the concepts and techniques of binary arithmetic and radix conversion. This material is a prerequisite to understanding parts of Chapter 4. Appendix B explains

---

† It has, in fact, been said that they are written in a foreign language—one whose words appear to be English but whose sentence structure, grammar, and meaning are totally different.

the symbols used for digital logic. It is primarily background for chapters 10 and 11.

The order of the chapters represents a logical progression and development, but it is certainly not the only one possible. Assembly language programming is presented early enough to permit practical experience with homework projects. Sufficient time must be available to cope with the slow turnaround time, down time, and machine unavailability that the student is likely to encounter. This book in its entirety is a suitable text for a two-term course. When only one term is available, the instructor will have to select appropriate material.

There are many ways to organize this material, both when writing a text and when teaching a course. Ideally, two full semesters should be devoted to the subjects covered here, perhaps with reference to supplementary material such as that listed at the end of each chapter. If assembly language for a specific computer is taught in the course or as an associated laboratory course, the manufacturer's manuals are required even though they may not be adequate for classroom use.

Electrical engineering students and computer science students will be able to partly tailor their use of this book by emphasizing the hardware (Chapters 10-12) or software (Chapters 13, 14) portions of Section 3. This division should be effected only if the time available is limited, since to devote undivided attention to one at the expense of the other is contrary to the philosophy of the book—that understanding of both hardware and software (on this level) is of prime importance to both computer scientists and electrical engineers.

If only one semester is available, emphasis should certainly be placed on the teaching of the assembly language for the computer available to the students. Section 1 serves as a good introduction to the subject, and Chapter 7 may be used in parallel with the computer manuals. A significant trade-off here is that a valuable but limited subset of a particular assembly language can be taught in a restricted period of time. More time permits greater depth, more varied applications, and more attention to subtleties. In any time that remains, portions of section 3 may be selected, depending on the orientation desired.

Much of the material covered here will in fact appear in other courses in a full curriculum. Anything that has been previously covered (such as the simplified discussion of electricity in Chapter 10) should be omitted, of course. If the material will be treated more fully in a later course, it should probably be included here simply as background information.

This text is designed to be used for a course such as the one designated B2 by the Association for Computing Machinery's Committee for Curriculum in Computer Science. The book goes much further than B2, and therefore a deeper course may be taught from it if time permits. From an electrical engineering standpoint, the advent of the minicomputer has made general-purpose digital computers economically attractive as system components. It is no longer necessary and may not be economically justifiable to design and manufacture limited-production special-purpose digital equipment. Rather, an off-the-shelf digital computer may be made part of a system as simply as an amplifier or a motor. This book will provide sufficient introduction to many of the relevant aspects of computer systems which an engineer will need to know in order to incorporate a computer within a larger system.

Certainly no one will become an expert from just reading this book. If interest is stimulated, the student may pursue further courses in his area of interest. If knowledge of the computer remains only one of a set of tools rather than a speciality, the student will have gained the background and vocabulary to talk to the experts when he needs to.

We wish to thank our friends and colleagues who have taken the time to read this manuscript and comment on it: A. B. Marcovitz, E. P. O'Grady, J. H. Pugsley, and D. E. Rippy. We especially wish to thank E. F. Miller, Jr., who also contributed his experience with commercial large-scale systems. S. D. Campbell and E. Penniman performed yeoman service in typing the manuscript.

College Park, Maryland M. D. A.
Gaithersburg, Maryland P. G. S.
December 1972

CONTENTS

SECTION 1
INTRODUCTION TO THE DESIGN AND FUNCTION OF COMPUTER SYSTEMS

1 WHAT IS A COMPUTER?

1.1 THE HARDWARE/SOFTWARE RELATIONSHIP

A computer may be described from two different viewpoints: what it does and what it is made of. Both are valid. It is our purpose to explore both viewpoints, thereby seeing the whole picture in perspective.

The design of the hardware must necessarily begin with a statement of the functions to be performed and the constraints on them. From this beginning the implementation follows naturally. A computer is a piece of electronic and electromechanical equipment, composed of a very large number of transistors, resistors, and so forth. Its circuits are designed to treat electrical signals as if they represented numbers, and to follow a sequence of instructions dealing with the processes to which these numbers will be subjected.

A computer, the hardware, is simply dormant capability. Computer hardware can do nothing without programs, the software. But software as well is dormant; a computer program must be executed by a computer. The capability so often referred to as a computer or data processing system is neither the physical equipment nor the invisible software—it is the two together.

Since, ultimately, the only computer system of any significance to the outside world is the system composed of hardware *plus* software, one cannot claim to have truly designed a computer system without having designed both. Integral hardware/software design requires not only a knowledge of machine design and program design, but a thorough understanding of the total interdependence of hardware and software in a working system.

1.2 COMPUTER SYSTEM DESIGN

A computer designer has open to him a flexibility denied other engineers: Most functions may be implemented in either hardware or software! An intelligent design can make the most of both.

Design criteria will vary with the system application and with implementation cost. Some considerations that might be appropriate follow.

a) The system must be able to perform required computations with great speed.

b) Symbols used for data representation must be such that there is little chance of confusing them with one another or with other symbols.

c) The system must have the capability of handling numbers of sufficient precision (i.e., enough digits) for the problem.

d) Components and circuits used should be economical, reliable, few in variety though large in number, and easily replaced.

e) Front-panel or maintenance-panel indicators should accurately reflect the state of the program being run.

f) The system should be expandable in the field to other configurations of equipment, so that performance can be changed as needs change.

g) There should be program compatibility with other machines in the same product line.

h) Software should be compatible with recognized standards, such as FORTRAN, and with the requirements of other machines.

We discuss hardware and software together in this book because their interconnection and interdependence is inescapable. One cannot design good hardware without knowledge of how it is going to be used, nor can one do good programming without some knowledge of the hardware with which he is operating. Once the essentials of a computer system have been provided, it is generally true that any function can be performed either in hardware or in software or in a combination thereof. The selection of hardware or software implementation depends on the demands and the costs of performing the function. A frequently performed function would probably be better done in hardware, an infrequently performed one in software.

1.3 ALGORITHMIC METHODS

As a starting point, we will assume that the reader is a competent programmer in some algorithmic language, such as MAD, FORTRAN, ALGOL, or PL/1. We do not expect him to be an expert programmer, knowing all the tricks and subtleties which are available in these languages, but we do assume that he has been exposed to the algorithmic method. This method is summarized below for the purposes of review.

The algorithmic method may be defined as a complete plan for the solution of a problem that takes into account all contingencies that may occur. There are three basic characteristics of an algorithm: it is finite, definite (i.e., unambiguous), and complete (i.e., terminating). The algorithm will certainly include formulas and procedures, but it will also include such things as the flow of control from one part of the solution to another, the selection of the formulas and procedures to be followed, the methods for inputting and outputting the data on which the formulas and procedures are to be exercised and the handling of errors. The algorithm may be described independent of the language in which it is implemented by means of a flowchart.

A flowchart uses boxes of specific shapes to represent specific operations being performed by the digital computer. For algorithmic language the flowchart boxes that appear in this book are closely related to those of many systems in common

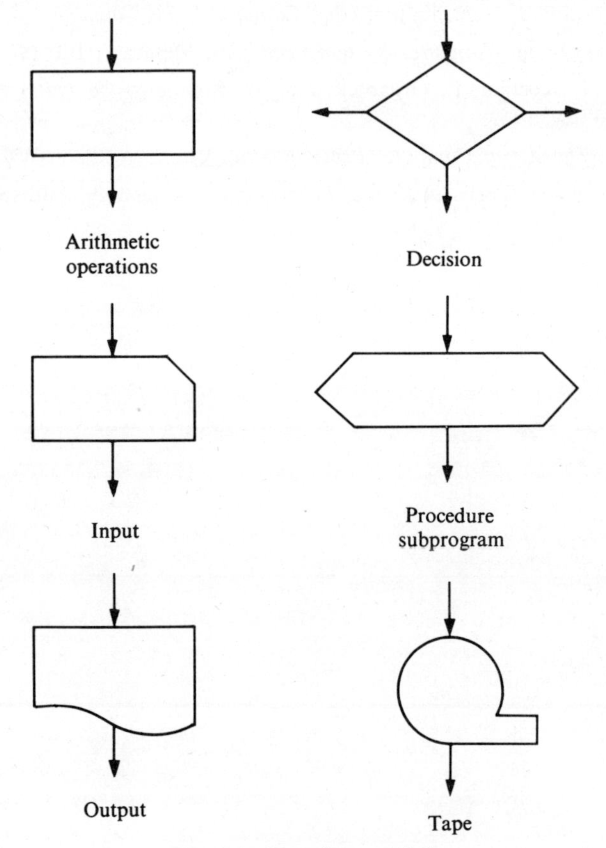

Fig. 1.1. Flowchart symbols

use, as shown in Fig. 1.1. The arrows entering and exiting from these boxes indicate the flow of control through the program. Note that the diamond-shaped box is used for a decision. Depending on the results of the comparison indicated within the box, various exit branches may be followed. These diamond-shaped boxes control the flow through a program by providing alternative procedures that can be followed.

Each box of the flowchart may be thought of as representing a group of machine operations. For instance, note that there are boxes for arithmetic operations, for input, for output, for subroutine linkage, and for decision making. The arrows connecting the boxes represent the logical order in which operations are carried out.

REFERENCES

ANSI Flowchart Standards, *Computing Surveys*, Vol. 2, June 1970.

R. R. Fenichel and J. Neizenbaum, eds., *Computers and Computation*, Freeman, 1971.

H. Hellerman, *Digital Computer System Principles*, McGraw-Hill, 1967.

A. Forsythe, T. Keenan, E. Organick, and W. Stenberg, *Computer Science: A First Course*, Wiley, 1969.

D. E. Knuth, *The Art of Computer Programming*, Vol. 2, Addison-Wesley, 1968.

B. A. Galler and A. J. Perlis, *A View of Programming Languages*, Addison-Wesley, 1970.

2
WHAT THE HARDWARE DOES

2.1 ORGANIZATION OF THE EQUIPMENT

The organization of a simple computer is presented in the block diagram of Fig. 2.1. The *logical* functioning of the hardware is broken up into several *modules*, each of which is charged with a discrete set of functions. The physical division of the equipment into cabinets and chassis very often follows the logical structure, but this is not actually necessary.

The functions provided for are *control*, *memory*, *arithmetic*, and *input/output*. Their purposes follow. The memory stores both the data to be processed and the instructions for processing them. The control unit supervises the sequence and manner in which the data are operated on, and it does so as specified by the instructions. Both the instructions and data currently being operated on are stored in

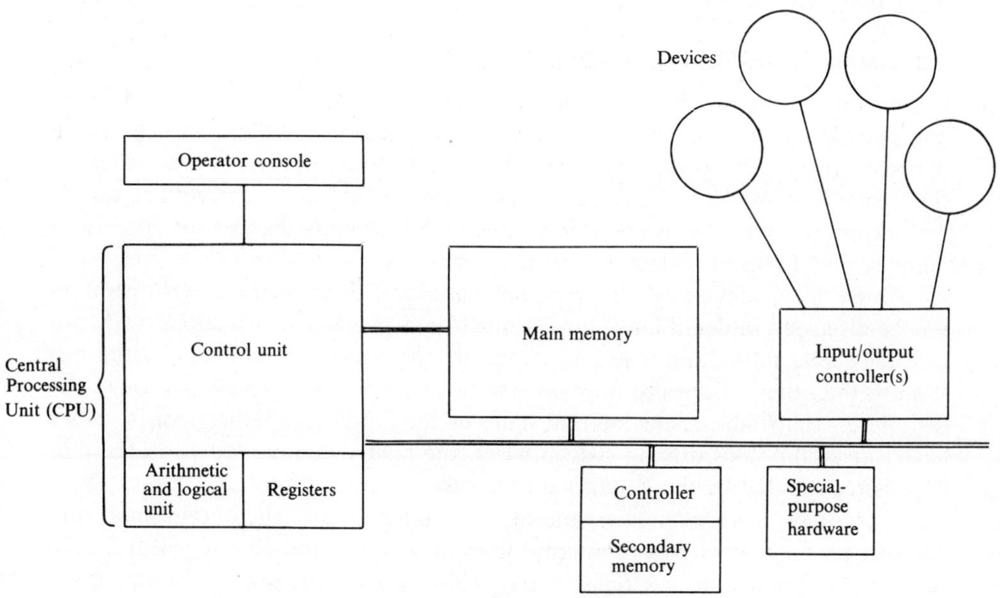

Fig. 2.1. The computer hardware

main memory during program execution. Most of the data processing operations are carried out by the arithmetic module, again with the supervision of the control unit.

Transfer of data and instructions between the human user and the memory is accomplished by the input/output section. A special input/output device, the control panel, serves the operator as a means of communication and enables him to initiate programs manually.

Each of these functions is discussed in detail below. First, however, it is important to understand how these pieces of equipment actually handle and transfer data among themselves. To accomplish that end, we will introduce some concepts common to all of them.

2.2 USE OF THE BINARY NUMBER SYSTEM

Digital computers, in general, are constructed of electronic circuits with only two operating states. This fact results in economies of manufacture and other advantages, which will be discussed later when the circuit details are covered. The decision to use two-state circuits, which is essentially a hardware restriction, has wide ramifications throughout the machine organization and the software.

Specifically, the availability of two-valued logic circuits leads naturally to the use of the binary number system. In this system, the radix, or base, is 2; and the only permissible digits are 0 and 1. A given logic element may therefore represent 0 or 1 with its two states, and representation of large numbers is possible simply by using many elements.

2.3 LOGIC, FLIP-FLOPS, AND REGISTERS

In general, binary numbers, whether commands or data, are represented in the computer by voltages, called *logic levels*. One of the states of the device can represent the binary value 0 and the other state the binary value 1. A wire carries a single binary digit (*bit*), the value of the bit corresponding to the voltage on the wire. Only two logic levels are needed, one each for binary 0 and 1. A sixteen-bit number would require sixteen wires in this scheme.

An electronic device called a *flip-flop* is used for binary storage. Its operation will be discussed in detail later, but its function is simple. It is a circuit with two possible stable states, and it has a mechanism that enables it to change from one state to the other. Each flip-flop has one or more output wires, which carry the logic levels that indicate the internal states of the flip-flop, whether 1 or 0. If we have sixteen flip-flops driving sixteen wires, the binary number stored in the flip-flops is represented by the voltages on the wires.

A group of flip-flops thus organized is called a *register*. The output lines from one register may be wired to the input lines of another, and on a signal from the control unit, the data in one register may be transferred, or copied, into the other. The control unit may also have other signals that cause the register to be cleared (all flip-flops set to 0), to be complemented (0's change to 1's, 1's to 0's), or to have all its bits shifted left or right. In a left shift, for example, the eighth bit is copied into the seventh position, the ninth bit replaces the eighth, and so on. All these functions are performed directly by the electronic circuitry and constitute the basic mechanism by which data are moved from place to place within the computer.

It is easy to see that registers are very useful, and there are many of them in the machine. Their contents can be changed, cleared, or transferred into much faster than main memory. Therefore many register operations may take place during one main memory cycle.

2.4 MAIN MEMORY

The first functional module to be discussed is called *main memory*. Since the memory function in many machines is performed in tiny magnetic toroids called *cores*, the main memory is often referred to colloquially as core.

The function of main memory is to store data and programs in the form of binary numbers. Each number is kept in a specific *location*, which has an identifying number, called an *address*. The memory is used in two ways. In a *read* function, the number previously stored at a given address is fetched. In a *write* function, a new number is placed at the address, destroying the previous contents. Note that each location has *two* numbers associated with it, the address and the contents. These numbers should not be confused.

The memory module itself, like all other modules, communicates with the rest of the machine by means of registers and control lines. Figure 2.2 is a block diagram of a much simplified memory module. The two registers shown are the *memory address register* (MAR), which stores the address of a location, and the *memory buffer register* (MBR), which stores the contents of a location. The control lines are "read command," "write command," and "busy."

When a read function is performed, the address of the cell to be read is loaded (transferred) into the MAR, and the read command line is energized with a voltage pulse. Upon completion of some fixed period of time, the contents of memory at the address named will have been loaded into the MBR. The previous contents of the MBR are destroyed. In most memory systems, the contents of the memory

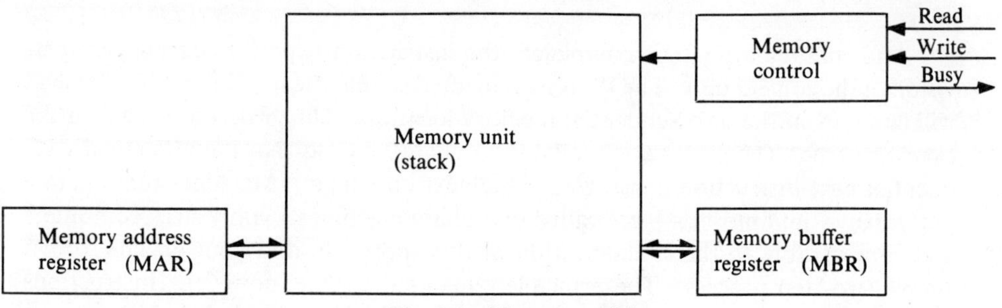

Fig. 2.2. Memory module

at the address named are not changed. If they were destroyed in the process of reading, they are automatically restored by the internal logic of the memory module.

When a write function is performed, the address of the cell to be written into is loaded into MAR, and the item of data to be placed in memory at that address is loaded into MBR. The write command line is then pulsed. Upon completion of some fixed period of time, the contents of MBR will have been placed in memory at the address named. The previous contents of memory at that address are destroyed, and the contents of MAR and MBR are unchanged. The memory busy signal is a logic level that indicates whether the memory is free to accept a read or write command.

This type of memory is called *random access* because any address may be read or written in some fixed time. This might not be true of *sequential access* memories.

2.5 CONTROL AND THE CONTROL UNIT

In general, all functional modules will operate in much the same way. They will receive data for or transmit data from their internal registers, receive commands from other modules, and issue busy signals to other modules. This combination of activities is called *asynchronous* operation, since each module is independent of the whole for the timing of its internal function.

One module for which the description above is incomplete is the *control unit*. This piece of hardware interprets the instructions written by the programmer and causes them to be performed. The actual output of the control unit is a series of commands in sequence to various modules to effect the instructions. The control unit circuitry generates these command sequences in response to the contents of each instruction.

Instructions are stored in memory as binary numbers. In order to execute them, the control unit must retrieve them. One of the processor registers, the *program counter* (PC), contains the memory address of the next instruction. When the current instruction is finished, the PC is transferred to the MAR, and a read operation is performed. This places the binary number which is the next instruction in the MBR. The MBR is then transferred to yet another register, the *instruction register* (IR), where its meaning is decoded. This procedure is called the *fetch phase*.

After the fetch phase is completed, the instruction is performed under supervision of the control unit. The PC is incremented so that the next instruction fetched will come from the next sequential memory location. This procedure is called the *execute phase*. (Facilities are available to change PC so that the numerically sequential next instruction is not *always* the next one done.) Machines that operate in this way—and most do—are called two-phase machines. Many large computers add considerably to the sophistication of this approach, and some deviate from a simple two-step process. The principle remains the same, however. Instructions are fetched from memory, decoded, and executed. The only facility in most machines for differentiating data from instructions is the PC. When the PC points to a binary number, that number is assumed to be an instruction, and the programmer is usually the one responsible for ensuring that it is.

2.6 ARITHMETIC AND LOGICAL UNIT

Another module of interest is the *arithmetic and logical unit* (ALU), which performs the additions, logical combinations, and other operations asked for. It usually contains one or more accumulators. An *accumulator* is a register in which the machine forms the result of an arithmetic operation. Typical instructions load an accumulator from memory, add to it, store it in memory, clear it, shift it, complement it, and increment it. Most programs concentrate on the accumulators, and most of the action takes place there. Some of the functions mentioned above are accomplished by hardware connected directly to the accumulator flip-flops. Others are done by special circuits, such as adders, multipliers, etc. These special circuits are never directly accessible to the programmer, but he may use them by sending the contents of an accumulator or of some memory location to them to be added, multiplied, or whatever. The results of such operations usually are returned to the accumulators.

2.7 CENTRAL PROCESSING UNIT

The control unit, the ALU, and the registers are normally grouped together, both logically and physically, to form the *central processing unit* (CPU). This organization collects together in one module all the hardware normally associated with the execution of instructions. Physical proximity of the components often makes greater speed possible and reduces the number of interconnections. Despite the great complexity of the CPU, it is usually the least expensive part of a computer system, owing to better technology in logic circuits than in memory devices and peripheral equipment.

Some computers incorporate more than one CPU, usually at great expense for more complicated software. Such computers constitute an exception to standard configuration, and they are employed only in sophisticated systems.

2.8 INPUT/OUTPUT

Other specialized pieces of hardware for transferring data into and out of the machine are available with many different organizations. They all operate on the principle of transfer of data, commands, and signals. Input to the computer takes the form of punched cards, magnetic tape, paper tape, typewriter-like devices, and specially prepared (paper) documents. Output from the computer may be directed to essentially the same media as those that serve as input or to other media, such as television-like displays. Input and output are usually discussed together since these are the functions by which the computer communicates with the outside world. Human factors involving convenience, reliability, and suitability must be considered part of the input/output process along with accuracy and reliability of the data themselves.

The main function of input/output equipment is to transform information from a human-readable form to a machine-readable form, and vice versa. For input, the human being depresses keys on some sort of keyboard device to indicate the information he wishes to transmit. The pressing of the key is converted into an electrical signal, which may go through a number of intermediate steps or media, such as punched cards, but is eventually fed into the memory of the computer system. The usual device for output is a printing mechanism, which takes the electrical signals being stored in the computer and converts them electromechanically to printed characters.

Another input/output function is to exchange data with such *secondary memory* devices as bulk core, disk, or magnetic tape. These devices are used to greatly increase the useful storage available in main memory. Programs and data may be swapped in and out of main memory as they are needed. The exchange usually takes place at high speeds, although the time required to find the desired swap area may be considerable. Main memory is almost always organized so that the time it takes to access any particular location is the same as for any other location (*random access*). This situation will not necessarily prevail in secondary memory, where the access time may be variable and may also be large. For this reason, it is uneconomical to program directly with data or instructions located in secondary memory. Large blocks of programs and data are brought into main memory and used there.

REFERENCES

Burroughs Corporation, *Digital Computer Principles*, McGraw-Hill, 1969.

R. R. Fenichel and J. Neizenbaum, eds., *Computers and Computation*, Freeman, 1971.

I. Flores, *Computer Organization*, Prentice-Hall, 1969.

J. K. Iliffe, *Basic Machine Principles*, American Elsevier, 1968.

QUESTIONS

1. What is the basic function performed by the fetch phase of a computer instruction?
2. What determines the function performed by the execute phase of a computer instruction?
3. True or false: All data going to and from main memory must pass through the memory address register. Explain.
4. Given that data can be copied from one position in a register to another position in the same register, as in a shift operation, explain how it would be possible to transfer data from one register to another. Would the data in the source register (the one being transferred from) necessarily be lost or destroyed in the process?

5. Would it be possible to use large numbers of flip-flops in place of cores for main memory? Would it be practical? Explain.
6. In the following sequence of memory operations, the contents of MAR and MBR are shown before the indicated command is given. Show them after the command has been executed.

MAR before	MBR before	COMMAND	MAR after	MBR after
0100	0000	write		
0101	3277	write		
0102	4440	write		
0101	9277	read		
0103	7116	write		
0100	7116	read		
0103	0000	read		
0102	7116	read		

3
WHAT IS SOFTWARE?

Software, by definition, is any program written for the computer. Software may be divided into two broad general categories: applications programs, which are written to solve users' problems, and systems programs, which are concerned with operating the computer service.

3.1 PROGRAMMING CRITERIA

The minimum acceptable level of performance of any program requires that it produce correct outputs, given proper inputs. Additional criteria distinguish good programs from merely adequate ones. To a certain extent these criteria may be mutually exclusive; if so, you will have to assign priorities to them in terms of their importance. An engineering trade-off may have to be made among objectives.

Some criteria other than the bare necessity of doing the job are:

1. *Storage minimization.* Computer memory is an expensive resource, and its use should be minimized by reducing the total number of instructions and reserved data storage areas.
2. *Rapid execution.* Most computer accounting systems charge for execution time. This is one simple measure of resource utilization. The program that does its job quickly is to be preferred.
3. *Safeguards.* It is naive to assume that other people using your program will always do so correctly. A set of tests and messages to detect and identify erroneous inputs or other conditions will certainly make a program more "idiot-proof."
4. *Documentation.* A program's utility is enhanced by the ease with which other people can use it. Documentation exists at many levels: for the person who wishes to know how to use the program; for the person who wishes to know the methodology implicit in it; for the person who wishes to follow the coding; and for the person who wishes to modify the program.
5. *Creation effort.* The law of diminishing returns applies to programming. It is quite possible to put too much time and effort into a program while pursuing the preceding criteria.

3.2 TRANSLATION

Most programming is done in a *higher-level language*, one that is more or less well suited to the task to be performed. Higher-level languages are usually designed with human convenience in mind. Although they may not completely succeed, these languages should facilitate the expression of complex algorithmic, mathematical, and information processing procedures.

Machine language is an intrinsic property of any computer. It consists of a vocabulary of binary numbers, each of which the machine interprets directly as an instruction. The computer fetches and decodes the instructions, as indicated in Chapter 2, and then performs the desired operation.

Thus far we have mentioned two languages for describing the program which the computer executes in solving a problem. What we lack is a means to transform information in a higher-level language to machine language. This transformation, known as *translation*, is an algorithmic process well suited to machine implementation. The syntactic rules of higher-level languages are designed to facilitate translation.

3.3 SOURCE AND OBJECT PROGRAMS

When a program is expressed in a form which will be translated into machine language, it is called a *source program*. It is translated into the *object program*. The source may be essentially machine-independent and therefore portable. The object program is in a very machine-dependent form, suitable for the computer on which it will be run. The very purpose of the translation process has been to produce this object program in the machine language of a specific computer.

3.4 THE SUPERVISOR

Except in the smallest and most primitive computers, there is some software which interfaces between the user program's source or object and the hardware. The various names applied to this software include *system*, *executive*, *monitor*, and *supervisor*. To some extent these names differentiate the degree of service provided; we shall use the most general term, *supervisor*.

A supervisor is provided to increase the cost effectiveness of the computer system. The price of the supervisor includes the direct cost of writing and debugging it and the indirect cost of the memory space it occupies and of the execution time spent in the supervisor instead of in a user program (called *overhead*). The advantages gained from the use of a supervisor include: orderly transition of control from one program to another; efficient handling of input/output, secondary storage, and equipment for the user; performance of bookkeeping functions; software services, such as libraries and loaders; and protection of the operating state of the system from the users and the users from one another.

The supervisor must intimately deal with the hardware organization. One may view the hardware as the innermost core machine and the supervisor as a covering all around the core. If the cover is intact, the combination of hardware and supervisor can be viewed as a new machine. This machine is easier to use than the bare hardware, but it may be less amenable to manipulation by a sophisticated user.

3.5 ASSEMBLY LANGUAGE

Between procedure-oriented higher-level languages and binary machine language, there is another level, known as *assembly language* or sometimes *assembler language*. It is characterized as a *low-level language* because it is only one step removed from the 1's and 0's of machine-language programming. In basic assembly language, there is a one-to-one correspondence between a single line of program and a machine-language instruction word. Through the convenience of *mnemonic instructions* and *symbolic labeling*, the assembly language provides an improved interface between the machine-language object program and the programmer. The objective of these two convenience features is to permit the use of names instead of numbers for telling the computer what instructions to follow and for referring to memory locations.

In machine language, operations and operands are specified numerically. The instruction is subdivided into *fields*, with the contents of each field specifying or modifying an operation or an operand. An example of this procedure is shown in Fig. 3.1. The contents of each field consist of a number. The intent of the number, implied by its location in the machine word, is interpreted by the hardware. As we have mentioned previously, the use of binary numbers for the machine-language instruction makes it very easy for the machine to understand but quite difficult for the human user of the computer.

Correspondence between assembly-language and machine-language programs

There is a unique relationship between the instruction word fields in assembly language and the various fields within an instruction word in machine language. In the machine-language instruction word, these fields are not necessarily in the same order as they are in assembly language, simply because the machine word is formatted for the convenience of the instruction decoder of the CPU, whereas the assembly language word is formatted for the convenience of the human programmer. Furthermore, the assembly-language fields may be coded symbolically, but in machine language these fields are filled with appropriate patterns of 1's and 0's.

It is not necessary for the programmer to know anything about the format of the machine instruction word. He simply writes his program in assembly language. It is processed by the assembler and translated into machine language for execution. The practical problem of debugging, however, makes it advantageous for the programmer to be able to read a machine-language program. He can do so, of course, by referring to the binary instruction format for each computer. The construction and operation of the *assembler*—the program which performs the translation from assembly language to machine language—will be covered in a subsequent chapter.

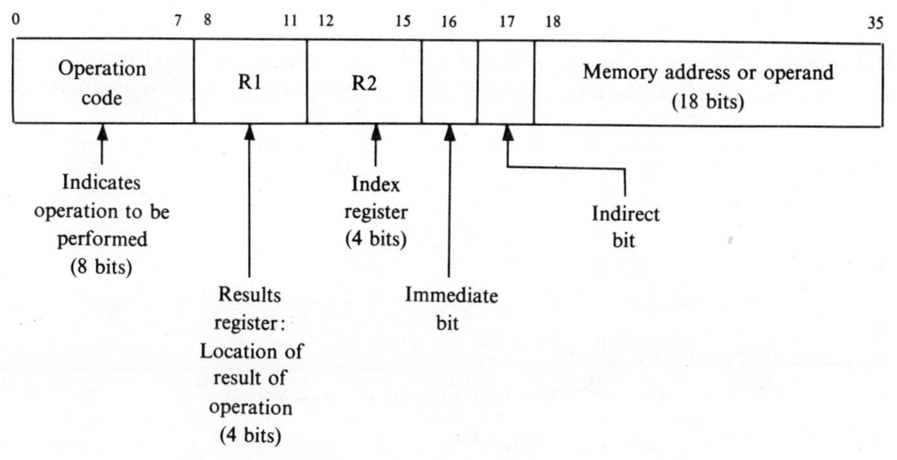

Fig. 3.1. Fields in an instruction for a typical 36-bit machine

Assembly language field delineation

From the standpoint of the user, the most convenient set of rules for field delineation in assembly language is what is known as *free-field format.* The order of the fields is specified, but not the column position on the coding form, punched card, or typewriter keyboard. The termination of a field is indicated by a special character that is not otherwise permitted within a field. This character will often be a space or a comma. The label field customarily begins in column 1. If column 1 contains a space, then no label is being used. The rightmost fields, if not used, may be empty. However, if an intermediate field is not used for some reason, this information must be transmitted to the assembler by the use of such a combination as ", ," to indicate a null field.

In an alternative format, known as *fixed field*, certain columns are designated for each field. If a given field is to be nonempty, the information for it must be entered in the columns specified. Fixed-field formats are relatively inconvenient with punched cards and extremely inconvenient when operating from a conversational terminal such as a teletypewriter. For the purposes of illustration in this text, it is convenient to use a fixed-field format. We can then show that a field is not used simply by leaving it blank. Figure 3.2 shows some typical assembly-language statements in fixed-field format.

The operation field

The operation field contains the directive telling the machine what to do. It is usually a *mnemonic* abbreviation referring to a well-defined operation. Each operation the computer can perform has a mnemonic operation code, or *op code*, for use by the programmer.

Statement label field	Operation field	Register designation field	Operand address field	Index register designation field
START	LOAD REGISTER	REGISTER ONE	LOC	XR1
	ADD TO REGISTER	REGISTER ONE	CONST1	
	STORE REGISTER	REGISTER ONE	LOC2	

Fig. 3.2. Fixed-field assembly-language format

Symbolic addressing

A major convenience of an assembler—perhaps the most important one—is the ability to assign arbitrary symbolic names to individual memory locations in the computer. Since a memory location can store either an instruction or a piece of data, the symbol is used to refer to either. The symbol refers to the *address* at which the data or the instruction is stored, not the contents of memory at that address.

A symbolic name consists of a string of printable characters of some specified maximum length. Other constraints on names, such as forbidding the use of the dollar sign ($) or the use of a number as the first character of the string, are common. Each discrete symbol is assigned to a unique numerical memory location (perhaps relative to some arbitrary origin), and it always refers to only that one location during that running of the program.

The label field

The use of a *statement label* makes possible symbolic reference to an instruction location. Without the use of a statement label, such a location could be referenced only by its numerical address. Reference by numerical address is extremely inconvenient because it involves the manual counting of each statement in order to obtain an address. It also requires knowledge of the location in main memory where the program will reside. When a symbolic name is coded into the statement-label field, an association is made between that label and the instruction word. A reference to the statement label is equivalent to a reference to the address where that instruction word is stored.

The operand field

Many computer operations have an associated operand which is the subject of the operation being performed, such as an addend. An assembly-language field is reserved for the specification of the address of this operand or, sometimes, the operand itself.

The most common way of specifying the operand is to insert the symbol for the *operand address* in the operand field. The operand address is, in turn, specified as a symbolic statement label by its occurrence in the statement-label field of another line of code, a data declaration, or an operand declaration.

Indirect Addressing

A very powerful tool is the use of the operand field as the location where the address of the operand is to be found. The operand-address field is used as the *address of the address* on which the operation is to take place. This process is known as *indirect addressing*.

In assembly-language programming, indirect addressing is often specified by the addition of a special symbol to the line of code (e.g., an asterisk before the operand address). As an example, say that indirect addressing is to take place, and we have specified that the operand address has the symbolic name BOX2. When the instruction is executed, the machine will go to the address specified by BOX2. It will consider the contents of that location to be the address on which the operation is to take place (see Fig. 3.3). Depending on the organization of the computer, it is quite possible that the address in the word referred to indirectly will in turn indicate further indirect addressing. Some machines permit only a specified number of such indirect address references, some permit none, and others impose no limit at all.

Immediates and Literals

It is sometimes convenient to use the operand field not for the address where the operand is to be found but for the operand itself. An integer operand that is small enough can be stored directly in the operand field. The operand is then referred to as an *immediate*. It is of course necessary to inform the computer that the content of the operand field is the operand itself. Specification of an immediate varies among assembly languages. In our treatment we will spell out the entire word **immediate.** In the machine instruction word, a single bit may indicate the presence of an immediate, or an entirely different op code may be used.

Alternatively, a constant may be designated as a *literal*. Like any other data word, a literal must be stored in a memory location, but it may be referred to in the instruction by its value rather than by its symbolic location. In execution there is no distinction between a value that was entered as a literal and one that was simply stored in a memory location. The assembler will take a constant defined as a literal, reserve a memory location for it, place the constant in that location, and place the address of the constant in the operand-address field of the instruction. If the same literal is used by more than one instruction, only one location is assigned, and it is referred to by several instructions. For our purposes, when we use a variable or a constant as a literal, we shall designate it by explicitly stating the word **literal.**

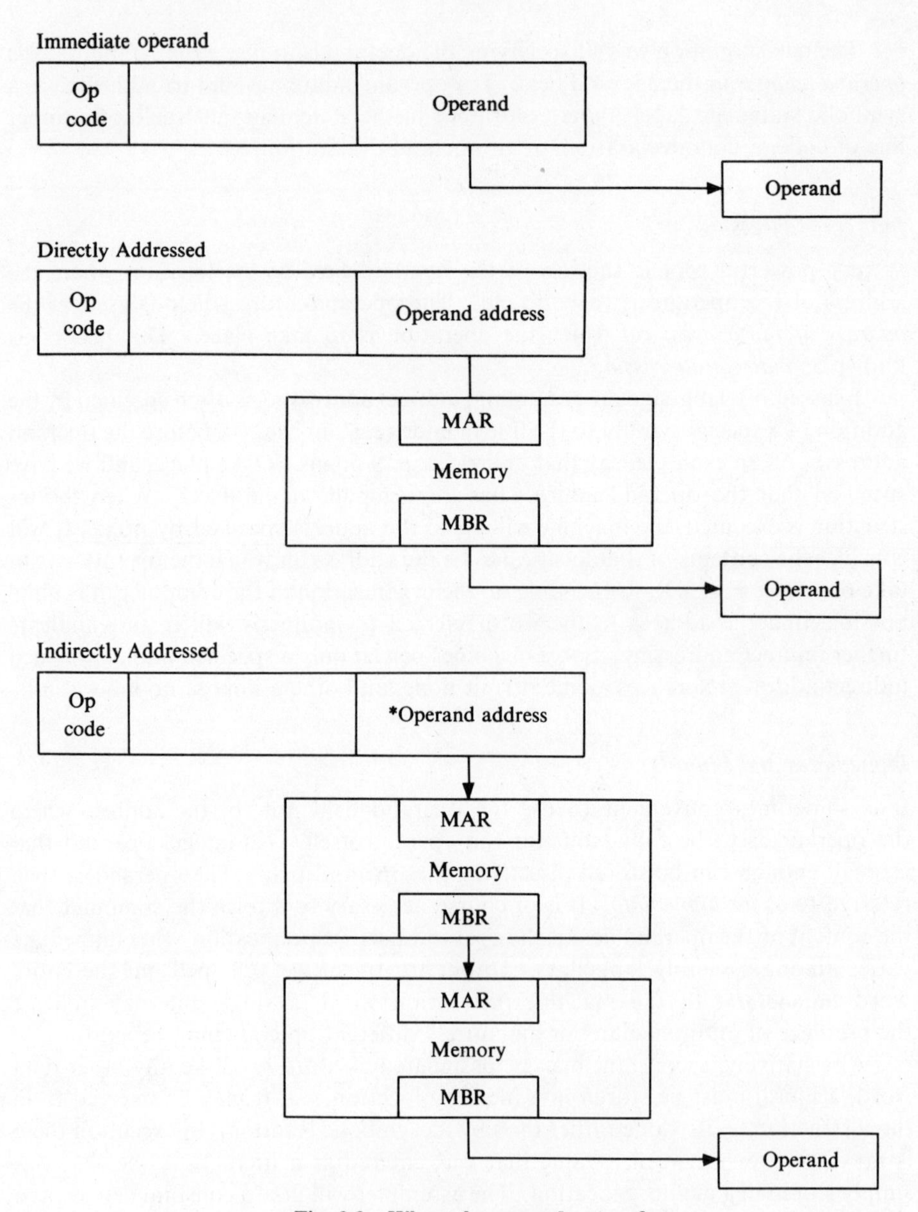

Fig. 3.3. Where the operand comes from

Index register address modification

It is frequently desirable to change the effective operand address by adding a variable to the operand address specified in the instruction. In assembly-language programming, the instruction word will contain a field for the specification of the index register to be used. When an index register is specified, the contents of that register are added to the operand *base address* (the one in the instruction word) to form the *effective address* (the actual address to reference in memory). If no index register is specified, address modification does not take place.

Index register address modification may be included in a command in which indirect addressing takes place. In this case the contents of the index register are added to the operand address specified. The effective address calculated is then used for the indirect addressing. The computer will go to the address calculated and from there take the address on which the operation will take place (pre-indexing). It is also possible in some machines to do the indirect step before indexing (post-indexing).

3.6 FUNDAMENTAL ASSEMBLY-LANGUAGE PROGRAMMING TECHNIQUES

When programming in assembly language, you can perform all the programming operations available in a higher-level language. In addition, you can take full advantage of the hardware instruction set of your computer.

When programming in assembly language a task that could also have been programmed in higher-level language, you may be able to write a better program, but you will do so at the expense of more detailed programming. In assembly language programming you must pay attention to details.

Linear arrays

The *linear array*, a list of data, is named by the symbolic label assigned to its first location. Only the first data storage location of the array has a symbolic label. We call it the array reference location. All other data items in the array are stored in sequential locations following the array reference location, and they are addressed relative to the reference location. For example, consider an array, the first element of which has the symbolic label RAY. The second element in the array will be stored in the location that is numerically one greater than the location symbolically referenced as RAY. The third element of the array will be stored in the location two greater than the location symbolically referenced by RAY, and so forth. Figure 3.4 shows a section of memory indicating the symbolic location RAY and the location of the subsequent elements of the array RAY.

Relative Addressing

In assembly language, there are two ways to reference the elements of RAY. The first is called *relative addressing*. In the operand-address field we could say RAY

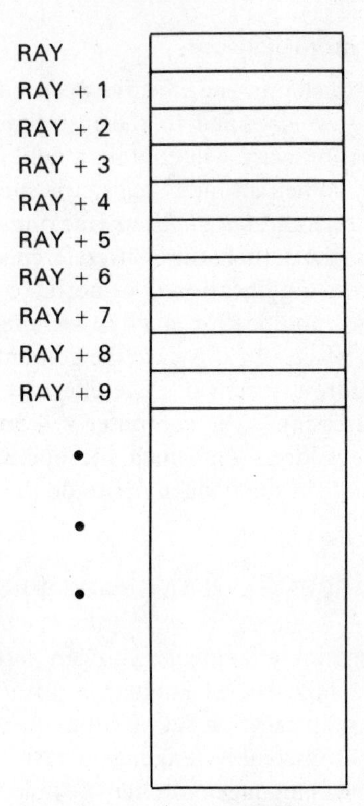

Fig. 3.4. Relative addressing

plus or minus a constant. For example, to address the fourth element of the array in the operand field, we would simply write RAY + 3. Relative addressing is convenient when the element of the array to be addressed is constant. However, most assembly languages will not permit arithmetic operations with variables for relative addressing, as in RAY + X.

Indexing

The second technique for addressing the elements of an array makes use of the index registers. We place the symbolic name of the reference location in the operand-address field of the instruction and some integer value in an index register. We then calculate the effective address by adding the contents of the index register to the operand address, thus pointing to an element of the array.

This procedure for obtaining the address of an element in a linear array is called *indexing*. It is directly analogous to the use of a single subscript in a higher-level language. However, there is a complication in that some higher-level languages permit a 0 subscript and others do not. If the 0 subscript is permitted, we simply load the symbolic reference of the array into the operand field and the integer subscript into the index register. If the index register is 0, then nothing is added to the operand address, and we have the 0 subscript. If the index register has a 1 in it, 1 is added to the contents of the operand address which then points to the first array element. Indexing follows directly, producing the effective address.

If we do not choose to allow the 0 subscript—perhaps only because we are unaccustomed to counting from 0—we then combine the techniques of relative addressing and index register address modification. In the operand-address field we place the array reference location minus 1. We then load our subscript into the index register exactly as we did before, but with the restriction that the smallest integer subscript is 1. Consider what happens when the integer subscript is indeed 1. We have as the direct address RAY − 1. To that we add the contents of the index register, namely 1, producing RAY − 1 + 1, which is simply RAY. We have reduced the amount of space that will be reserved for the array by one location, the 0 subscript.

Loops

A *loop* is an instruction or set of instructions which is repetitively executed, perhaps with some instruction modification. One type of loop calls for the repeated execution of a block of statements for sequential values of some indexed variable. A second type of loop is condition-controlled. It is executed until some predetermined condition occurs. The features common to all loops of the first type, apparent in the flowchart representation of Fig. 3.5, are the initialization of the index, its modi-

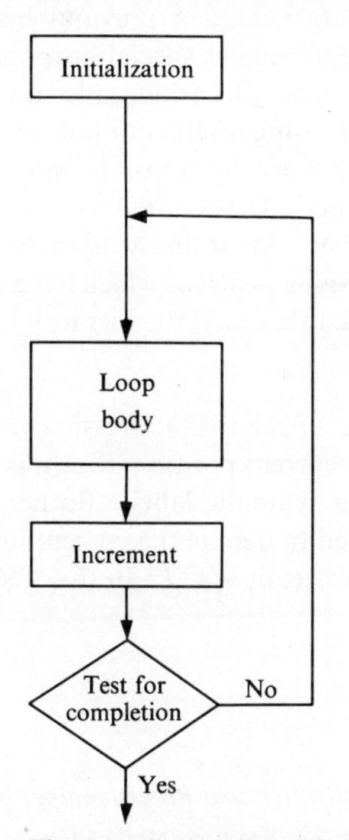

Fig. 3.5. A programmed loop

fication, and the testing of the index for completion. The indexing value is generally stored in an index register so that it may be used for address modification. The operation of the loop therefore depends on loading an arithmetic value into an index register, incrementing or decrementing the contents of that register by a specified amount, and performing an arithmetic comparison on the resultant contents. The use of indexing in loops is so common that machine languages generally contain a command that will perform the increment and comparison in one operation. (We shall discuss such commands in Chapter 7.) In the execution of a loop it is frequently immaterial whether we count up from the lower limit or down from the upper limit. Since it is always possible to make the lower limit 0 by the use of relative addressing in the operand-address field, it is sometimes convenient to count down from the upper limit to 0, making the test for completion simply the determination of when the index passes through zero. Then we do not have to store a final index value for comparison to see if the loop should be terminated. However, if we wish to count upward, it will be necessary to store an upper limit somewhere so that the comparison can be made. To perform the counting-up operation with a single command, we need to specify three quantities: the index register to be used, the amount by which it is to be incremented, and the upper limit.

Software switches

In the execution of a program, it is often desirable to follow alternative paths of execution, depending on data values or previous history. The indication as to which path to follow may be stored in a special-purpose or general-purpose memory location. Since the special-purpose location, called a *software switch*, can store only a 1 or a 0, it is devoted to indicating whether or not some specific event has occurred. In fact, these software switches are flip-flops, the same hardware devices mentioned earlier as register components. These flip-flops may be individually set, reset, or tested by software instructions. An ordinary memory location may be used to store a wide variety of numerical values, each of which has a different assigned meaning to the programmer. When used this way, the memory location is known as a *switch*.

Reflexive addressing

Closely akin to the process of relative addressing is that of *reflexive addressing*, or addressing relative to the current position. For this procedure it is not necessary that the current line have a symbolic label. Rather, a special symbol, often an asterisk or dollar sign, is used to denote the current line, and an arithmetic plus or minus indicates the direction, as in $* + 7$. In this treatment we shall spell out the words **present position.**

REFERENCES

C. W. Gear, *Computer Organization and Programming*, McGraw-Hill, 1969.

I. Flores, *Computer Organization*, Prentice-Hall, 1969.

Burroughs Corporation, *Digital Computer Principles*, McGraw-Hill, 1969.

D. E. Knuth, *The Art of Computer Programming*, Vols. 1 and 2, Addison-Wesley, 1968, 1969.

P. Wegner, *Programming Languages, Information Structures, and Machine Organization*, McGraw-Hill, 1968.

W. D. Maurer, *Programming: An Introduction to Computer Languages and Techniques*, Holden-Day, 1969.

F. K. Walnut, *Introduction to Computer Programming and Coding*, Prentice-Hall, 1968.

J. K. Iliffe, *Basic Machine Principles*, American Elsevier, 1968.

G. M. Weinberg, *The Psychology of Computer Programming*, Van Nostrand Reinhold, 1971.

QUESTIONS

1. Draw a flowchart relating source program, object program, translator, and supervisor.
2. If the supervisor causes overhead, why introduce it at all?
3. Are the contents of the program counter ever modified during the fetch cycle of an instruction? If so, how?
4. Are the contents of the program counter ever not modified during the execute cycle of an instruction? If not, how does this come about?
5. How would you design an accounting system that charges computer users fairly for the portion of the resources they employ in their programs? Specifically, determine a list of resources and a charge schedule for their use on your computer.
6. If a linear array is basic, how are higher dimensional arrays possible? Do multiple subscripts employ multiple index registers?
7. Why might it be difficult to provide for relative and reflexive addressing in higher-level language?
8. When using index registers, how would you refer to a memory location whose actual address is the same as the contents of an index register?
9. Describe in detail how the effective address of a memory reference instruction is computed from the contents of the instruction and the contents of the index register.
10. How would it be possible for the effective address of an instruction to be the same as the operand address in the instruction?
11. If a negative number is loaded into an index register and that index register is then used in a memory reference instruction, what will be the result in the computation of the effective address?
12. Would it be possible to use the same adder circuitry for the arithmetic computations programmed by a user and for the address computations that produce an effective address in an index instruction?
13. Write out the contents of several memory locations, one of which contains an

instruction that refers to some of the others. The instruction we are concerned with contains both indirect addressing and indexing. Find the effective address of the instruction and show whether it is different for pre-indexing, where the indexing is done before the indirect reference, and post-indexing, where the indexing is done after the indirect reference.

14. Memory location 1000 contains the number 3627. In a memory reference instruction, the operand address is 1000, and indirect addressing is specified by setting the indirect bit. What is the effective address?
15. Design a machine organization in which an infinite number of levels of indirect addressing are allowed. In this machine, where would the indirect address indicating bit have to be in order to allow an arbitrary number of levels?
16. Execution of a normal memory reference instruction in an ordinary two-phase machine takes two memory cycles, one for fetch and one for execute. Contrast this with an instruction having indexing.
17. If execution of a normal memory reference instruction requires two memory cycles, how many memory cycles are required for an instruction with one level of indirect addressing?
18. How many memory cycles are required for a memory reference instruction with n levels of indirect addressing?
19. Is it meaningful to have an immediate register-to-register instruction?
20. Is it meaningful to have an indexed immediate instruction?
21. Is it meaningful to have an indirectly addressed immediate instruction?
22. Index register 7 contains the number 2. An immediate instruction which loads data into the accumulator is indexed by register 7. The immediate instruction calls for the number 17 to be loaded into the accumulator. What number will actually be loaded?
23. Memory location 1000 contains the number 3124. An immediate instruction calls for the loading of the number 1000 into the accumulator, and indirect addressing is asked for. What number will actually be loaded into the accumulator?
24. A conditional jump is a jump whose occurrence depends on the result of a certain test. Describe some of the tests you think might serve this purpose.
25. What class of instruction would you use to test an input/output device to see if it was ready to receive a data transfer?
26. Describe an instruction or an instruction sequence which would test an input/output device for ready and suspend processing until the device was ready for the data transfer.
27. List some advantages and disadvantages of using assembly language rather than machine language for writing programs.
28. Describe a typical way in which the instruction fields in an assembly-language instruction correspond to the instruction fields in a machine-language instruction.
29. Which field, if any, in a machine-language instruction corresponds to the statement-label field in an assembly-language instruction?

30. How can the assembler make it easier to program a machine that does not have immediate instructions?
31. Describe in detail why it would be desirable to have a single computer instruction which would increment or decrement the contents of an index register, compare the new contents with a stored constant and conditionally jump, depending on the results of the comparison.
32. Why is it possibly more efficient to count backward from the top of an array to the zeroth element than to count from the first element to some terminating value?
33. A computer with 8-bit words in main memory has the instruction word divided into two fields. The field containing the operation has three bits, and the field containing the storage address has five. How many different instructions will this machine perform? To how many memory locations can a single instruction refer?
34. In a machine with multiple accumulators, would there be an instruction such as SEND ACCUMULATOR #1 TO ADDER?
35. A main memory has 8192 words of storage (sometimes described as 8K), and each word is 36 bits long.

    a) How many bits does the entire memory hold?
    b) How many bits of *register* storage are required for the MAR?
    c) For the MBR?
36. Why must the program counter be transferred to the MAR at the beginning of each fetch phase?
37. What is the effect of executing an instruction that adds one to the program counter?
38. What is the effect of executing an instruction which subtracts one from the program counter?
39. Why are instructions which modify the contents of the program counter necessary?
40. Consider a machine with a program counter. Describe three instructions that might operate on it. What would be the effect on the flow of control of the program from each of these instructions?
41. Is it possible to fetch a piece of data and interpret it as an instruction? Is this a good idea, or is it dangerous? Elaborate.
42. When you normally give instructions to another person in English, are you using "elemental" or "higher-level" language? Give examples of both kinds of instruction when telling someone to bring you a glass of water.
43. One small commercial computer does not provide index registers but rather reserves eight locations in memory for special use. In particular, an indirect reference to any of these locations not only provides the indirect addressing but also increments the contents of that location. This procedure is called auto-indexing. For example, if location 3 contains the number 164 and the instruction is ADD * 3, where * means indirect addressing, then the effective instruction is ADD 164, and the contents of location 3 become 165.

    Discuss the advantages and disadvantages of this scheme in comparison with an index register organization for computers with short word lengths.

4
DIGITAL REPRESENTATION OF INFORMATION

What information do we want to be able to represent in the computer? Everything! A computer is an information processing machine, and we do not wish to limit or restrict its capabilities. Information may be arbitrarily classified into several categories.

Positive and negative integers
Numbers in scientific notation (e.g., 7.3×10^5)
Alphanumeric character strings
Instructions
Events

All these different types of information must be represented in the machine in binary, and therefore a coding scheme or plan must be designed for each.

4.1 CONTINUOUS AND DISCRETE REPRESENTATIONS

Most of the information with which we deal is continuous in nature. All measurements, whether in the metric system or the English system of units, can be expressed in finer and finer gradations, and if we had the necessary equipment, we could obtain measurements to any degree of precision required. The only noncontinuous system with which we are likely to have had experience is the set of integers. We use the integers for counting, and we probably learned them as our introduction to numbers. The symbols we use to represent numerical and nonnumerical information (the numerals in our decimal system and the letters in our alphabet) have evolved over a long period into forms that are convenient and natural. The system is not always logical, but it *is* the system we have grown up with and therefore feel most comfortable with. When we attempt to put information into a digital computer we find that the computer's organization is significantly different from that of human beings. The forms convenient for human use are not well suited to the computer. In the future, forms convenient for the computer may become accepted and used by people, but any discussion of whether such a development is desirable is beyond the scope of this text. For the present, at least, com-

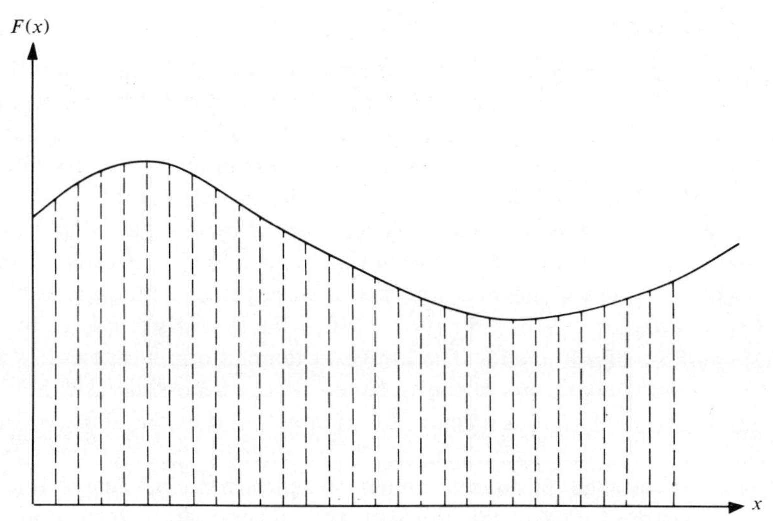

Fig. 4.1. Discrete sampling of continuous information

puters serve people—not the reverse—and efforts must be made to improve the computer's capability of operating on human terms.

One difficulty between human and computer is encountered, for example, in the representation of numerical information. The digital computer, by its very nature, is capable of representing only discrete quantities of information. One of the problems the computer user faces is the conversion of information available to him in his continuous world to the discrete world of the digital computer. The solution is not difficult, however. If his data are expressed in the form of some continuous curve, he need only take his data samples at sufficiently close points along the curve to make it contain all the significance expressed by the curve. An example of this sampling process is given in Fig. 4.1. In making those measurements, the user assigns the closest *digital* value he can resolve to each *analog* quantity he measures.

When there is a continuous relationship between x and y, we may write simply

$$y = F(x),$$

meaning y is a function of x; for every value of the independent variable x there exists a value of the dependent variable y. The rule which assigns a y to each value of x is the functional relationship F. When we deal with discrete sampling, the relationship must be rewritten

$$y_i = F(x_i),$$

meaning that for each discrete (or sampling) value of x, namely x_i, there exists a value of y, namely, y_i. Each of these x_i, y_i points must lie on the continuous function F, within the limits of precision chosen for the sampling process.

4.2 REPRESENTATION OF SYMBOLIC INFORMATION

In many cases the information to be represented within the digital computer is not numerical in nature but symbolic. It can even be argued that the numbers represented within the digital computer are also symbols. (This depends on your viewpoint. There are physically countable voltages representing the integers within the machine.) Information is represented symbolically in the computer. Since each distinct number or character is internally represented by a sequence of 1's or 0's, the grouping of 1's and 0's forms a symbol that stands for some external state or event. Depending on the meaning attached to the symbols, rules are established for their manipulation and combination. When the symbols represent numbers, the rules are those of arithmetic. The simplest examples of non-numeric symbolic information are the characters of the alphabet. It is not possible to write a letter of the alphabet directly into a memory location within the computer; we have to use a binary number to represent it. Each character of the computer's external character set is assigned an internal numerical representation. The character set usually includes the letters of the alphabet, the numerals from zero to nine, and some special characters, such as arithmetic operators, common punctuation marks, etc.

The American Standard Code for Information Interchange (ASCII)

During the early years of computer development, various codes were established to represent the symbolic character set. In all these codes the possible number of different symbols was limited by the number of binary digits, called *bits*, allocated to represent any character. Each bit can take on either of two states: 0 or 1. Therefore a group of n bits can take on 2^n states. If 6 bits were devoted to each symbol, there could be $2^6 = 64$ different symbols; if 7 bits, $2^7 = 128$ different symbols; if 8 bits, $2^8 = 256$ different symbols.

The lack of standardization of early symbol codes made it extremely awkward to transfer information from one computer system to another, but in recent years a standard has become established. Although officially intended only for use in communication between computers, this external standard is expected to become a *de facto* internal standard. Its formal name, American Standard Code for Information Interchange, is commonly abbreviated to ASCII (Fig. 4.2). Note that ASCII is a 7-bit code representing 128 different symbols. The bit positions are shown with bit 1, the least significant bit, in the rightmost position. Only 96 of the symbols are printable; 32 code combinations are reserved for control functions. The codes for the control functions have bits 6 and 7 set to 0. The meanings of the control functions are given in Fig. 4.3.

In the implementation of the ASCII code on punched paper tape, the perforations are arranged in eight longitudinal tracks, one for each of the seven information bits and one for error checking. There are smaller sprocket feed holes punched in the tape for mechanical feed and for frame reference. Bits 1, 2, and 3 appear to one side of the feed holes, as shown in Fig. 4.4.

Bits												
b_7					0	0	0	0	1	1	1	1
b_6					0	0	1	1	0	0	1	1
b_5					0	1	0	1	0	1	0	1
b_4	b_3	b_2	b_1	Column → / Row ↓	0	1	2	3	4	5	6	7
0	0	0	0	0	NUL	DLE	SP	0	@	P	`	p
0	0	0	1	1	SOH	DC1	!	1	A	Q	a	q
0	0	1	0	2	STX	DC2	″	2	B	R	b	r
0	0	1	1	3	ETX	DC3	#	3	C	S	c	s
0	1	0	0	4	EOT	DC4	$	4	D	T	d	t
0	1	0	1	5	ENQ	NAK	%	5	E	U	e	u
0	1	1	0	6	ACK	SYN	&	6	F	V	f	v
0	1	1	1	7	BEL	ETB	’	7	G	W	g	w
1	0	0	0	8	BS	CAN	(	8	H	X	h	x
1	0	0	1	9	HT	EM	)	9	I	Y	i	y
1	0	1	0	10	LF	SUB	★	:	J	Z	j	z
1	0	1	1	11	VT	ESC	+	;	K	[	k	{
1	1	0	0	12	FF	FS	,	<	L	\	l	¦
1	1	0	1	13	CR	GS	—	=	M	]	m	}
1	1	1	0	14	SO	RS	.	>	N	^	n	~
1	1	1	1	15	SI	US	/	?	O	——	o	DEL

Fig. 4.2. American Standard Code for Information Interchange (ASCII)

There is also a version of ASCII for the $3\frac{1}{4}'' \times 7\frac{3}{8}''$ rectangular-hole punched card. The standard Hollerith code for these punched cards specifies 256 hole patterns in 12 rows of punching. Figure 4.5 shows the patterns for the capital letters and the numbers.

NUL	Null	DLE	Data Link Escape (CC)
SOH	Start of Heading (CC)	DC1	Device Control 1
STX	Start of Text (CC)	DC2	Device Control 2
ETX	End of Text (CC)	DC3	Device Control 3
EOT	End of Transmission (CC)	DC4	Device Control 4 (Stop)
ENQ	Enquiry (CC)	NAK	Negative Acknowledge (CC)
ACK	Acknowledge (CC)	SYN	Synchronous Idle (CC)
BEL	Bell (audible or attention signal)	ETB	End of Transmission Block (CC)
BS	Backspace (FE)	CAN	Cancel
HT	Horizontal Tabulation (punched card skip) (FE)	EM	End of Medium
		SUB	Substitute
LF	Line Feed (FE)	ESC	Escape
VT	Vertical Tabulation (FE)	FS	File Separator (IS)
FF	Form Feed (FE)	GS	Group Separator (IS)
CR	Carriage Return (FE)	RS	Record Separator (IS)
SO	Shift Out	US	Unit Separator (IS)
SI	Shift In	DEL	Delete†

CC: Communication control
FE: Format effector
IS: Information separator

† In the strict sense, DEL is not a control character.

Fig. 4.3. ASCII control characters

Octal and hexadecimal

Although the binary numbering system exactly represents the internal symbolic code, machine instruction, or data word, it is an inconvenient system to use. Because there are only two symbols (and states), a large number of digits is required to represent a reasonably sized number. Instead of binary, it is common to employ another number system whose radix, or base, is a power of 2. (A detailed discussion of number systems, radix conversion, and positional notation may be found in Appendix A.) As the radix increases, so does the number of symbols (numerals), and consequently the number of digits required to represent a given quantity decreases.

The commonly used, practical choices are the octal number system (base 8) and the hexadecimal number system (base 16). The counting sequences in these two systems in correspondence with the decimal and binary systems are shown in Fig. 4.6. Note that hexadecimal requires 16 symbols; the capital letters A, B, C,

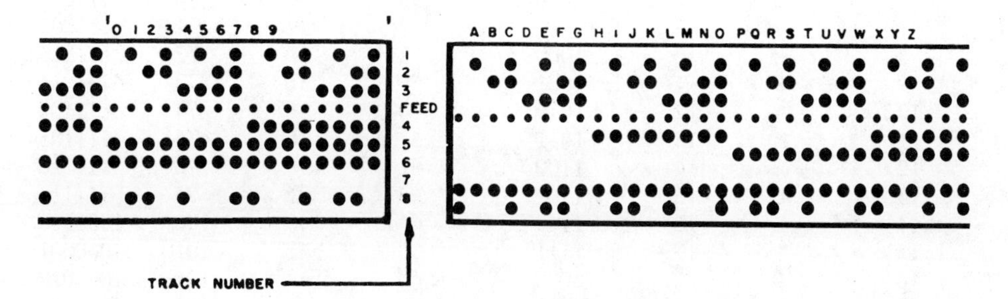

Fig. 4.4. ASCII code on punched paper tape

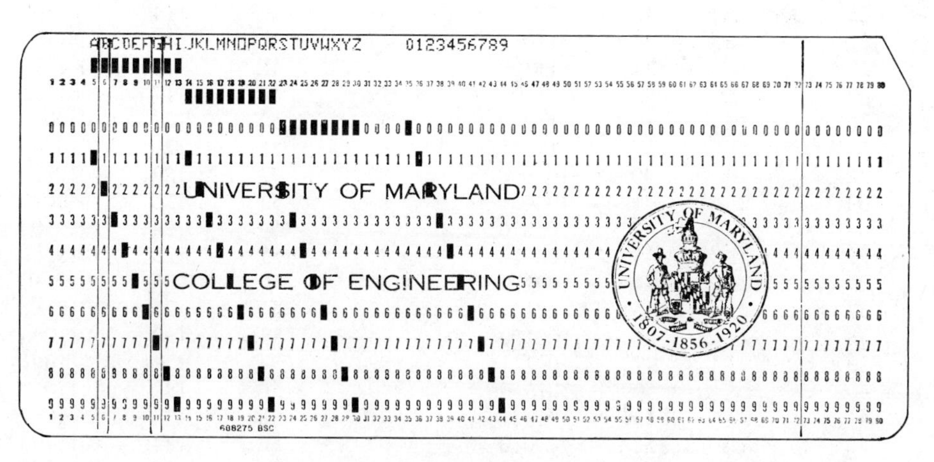

Fig. 4.5. Twelve-row Hollerith punch card

Decimal	Hexa-decimal	Octal	Binary	Decimal	Hexa-decimal	Octal	Binary
1	1	1	1	51	33	63	110011
2	2	2	10	52	34	64	110100
3	3	3	11	53	35	65	110101
4	4	4	100	54	36	66	110110
5	5	5	101	55	37	67	110111
6	6	6	110	56	38	70	111000
7	7	7	111	57	39	71	111001
8	8	10	1000	58	3A	72	111010
9	9	11	1001	59	3B	73	111011
10	A	12	1010	60	3C	74	111100
11	B	13	1011	61	3D	75	111101
12	C	14	1100	62	3E	76	111110
13	D	15	1101	63	3F	77	111111
14	E	16	1110	64	40	100	1000000
15	F	17	1111	65	41	101	1000001
16	10	20	10000	66	42	102	1000010
17	11	21	10001	67	43	103	1000011
18	12	22	10010	68	44	104	1000100
19	13	23	10011	69	45	105	1000101
20	14	24	10100	70	46	106	1000110
21	15	25	10101	71	47	107	1000111
22	16	26	10110	72	48	110	1001000
23	17	27	10111	73	49	111	1001001
24	18	30	11000	74	4A	112	1001010
25	19	31	11001	75	4B	113	1001011
26	1A	32	11010	76	4C	114	1001100
27	1B	33	11011	77	4D	115	1001101
28	1C	34	11100	78	4E	116	1001110
29	1D	35	11101	79	4F	117	1001111
30	1E	36	11110	80	50	120	1010000
31	1F	37	11111	81	51	121	1010001
32	20	40	100000	82	52	122	1010010
33	21	41	100001	83	53	123	1010011
34	22	42	100010	84	54	124	1010100
35	23	43	100011	85	55	125	1010101
36	24	44	100100	86	56	126	1010110
37	25	45	100101	87	57	127	1010111
38	26	46	100110	88	58	130	1011000
39	27	47	100111	89	59	131	1011001
40	28	50	101000	90	5A	132	1011010
41	29	51	101001	91	5B	133	1011011
42	2A	52	101010	92	5C	134	1011100
43	2B	53	101011	93	5D	135	1011101
44	2C	54	101100	94	5E	136	1011110
45	2D	55	101101	95	5F	137	1011111
46	2E	56	101110	96	60	140	1100000
47	2F	57	101111	97	61	141	1100001
48	30	60	110000	98	62	142	1100010
49	31	61	110001	99	63	143	1100011
50	32	62	110010	100	64	144	1100100

Fig. 4.6. Counting sequences

D, E, and F represent the quantities 10 through 15. Each octal digit, being a multiplier of a power of 2^3, represents and is equivalent to 3 bits. Each hexadecimal digit, being a multiplier of a power of 2^4, represents and is equivalent to 4 bits. Thus 36 bits may be represented by 12 octal digits or 9 hexadecimal digits.

Representation of events

If we are interested in representing neither numerical nor alphabetic information but events, we can still use symbolic representations. For example, let us assume that our computer system is involved in the monitoring of the operation of a steel mill. We may want to record and take some action on the speed at which the rollers are turning, the pressure the rollers are exerting, the temperature, thickness, and perhaps the color of the steel, the atmospheric conditions, the time of day, and maybe even the time remaining before the shift changes. If there is a discrete set of events, we can assign some numerical representation to each event. We can then convert the occurrence of an event into its numerical representation and proceed as if the numerical representation and the event were synonymous. So far as the computer is concerned, there is only the numerical representation. So far as the human observer is concerned, there is a one-to-one relationship between the numerical representation and the event itself, and he therefore has no trouble in thinking of the event as being represented in numerical form.

How many numbers?

Since events, symbols, and numbers have a one-to-one relationship, we can henceforth restrict our discussion to numbers. How many numbers can be represented within the digital computer? How many elementary storage locations are required within the digital computer to represent a given number? The answer depends on how much information an elementary storage location can accommodate. The most common storage devices are those in which each elementary storage location is capable of maintaining one of two possible states. Therefore, as already noted, 2^n numbers can be represented by n bits. The grouping of these n binary digits can be called an n-tuple, or a collection of n bits.

4.3 ADDRESSABLE STORAGE

Since practical computer memories may contain a very large number of bits—often as many as 10^5 to 10^8 bits—it is virtually impossible for the user to keep track of the meaning attached to every bit in the memory. At the very least, it would be something of an intellectual strain to do so. On the other hand, it would be wasteful in terms of hardware cost for the computer to provide explicit addressing of every bit in memory, although there are cases in which doing so is desirable. Computer systems employ various techniques to get around this problem.

One alternative is to arrange the memory so that groups of n bits together form a *word*. The computer memory can then be organized to store or retrieve (fetch) n bits at a time. In many typical computer systems, n, the number of bits in a word, is between 12 and 64. Each word in the memory is assigned a unique *address* or serial number. Thus any word of n bits may be accessed by referring to its address.

For example, for the solution of an arithmetic problem, one location, say address 0264, is assigned to store the value of a variable, X. The value of X will then always be found in cell 0264, although that value (the contents of the cell) can vary according to the results of the arithmetic computations performed.

A word may not be the only storage unit to which an address is assigned. The smallest such unit is called the *smallest addressable increment* of memory. In the example above, the smallest addressable increment is a word. Most computers—including the one we will be working with in this book—operate in this fashion, that is, with a memory word as the smallest addressable increment.

In some computer systems, the smallest addressable increment of the memory and a memory word are not the same quantity of information. In particular, for the representation of certain information, especially alphanumeric information, the use of an entire computer word for just one element may be wasteful. Examination of the ASCII character set in Fig. 4.2 shows that six characters (each 7 bits long) can be stored in a 42-bit computer word. When manipulation of alphanumeric characters is important to the problem being solved, it is valuable to be able to address each character in the machine word. Often each character within a computer word is called a *byte*, and computers with instructions which can address a byte within a word are called *byte-addressable.*

The relationship between the width of a memory word, the number of bits stored or retrieved in a single operation of the memory hardware, and the smallest addressable increment in the memory is relatively simple. Figure 4.7 shows a portion of the memory of a computer system which is byte-addressable and in which a memory word has 4 bytes; each byte is 8 bits long, and so the machine word is actually 32 bits long. Since computers operate in binary notation, we have shown both the binary and the decimal values of the addresses. Note that the full-word decimal addresses increase by 4 for each computer word; similarly (and for the same reason) the lowest-order two bits of each word's binary address always have the value 00.

In this memory system, all but the lowest-order two bits of each address are used by the computer to select the indicated word from memory. That is to say, everything above the low-order two bits constitutes the word address. Each selection retrieves a full 4-byte (in this example, 32-bit) word. The remaining two low-order bits are applied to further select the byte within the word *after* the word is retrieved from memory. For example, consider the word whose decimal address is 20 and whose binary address is 010100. The address of the zeroth byte is 010100, of the first byte 010101, of the second byte 010110, and of the third byte 010111. Note that since the address of the zeroth byte and the address of the full word are the same, the instruction op code must specify whether full word or byte operands are involved. In Fig. 4.7 we address the first byte of this word; thus the decimal address is 21, and the binary address is 010101. The byte which is actually indicated

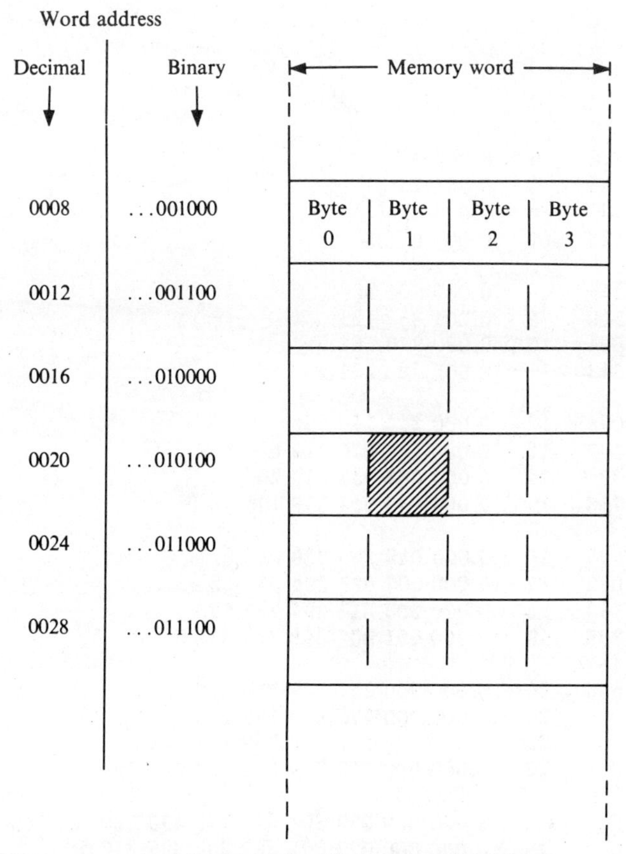

Fig. 4.7. Example of byte-addressable memory

with the complete address is the shaded one. Hence, in this computer organization, each byte is addressable, even though the memory is only word-addressable.

The reader should verify that we could extend this memory organization to include *bit-addressability*, in which each bit of each byte of each word would have a unique address, merely by adding another three bits to the address. In practical computers, however, full bit-addressability is usually more costly in additional hardware than its usefulness can justify. The choice of a smallest addressable increment is another hardware/software trade-off to be made by the machine designer.

4.4 BINARY ENCODING OF INTEGERS

We will begin our discussion of the encoding of numerical information within the computer by considering the integers. Familiarity with binary arithmetic is assumed, but readers wishing to review it may turn to Appendix A. As noted earlier, if we have n bits of information we can represent 2^n different numerical values. If

2^n	n	2^{-n}
1	0	1.0
2	1	0.5
4	2	0.25
8	3	0.125
16	4	0.062 5
32	5	0.031 25
64	6	0.015 625
128	7	0.007 812 5
256	8	0.003 906 25
512	9	0.001 953 125
1 024	10	0.000 976 562 5
2 048	11	0.000 488 281 25
4 096	12	0.000 244 140 625
8 192	13	0.000 122 070 312 5
16 384	14	0.000 061 035 156 25
32 768	15	0.000 030 517 578 125
65 536	16	0.000 015 258 789 062 5
131 072	17	0.000 007 629 394 531 25
262 144	18	0.000 003 814 697 265 625
524 288	19	0.000 001 907 348 632 812 5
1 048 576	20	0.000 000 953 674 316 406 25
2 097 152	21	0.000 000 476 837 158 203 125
4 194 304	22	0.000 000 238 418 579 101 562 5
8 388 608	23	0.000 000 119 209 289 550 781 25
16 777 216	24	0.000 000 059 604 644 775 390 625
33 554 432	25	0.000 000 029 802 322 387 695 312 5
67 108 864	26	0.000 000 014 901 161 193 847 656 25
134 217 728	27	0.000 000 007 450 580 596 923 828 125
268 435 456	28	0.000 000 003 725 290 298 461 914 062 5
536 870 912	29	0.000 000 001 862 645 149 230 957 031 25
1 073 741 824	30	0.000 000 000 931 322 574 615 478 515 625
2 147 483 648	31	0.000 000 000 465 661 287 307 739 257 812 5
4 294 967 296	32	0.000 000 000 232 830 643 653 869 628 906 25
8 589 934 592	33	0.000 000 000 116 415 321 826 934 814 453 125
17 179 869 184	34	0.000 000 000 058 207 660 913 467 407 226 562 5
34 359 738 368	35	0.000 000 000 029 103 830 456 733 703 613 281 25
68 719 476 736	36	0.000 000 000 014 551 915 228 366 851 806 640 625
137 438 953 472	37	0.000 000 000 007 275 957 614 183 425 903 320 312 5
274 877 906 944	38	0.000 000 000 003 637 978 807 091 712 951 660 156 25
549 755 813 888	39	0.000 000 000 001 818 989 403 545 856 475 830 078 125

Fig. 4.8. Decimal representation of powers of 2

we start with the numerical value of 0 and count upward, the largest integer that can be represented by n bits is $2^n - 1$. Human beings find working with powers of 2 hard because it is not our normal way of thinking. We have been taught to operate in a decimal number system and can easily conceive of powers of ten. Figure 4.8 gives the powers of 2 in decimal notation.

We must modify one statement above, however. With n bits we can represent up to $2^n - 1$ *positive integers*. The use of only positive numbers very much restricts our operations. Let us therefore consider how we can introduce negative integers into our set. The most common way to represent a negative integer is to preface or precede it with a special symbol. The common symbols for positive and negative representations are the plus sign (+) and the minus sign (−), respectively. The words "plus sign" and "minus sign," of course, are merely designations that have been attached to these symbols in accordance with their applications. The symbols for the plus sign and the minus sign are carried over into the binary number system as well. However, when we are representing information within a computer, we have no way of drawing a plus sign or a minus sign into a bit. Thus we are again faced with the problem of symbolic encoding, but that problem is easily solved by convention. The value of 0 is applied to either plus or minus, and the value of 1 is then applied to the other. By generally accepted convention, the binary value of 1 represents minus. If we are going to devote a binary digit to the representation of the sign, then for n bits we will have only $n - 1$ bits available for the representation of the number itself. Since the $n - 1$ bits are said to represent the magnitude of the number, the usual name of this representation is *sign-magnitude*. Since only $n - 1$ bits are now available for the representation of the magnitude of the integer, our range of integers is $\pm(2^{n-1} - 1)$.

Complement Representation of Negative Numbers

Sign-magnitude representation of signed numbers seems most natural since it is the way humans work with decimal numbers. The rules and procedures for calculation with decimal numbers are well suited to the sign-magnitude representation. These manual rules and techniques, with suitable modification, of course, can be used for binary arithmetic as well, but they are not necessarily the best in terms of computer design. For example, subtraction, by use of computer circuits, is *not* most easily done in a sign-magnitude representation.

Another method for representing negative numbers may be employed when the number of places that represent a number is fixed. Known as the *complement* form or representation, it is a result of modulus arithmetic.

If we have a modulus of 10,000 in decimal, 9999 is the largest number we can represent. If we tried to put in 36,492, we could hold only 6492. If M is the modulus, and D is the number we wish to represent, then

$$\frac{D}{M} = \text{Quotient(throw away)} + \text{Remainder(answer)}.$$

So we would say 36,492 modulo 10,000 = 6492. If you have only n bits, then you can represent a number only as large as $2^n - 1$, as we have discussed above. If you add 1 to $2^n - 1$, you obtain the number 0. All of the places will be zeros. You have generated a carry, called a *carryout* because it is a carry beyond the number of places you have and is therefore lost. This result is illustrated in Fig. 4.9. The only way you can express numbers in this system is modulo the largest possible number —binary, decimal, or any other number base.

```
    111...11     n bits, all 1's
          +1
 -----------
 (1)000...00     n + 1 bits
                 n bits, all 0's
                 n + 1 bit is 1
```

Fig. 4.9. n-bit carryout, arithmetic modulo 2^n

Let us consider the process of obtaining the negative of a number, A, in a normal unrestricted system. If we write $A + B = 0$, we imply that B is the additive inverse of A, and solving this algebraic system, we obtain $B = -A$. However, if we write that $A + B = 0$ (modulo n), when we solve the system algebraically, we obtain $B = kn - A$. If we let $k = 0$, we are then back to the unrestricted case and not working in a modulus form. However, let us consider the more interesting case where k is equal to 1. If $k = 1$, we have that $B = n - A$, and thus B is represented as a positive number. If we had to go through the work of subtraction in order to find B, we would have absolutely no advantage over representing it in the sign-magnitude form as $-A$. But as we shall soon see, there are ways of obtaining $(n - A)$ without performing the subtraction.

Assume for the moment that we have $(n - A)$, and consider the possibility of performing a subtraction. For instance, let us subtract A from X. For $D = X - A$ we can use $D = X + B$ (where B is the negative representation of A). In this case we expand to obtain $D = X + (n - A)$. Thus, using the complement form, we can perform subtraction by actually doing addition. The advantage within the computer is immense. We need only one circuit to perform the operations of addition and subtraction. If we wish to perform subtraction, we first complement the subtrahend.

Now let us turn to the problem of finding the complement without performing subtraction. This procedure will uncover one of the advantages of the binary number system. Consider finding the complement by performing subtraction, as shown in Fig. 4.10. We always subtract from a number composed of all 1's, as shown in Fig. 4.10(b), just as in decimal we would subtract from 9999. Thus the result of the subtraction is that every zero has become a one and every one has become a zero—a form known as the 1's complement. The 1's complement of a binary number is therefore found by inverting all the bits. This procedure is easily done by electronic circuits; we can perform the operation of finding the complement without subtracting. Even when using the 1's complement form for representing negative numbers, we note that the leftmost bit still represents the sign. A 0 denotes a positive number; a 1 denotes a negative number. This is not sign-magnitude but 1's complement form; in order to find the magnitude of this negative number we need to recomplement it.

(a)

(Employed in complementing)		(Not employed in complementing)	
1	1	0	0
−0	−1	−0	−1
1	0	0	1 (borrow 1)

(b)

111111...111	
−110010...010	(number)
001101...101	(1's complement)

Fig. 4.10. Binary subtraction: (a) Subtraction table; (b) Finding the complement.

Having now introduced the 1's complement to represent negative numbers, we may proceed to do both addition and subtraction by addition. As an example, let us consider adding +3 and −2. The binary representation of 3 is 011, and minus 2 is represented as 101 modulo 2^3, as shown in Fig. 4.11. When we perform the addition, we note that for the three digits representing our numbers we obtain 000; we also obtain a digit 1 in the fourth position as a result of a carryout. The three digits of significance, 000, are not the true result of adding 3 and minus 2. However, we note that if we add the carryout to the resultant sum, we obtain 001, which is indeed the correct result, 1.

Decimal	Binary	
3	011	
−2	101	
1	(1)000	
	└→1	(end-around carry)
	001	

Fig. 4.11. Addition/subtraction with 1's complement

In the representation of numbers in a 1's complement form, the carryout is to be added to the resultant intermediate sum. The procedure of adding the carryout to the sums is known as an *end-around carry*. (A proof of the correctness of the end-around carry is beyond the scope of this book but may be found in Flores, 1963.)

An alternative method of representing negative numbers is the *2's complement*. It is formed by subtracting the number from a binary value equal to 2^n (where n is the number of digits in the original number)—like subtracting from 10,000 in decimal. Figure 4.12 indicates the procedure for finding the 2's complement of the four-bit signed number $0101(5_{10})$. Recall that the highest number that can be expressed with n binary digits is $2^n - 1$, whereas the number of possible combinations is 2^n. However, *the 1's complement is the number which must be added to form the binary sum $2^n - 1$, whereas the 2's complement is the number which must be added to form the binary number 2^n*. In short, the 1's complement is one less than the 2's complement. Therefore, all that is necessary to form the 2's complement is to add 1 to the 1's complement of the number. Computer hardware may be designed to handle correctly any *one* of the three types of notation. Provided that all calculations are done consistently with one type, the answer is correct.

(a)

$$\begin{aligned} 2^4 &= 10000_2 \\ 5_{10} &= \underline{\ 0101_2} \\ -5_{10} &= \ 1011_2 \text{ (2's complement)} \end{aligned}$$

(b)

$$\begin{aligned} 5_{10} &= 0101_2 \\ -5_{10} &= 1010_2 \text{ (1's complement)} \\ & \quad \underline{+1} \\ -5_{10} &= 1011_2 \text{ (2's complement)} \end{aligned}$$

(c)

$$\begin{aligned} 5_{10} &= 0101 \\ & \quad \downarrow\downarrow\downarrow\downarrow \\ -5_{10} &= 1011 \text{ (2's complement)} \end{aligned}$$

Fig. 4.12. Finding the 2's complement: (a) by subtractions; (b) from 1's complement; (c) by right-to-left sequence.

You can find the 2's complement in a single computer process, but that process must be sequential rather than combinatorial. That is, the answer may not be expressed simply as a one-shot combination of the bits; rather, the bits must be treated one at a time as the process takes place. The process is to examine the number to be complemented bit-by-bit from right to left. For each 0 bit in the original number, the complement will also have a 0 bit until the first 1 bit is reached in the original. At that point, you place a 1 in the corresponding bit position in the complement. Thereafter, invert the bits in going from the original to its complement. This process is illustrated in Fig. 4.12(c). When performing addition with 2's complement representation, you ignore the carryout. This rule may be proved by the reader. Although we have discussed complement form for representing integers, there is no reason why it could not also represent fractions.

Positional notation is used with any radix system. A period, called the radix point, separates those digits which multiply positive powers of the radix from those digits which multiply negative powers of the radix. If the binary point is located at the left end of the word, the result is a binary fraction between 0 and 1. If it is at the right end, the result is an integer between 0 and $2^n - 1$. As before, the results of all arithmetic are correct so long as one such convention is followed consistently. It is usual to adhere to the integer convention. Fractional numbers are usually represented in the floating point form, which is discussed below.

Sign-magnitude, 1's complement, and 2's complement are three different schemes for assigning symbols to represent negative integers. All these schemes may be thought of as producing binary integers, but according to different rules.

Decimal number	Sign-magnitude binary number	1's complement binary number	2's complement binary number
+7	0111	0111	0111
+6	0110	0110	0110
+5	0101	0101	0101
+4	0100	0100	0100
+3	0011	0011	0011
+2	0010	0010	0010
+1	0001	0001	0001
+0	0000†	0000†	0000
−0	1000†	1111†	0000
−1	1001	1110	1111
−2	1010	1101	1110
−3	1011	1100	1101
−4	1100	1011	1100
−5	1101	1010	1011
−6	1110	1001	1010
−7	1111	1000	1001
−8	—	—	1000

† Note that both sign-magnitude and 1's complement have separate notations for +0 and −0. Only 2's complement does not. For this reason an extra number is available in 2's complement, which always represents the most negative number possible with the given number of bits.

Fig. 4.13. Four-bit binary number representations

It is interesting to compare the symbolic representation in Fig. 4.13, in which four bits are used to represent signed integers in each of the three schemes.

One can generalize the 1's complement and 2's complement representation of negative binary numbers to other number systems. The 2's complement corresponds to the radix complement, or true complement, and the 1's complement corresponds to the diminished radix complement. The diminished radix complement is found by the process of subtraction. The number of which you wish to find the complement is subtracted from an n-digit number, where all n digits are the largest number possible in the radix being used. For instance, in a base-10 system, decimal, take a number composed of all nines and subtract from it the number to be complemented. The result is the diminished radix complement. To find the true radix complement, add 1 to the rightmost (the least significant) digit of the diminished radix complement. In the decimal system the radix complement is called the tens complement.

Overflow

As a result of addition or subtraction, an attempt may be made to generate a number which cannot be represented in n bits. This condition, known as *overflow*, is a consequence of a restricted word size in the computer.

If overflow went undetected, the arithmetic would proceed, uninterrupted and incorrect. Therefore, it must be detected. Overflow can be detected by noting a change of sign where one is not expected. For example, the sum of two negative numbers should not be positive, and the sum of two positive numbers should not be negative. The hardware is usually equipped with special circuits which detect these conditions during the arithmetic operation. A hardware bit, or *flag*, is set if the operation was in error. The programmer must usually be responsible for the testing of this bit after every operation which might conceivably cause an error of this type. If such an error is detected, the programmer may decide to proceed, to halt and print an error message, or to recover the answer through other arithmetic operations.

Binary integer multiplication

The rules of multiplication are the same in all number systems, but in binary we have the advantage of a very short multiplication table (Fig. 4.14). Multiplication by 1 produces the same number; multiplication by 0 produces 0.

×	0	1
0	0	0
1	0	1

Fig. 4.14. Binary multiplication table

$$\begin{array}{r} 7_{10} \\ (\times)5_{10} \\ \hline 35_{10} \end{array} \qquad \begin{array}{r} 0111 \\ (\times)0101 \\ \hline 0111 \\ 0000 \\ 0111 \\ \hline 100011 \end{array}$$

Fig. 4.15. Multiplication in decimal and binary

Consider a process of "long multiplication" with multi-bit multiplier and multiplicand. When the multiplier bit is 0, no partial product is produced; when the multiplier bit is 1, the partial product is the multiplicand shifted left the proper number of bit positions to line up with the multiplier digit. At the end the partial products are summed. The multiplication of 0111 (decimal 7) by 0101 (decimal 5) is shown in Fig. 4.15.

Note that the product may contain more bit positions (be longer) than either the multiplier or multiplicand. Therefore the product register in a computer must be a double-length register if significance is not to be lost. Another point for computer implementation is the treatment of the partial products. The computer adder circuits we have previously discussed accept only two inputs, the addend and the

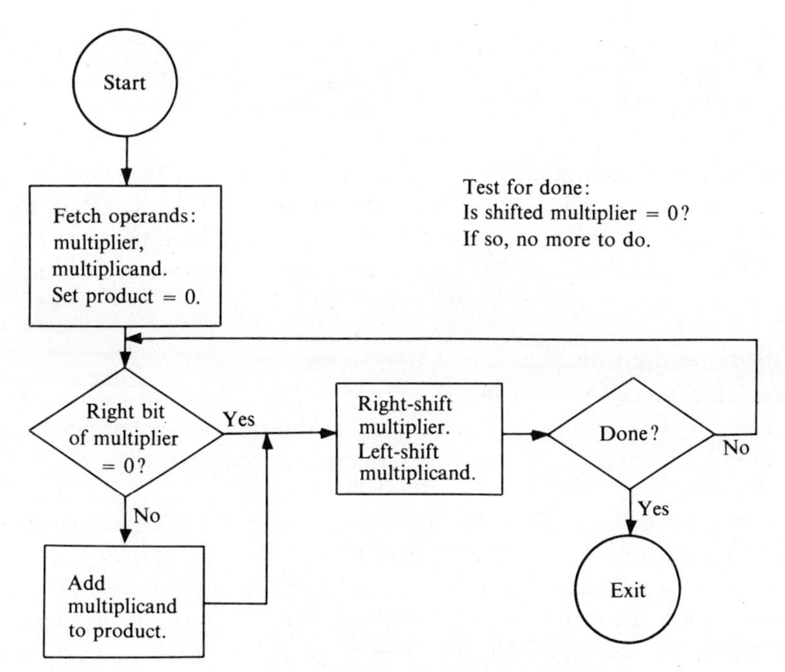

Fig. 4.16. Binary multiplication principles

augend; they cannot add a column of numbers. Therefore each partial product is added to the product being formed whenever it is generated. The product is thus available immediately at the end of the multiplication cycle. A flowchart of the multiplication process is presented in Fig. 4.16.

So far we have touched on multiplication of unsigned integers only. What happens when we deal with signed numbers, of course, depends on which of the three symbolic representations we are using for negative numbers. The multiplication of positive multiplier and positive multiplicand proceeds as noted above for all three forms, but when multiplier and multiplicand are not both positive, the multiplication algorithm branches.

The sign-magnitude representation poses the smallest problem. The magnitudes are always positive numbers that may be directly multiplied. The sign is determined separately from inspection of the signs of the multiplier and multiplicand. This task can be performed directly by computer circuitry.

When negative numbers are represented in the 1's complement form, we must inspect the sign bits of multiplier and multiplicand to determine how to proceed. We shall discuss only one pair of algorithms. If both numbers are negative the result will be positive; therefore they are both complemented before multiplication proceeds. When the sign differs, the answer will be negative, so we produce a product directly in complement form. The negative number is taken as the multiplicand. The rules of multiplication for positive numbers are followed except that as the multiplicand is shifted left, it is filled with 1's on the right, where empty space is created by the shifting. With a partial product, there is always an end-around carry.

For 2's complement multiplication we must also inspect the sign bits. If the signs are alike we proceed as above (complementing both if necessary). If the signs differ, the negative number is taken as the multiplicand. Multiplication then proceeds with no correction required; it produces a negative answer in 2's complement form.

Binary integer division

The division of signed positive integers proceeds independent of the method used to represent negative numbers. As in decimal arithmetic, division proceeds by trial subtraction. An unsuccessful trial is indicated by a change of sign of the partial remainder.

One popular division algorithm proceeds as follows: Load the dividend, left-justified, into a double-length register, zero-filling from the right. Similarly, make the divisor into a double-length word. Subtract the divisor from the dividend. If the subtraction is successful, enter a 1 in the quotient, replace the dividend by the partial remainder, and right-shift the divisor one place. If the subtraction is unsuccessful, resulting in a dividend sign change, enter a 0 in the quotient, discard the (incorrect) difference, leaving the dividend intact, and right-shift the divisor. Loop until the divisor has been completely shifted out, and then stop.

In integer division, both dividend and divisor are binary integers. Since the quotient will rarely be an integer, we must provide for both an integer and a fractional part of the answer. This problem may be avoided by considering the dividend and divisor to be binary fractions, with the implied binary point just to the left of the most significant bit, between it and the sign bit.

Preparatory to a fractional divide, the divisor and dividend are shifted so that the quotient will always be a binary fraction. We require that the divisor be greater than the dividend. When subtraction is completed, an appropriate shift of the quotient may be made to restore the binary point. The division algorithm is described by the flowchart of Fig. 4.17.

The algorithmic modifications required to handle the various representations of negative numbers may be determined by the reader with reference to those used for multiplication.

4.5 REPRESENTING LARGER AND SMALLER NUMBERS

The integer arithmetic we have covered so far is an exact process. A correctly operating computer, given the same input, will always produce the same results. The range of numbers handled by this representation is very limited, however. The numbers that can be represented in a 1's complement 16-bit computer word lie

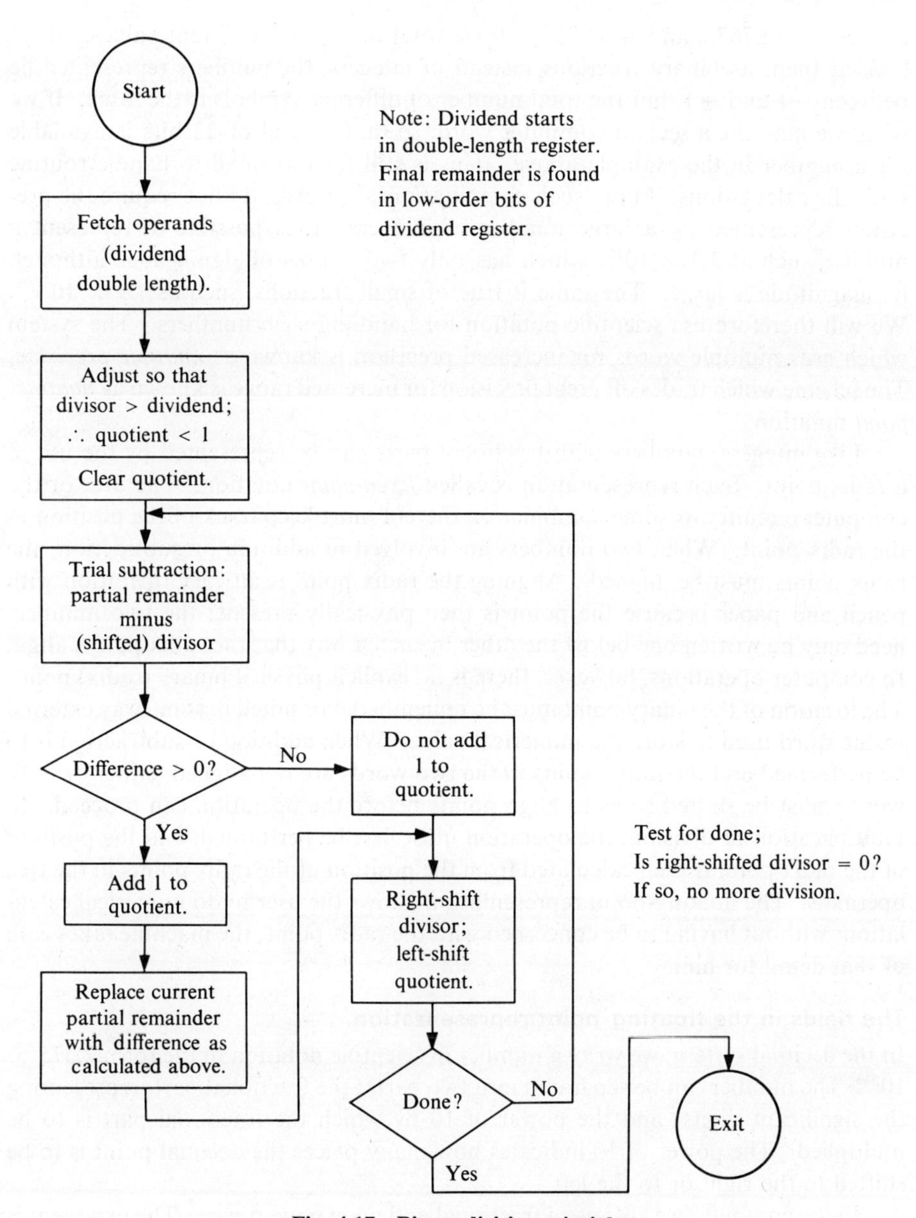

Fig. 4.17. Binary division principles

between $-32,767_{10}$ and $+32,767_{10}$, for a total of $2^{15} - 1$ different values. If we look at them as binary fractions instead of integers, the numbers represented lie between -1 and $+1$, but the total number of different symbols is the same. If we wish, we may use a second computer word, so that a total of 32 bits is available for a number in the example above. This is still far too small to handle routine scientific calculations. Many such computations, however, do not require the precision represented by a large number of integers. It is possible to represent a number, such as 3.3×10^{17}, which has only two figures of significance although its magnitude is large. The same is true of small fractions, such as 1.9×10^{-14}. We will therefore use scientific notation for handling such numbers. The system which uses multiple words for increased precision is known as *multiple precision.* The scheme which trades off great precision for increased range is known as *floating-point* notation.

Like integers, numbers with fractional parts can be represented by the use of a radix point. Such representation is called *fixed-point* notation. The user or the computer circuitry or some combination thereof must keep track of the position of the radix point. When two numbers are involved in addition or subtraction, the radix points must be aligned. Aligning the radix point is an easy operation with pencil and paper because the point is then physically present; the two numbers need only be written one below the other in such a way that the radix points align. In computer operations, however, there is no explicit physical binary (radix) point. The location of the binary point must be remembered or noted in some way external to the word used to store the numerical value. When addition or subtraction is to be performed and the radix points in the two words are not aligned, either or both words must be shifted so as to align points before the operation can proceed. In multiplication or division, the operation must first be performed, and the position of the radix point is then calculated from the position of the radix points in the two operands. The floating-point representation allows the user to do numerical calculations without having to be concerned with the radix point; the machine takes care of that detail for him.

The fields in the floating-point representation

In the decimal system, we write a number in scientific notation in the form 2.375×10^{21}. The number can be separated into two parts: the fractional part, representing the significant digits; and the power of 10 by which the fractional part is to be multiplied. The power of 10 indicates how many places the decimal point is to be shifted to the right or to the left.

In the computer we also use a fractional and an exponent part. The exponent is a power of some other base, usually 2, 8, or 16, instead of a power of 10. In floating-point notation, the fractional part is usually normalized; that is, the fractional part is always written as a radix point followed by a nonzero digit. One can always normalize a nonzero floating-point number by left-shifting or right-shifting the fractional part and correspondingly decreasing or increasing the exponent until a

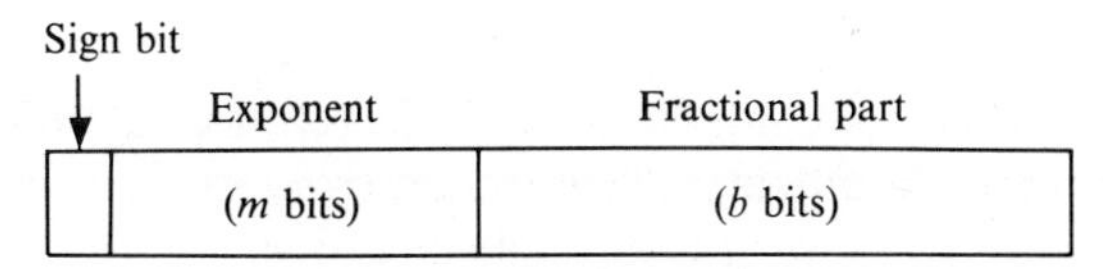

Fig. 4.18. Floating-point word

nonzero digit appears just to the right of the radix point. The fractional part is therefore between $(1/2)_{10}$ and 1 when normalized in binary representation. Similarly, the fractional part is between $(1/16)_{10}$ and 1 in hexadecimal representation.

When we represent a floating-point number in a computer word, we are faced with a question of how to treat the signs. Note that there are two signs to treat, that of the fractional part and that of the exponent. The most common scheme is to use sign-magnitude representation for the fractional part, possibly locating the sign bit of the number adjacent to the exponent rather than to the magnitude, as in Fig. 4.18. We thus have three parts, or fields, in the floating-point representation: the sign field, the exponent field, and the fractional field. One common arrangement of these fields, illustrated in Fig. 4.18, is discussed below.

A different scheme is used for representing the sign of the exponent. We simply add a constant to the exponent so that it is expressed as a positive number. This constant, known as the *bias*, or *offset*, is a binary number composed of a 1 followed by 0's to the field width of the exponent field. For example, in a machine with $m = 8$, the bias is $10000000_2 = 200_8$. The bias chosen is the middle of the range of numbers representable in m bits.

Floating-point field placement

In one common arrangement of these fields, shown in Fig. 4.18, the leftmost bit is the sign of the fractional part, the next m bits represent the exponent, and the remaining b bits represent the fractional part. Though not universal, this ordering of fields is very common.

Consider the situation in which information is stored as floating-point numbers in two computer words called X and Y. If we now choose to interpret the contents of X and Y as integer numbers, what property, if any, carries over? With the field ordering of Fig. 4.18, the equality or inequality relationship of X and Y is preserved, whether they are, in fact, floating-point numbers or integers. This result occurs because the sign bit is in the same place and the exponent is in a more significant digit position than is the fractional part.

The advantage of this preservation of ordering is that integer comparison instructions may be used with floating-point numbers. This particular field placement allows the same hardware instructions to be used for comparison of integers or floating-point numbers.

Multiple-precision floating point

In many applications, the word length of n bits for a floating-point word may be insufficient, because m bits cannot represent the necessary exponent of the floating-point number or 2^b cannot represent the desired precision of the floating-point number. Using more than one computer word is the solution. If only two computer words are used to represent a floating-point number, it is called double precision; if more than two words are used it is multiple precision.

4.6 DECIMAL ENCODING

In the organization of some computer systems, usually those for business applications, the direct representation of decimal digits within the computer has been desirable. Binary hardware is still used, however. Since there are ten decimal digits, four bits are needed to represent each numeral. Four bits can represent sixteen different numerals, so that when we represent a decimal digit, we waste six possible representations. There are a number of ways in which the decimal digits can be represented. The simplest, a binary representation that counts from 0 to 9, is called binary coded decimal, or 8421 BCD. It is shown, along with two other common schemes, in Fig. 4.19. The remaining binary numbers from 10 through 15 are simply not used. Each four-bit group representing a single decimal digit is called a *decade*.

Binary	8421 Decimal	2421 Decimal	Excess-3
0000	0	0	
0001	1	1	
0010	2	2	
0011	3	3	0
0100	4	4	1
0101	5		2
0110	6		3
0111	7		4
1000	8		5
1001	9		6
1010			7
1011		5	8
1100		6	9
1101		7	
1110		8	
1111		9	

Fig. 4.19. Three binary representations of decimal codes

The existence of unused code combinations presents additional complications in decimal arithmetic. For BCD addition, which is the simplest operation, the rules are as follows:

1. All the decades are added separately.
2. If the sum of any one decade is larger than 9, a carry must be propagated to the next higher decade.

3. If the sum in a decade is a binary number between 10 and 19 inclusive, a correction factor of 6 must be added to the decade.

4.7 WEIGHTED CODES

An alternative way of looking at the binary representation is to consider it to be weighted positional code. In the most common BCD code the weights of the positions are 1, 2, 4, and 8, counting from the right, corresponding to 2^0, 2^1, 2^2, and 2^3. However, the normal convention of positional notation places the largest power of the radix at the left (e.g., 2^3 2^2 2^1 2^0), thus giving rise to the name "8421 code."

One of the main problems with BCD arithmetic is the complexity involved in forming a complement. Techniques are available, but it is more desirable to have a method that permits complements to be formed by inverting each bit. If complement numbers can be formed by simply inverting each bit, the code is known as a self-complementing code. If a weighted code is to be self-complementing, the sum of the weights must be decimal 9. The 2421 code is self-complementing. The excess-3 code is also self-complementing, but it is not a weighted code. Note, however, that the normal BCD counting sequence has been altered by coding each decade as a binary number three units larger than the decimal number. That is, 0 is coded as 3, 1 as 4, and so on. This simple modification of the BCD code produces a self-complementing code and greatly simplifies the process of subtraction.

These codes have either unused states or redundant states. Redundancy of states occurs when more than one possible code combination can be used to represent a given decimal digit. These redundant states are not part of the proper code set, and the rules of arithmetic must be formulated to treat them properly when they occur.

4.8 UNIT DISTANCE CODES

When the information to be stored in the digital computer is input from some external physical medium, the use of a binary code has a number of pitfalls. Consider the case where the piece of data on the external medium changes from the value 3 to the value 4 and is then input to the computer. You will note in Fig. 4.20 that

Binary	Decimal
011	3
100	4

Fig. 4.20. Change from 3 to 4

such a change involves all three bits. If this change occurs continuously rather than at some discrete sample point, there is the possibility that all three bits may not change simultaneously (a property of the input device). In fact, any number from 0 to 7 may be generated during the change. To avoid this possibility of error generation, a different sort of code is sometimes used when going from continuous

8421
0000
0001
0011
0010
0110
0111
0101
0100
1100
1101
1111
1110
1010
1011
1001
1000

Fig. 4.21. Unit distance (Gray) code

to discrete. *Unit distance code* takes its name from the fact that any two successive digits differ by only one bit. A four-bit unit distance code to count from 0 to 15 is shown in Fig. 4.21. This scheme is also known as a Gray code after its inventor. Note that any two successive numbers differ in only one bit position. Note also the symmetry that has inspired another name for the code type, reflected codes. A typical transducer used to convert from rotating shaft position to a digital representation is shown in Fig. 4.22.

Fig. 4.22. Gray code cylinder analog-position-to-digital representation

REFERENCES

H. Hellerman, *Digital Computer System Principles*, McGraw-Hill, 1967.

C. W. Gear, *Computer Organization and Programming*, McGraw-Hill, 1969.

Burroughs Corporation, *Digital Computer Principles*, McGraw-Hill, 1969.

Y. Chu, *Digital Computer Design Fundamentals*, McGraw-Hill, 1962.

P. A. Ligomenides, *Information-Processing Machines*, Holt, Rinehart and Winston, 1969.

I. Flores, *The Logic of Computer Arithmetic*, Prentice-Hall, 1963.

QUESTIONS

1. What property do all *odd* binary integers have in common?
2. What happens to the value of a number if the binary point is shifted one place to the right? Two places? N places? One place to the left? Four places? N places? If the number is expressed in binary scientific notation, what happens to the value of the exponent in each of these cases?
3. Give two everyday examples where modulus arithmetic is encountered.
4. a) With N information bits, how many numbers may be represented?
   b) If we choose to start with the numerical value of 0 and count upward, what is the largest positive integer which can be represented in N bits?
   c) How does the answer to (b) change if the sign-magnitude form is used?
5. a) The waveform shown in Fig. 4.23 represents the output of an analog-to-digital converter.
   This signal is to be processed by a computer that uses the sign-magnitude representation for its numbers (integers).
   What must the *minimum* length of the computer word be in order to represent the discrete values (integers) in the waveform?
   What range of numbers should the computer word be able to represent?
   What range of numbers does the minimum or optimum word length actually represent?
   b) Repeat part (a) for a computer that uses the 1's complement representation for its numbers (integers).
   c) Repeat part (a) for a computer that uses the 2's complement representation for its numbers. (Be careful with this one!)
6. How are the diminished radix complement and true radix complement found? Use this procedure to find the diminished and true radix complement of a number in radix 16. How should a carryout be treated in both the reduced and true radix form?
7. a) What is negative zero? Give an example of its occurrence.
   b) Does negative zero occur during subtraction by addition of the 2's complement of the subtrahend? Explain.
8. Represent the following base-10 numbers in both 2's complement and sign-magnitude binary form. Then represent the negative of each of them in sign-magnitude, 1's complement, and 2's complement. Use 9 bits in all cases. It is possible that not all the representations requested will work with all the numbers given.
   0 10 72 254 255 256

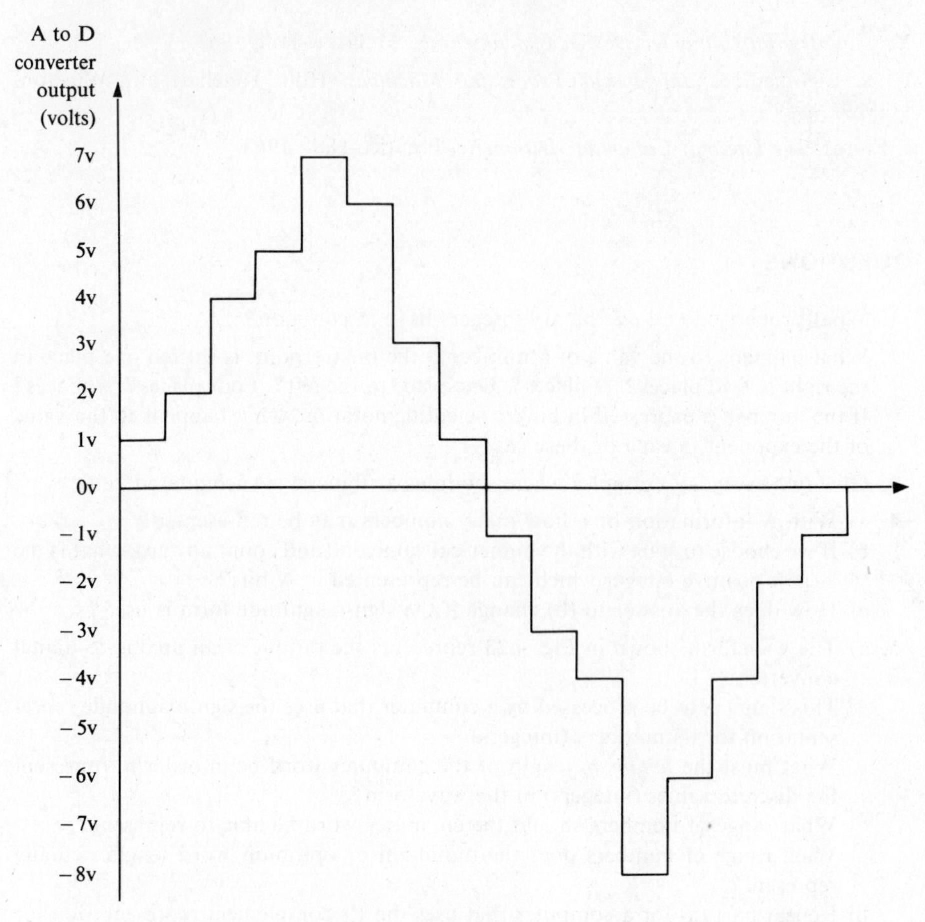

Fig. 4.23

9. In a machine that uses 12-bit binary numbers, what pattern of 12 bits represents -5 (in integer mode) if the representation is as indicated?
   a) sign-magnitude
   b) 1's complement
   c) 2's-complement

10. Perform the following exercises both in 2's complement and in 1's complement binary form, and convert the answer back to decimal as a check.

$$\begin{array}{r} 111_{10} \\ -\,66_{10} \\ \hline \end{array} \qquad \begin{array}{r} 42_{10} \\ -\,57_{10} \\ \hline \end{array}$$

11. a) Explain the advantage to the computer of using the complement form of a number to perform addition and subtraction.
    b) Describe the general method of expressing a number modulo n. Use the diminished radix complement.

12. Express the number 6372_{10} in each of the following forms.
    a) modulo 6×10^3
    b) modulo 2
    c) modulo 3186
13. What is an offset, or bias, and what is its function?
14. In a floating-point word 36 bits long in which 27 bits are reserved for the fractional part, 8 bits for the exponent and 1 bit for the sign, what are the largest and smallest numbers that can be represented (positive and negative)? How many significant figures (decimal) do they have?
15. In the representation of a floating-point word in the computer, there are two signs to treat, that of the fraction part and that of the exponent. Devise a scheme (other than the one described in the text) to represent these signs. Contrast your scheme with the one in the text.
16. Suppose that you are participating in the design of the floating-point hardware for a machine which uses 48-bit binary words. Given that 11 bits are used for the exponent, show how you would have the machine represent the decimal fraction $\frac{3}{8}$, giving particular attention to which bits are used for the exponent, the sign, etc., and defining "normalization" for floating-point numbers in the machine. Assume also that the machine already possesses integer arithmetic and integer compare and sign-testing instructions.
17. A certain digital computer has a 36-bit floating-point number format with a bias of 200_8 for its exponent field. The fractional field is 27 bits wide, and there is a sign bit, S, in the format.

S	Exponent	Fractional part
0	1 … 8	9 … 35

    a) Show in octal the largest and the smallest (nonzero) magnitudes that can be represented.
    b) If the exponent field is increased from 8 bits to 11 bits, what must the (approximate) bias be in order to express all exponent values as positive numbers?
    c) Represent the following octal numbers according to the given format.
       6.7420022   226677000   12.66
18. Given the following 5-bit binary integers:
    $A = 01001$
    $B = 11001$
    $C = 10111$
    a) What are the decimal equivalents if the 5 bits are used for unsigned integers?
    b) If the computer allows only 5 bits to represent signed integers, what are the decimal equivalents for each of the three representations—sign-magnitude, 1's complement, and 2's complement?
    c) Perform the indicated arithmetic and transform the results into decimal notation, explaining any (apparent) inconsistencies. Perform the operations for all three representations of signed integers.
       $S_1 = A + B$ $D_1 = B - C$
       $S_2 = B + C$ $D_2 = A - C$

19. Given the following 8-bit binary integers:
    A = 10110110
    B = 01110100
    C = 00011011
    a) If all 8 bits are used for unsigned integers, what are the decimal equivalents of integers represented internally in binary, 8421 BCD, and excess-3, respectively?
    b) Repeat question 18(c) for the numbers above as unsigned integers.
20. Because our alphabet is not directly understandable to a computer, a symbolic language is necessary to represent human instructions in the machine. How would CRUNCH be represented digitally according to ASCII specifications in a computer?
21. Register A, which has 12 flip-flops, contains 011001110011. If A represents a binary coded decimal integer (8421 BCD code) in 10's complement form, what is its value? Show the contents of register B if it contains
    a) the 10's complement of A,
    b) the 2's complement,
    c) the 1's complement.
22. Propose a four-bit weighted code for decimal digits that meets the following specifications. At least one of the weights is negative. The code is self-complementing in the sense that the 9's complement can be obtained by switching 1's to 0's and vice versa. Is the proposed code a unique representation?
23. Suppose that you are participating in the design of a computer with 42-bit words. Propose and defend briefly a word format for alphabetic information.
24. Given a 12-bit computer word in a machine employing 2's complement representation of negative numbers, what does 100000000000 mean?
25. Devise detailed algorithms for detecting integer overflow on machines using (a) sign-magnitude, (b) 1's complement, and (c) 2's complement representations of negative integers.
26. Does the double-length register required for multiplication and division have to be able to perform addition in both words? Explain your reasoning.
27. Draw a flowchart for signed integer multiplication when the representation of negative integers is (a) sign-magnitude, (b) 1's complement, and (c) 2's complenent.
28. The binary number 010 110 101 011 is stored in a 12-bit register.
    a) Convert the number to octal.
    b) Determine in both octal and binary the result of a left shift of one place,
    c) of three places,
    d) a right shift (of the original number) of two places,
    e) of eleven places.
    f) What is the effect of right and left shifts on the arithmetic value of the number?

5
ARCHITECTURE AND DESIGN OF COMPUTERS

5.1 INTRODUCTION

In the designing of a computer, there are many decisions to be made before the first wire is ever connected. The computer designer has available to him an exceptional range of flexibility. Most computing jobs can be done in a multitude of ways, and many approaches can provide viable solutions to the problems involved. Since it is possible to provide functions in either hardware or software, a good working knowledge of both is necessary if the designer is to make effective choices. We will look at the design process, with special emphasis on the available *trade-offs*, or compromises, and we will consider how these choices should be made.

5.2 THE SYSTEM DESIGNER'S JOB

Even though the daily work of computer engineers involves manipulation of rigorous logical concepts, there is no reason to believe that those engineers behave more rationally than other people. Consequently, computer design is not always what is should be. Section 5.5, covering traditional hardware design describes the situation in the real world. What we must do first, however, is to discuss a perhaps utopian ideal: How should a computer be designed?

Any proper systems design begins with two major considerations:

1. the capabilities of the equipment required to do the job(s) at hand, and
2. the interface between the equipment and the user.

A user—that is, anyone who interfaces with the system—will naturally consider the "computer" to be not only the contents of the electronic cabinets, but also the systems programs he routinely uses. The computer architect must therefore design a system in which the end-user interface includes procedure-oriented languages and other software functions. This is not to say that the designer must specify in complete detail the languages, compilers, and assemblers to be implemented on the system. It does mean that he must be fully aware of the hardware requirements implied by the fact that these languages form a major portion of what will be run on the system.

Having defined the jobs to be done, the capabilities required to do them, and the appearance of the hardware/software/human interface, the designer chooses a set of processes (job control system, arithmetic unit, bulk storage) which will provide both the services and the interface required. He also chooses a set of boundary conditions on these processes (minimum memory size, arithmetic speed, cost). These inputs—capability, interface, processes, and boundary conditions—constitute a set of data on which the designer does his creative work. He must then make a number of trade-offs to find a design which in his opinion best suits the input data.

5.3 HARDWARE/SOFTWARE REVISITED

The most obvious interrelationship between computer hardware and software is that they are interchangeable. Provided that the basic machine satisfies theoretical minimum requirements (i.e., storage, input/output), all other usual computer functions are realizable in programming.

From the standpoint of capability, hardware/software systems consist of a core of hardware functions on which are superimposed layers of software capability.

1. The nucleus is the hardware core. It is comprised of sufficient electronic equipment to implement the instruction set. Closely associated with the hardware are system-supplied subfunctions; software so general and so frequently used as to constitute near-hardware. Routines employed to extend the basic functional domain of the computer hardware or to serve as resident interfaces between all outsiders and the hardware are found at this layer.
2. The next layer is composed of support functions, primary software components that extend the overall convenience of the system. The monitor, supervisor, or executive may be a large important software component, but ultimately it is just a convenience extender, and its function of increasing job-to-job throughput is part of that convenience. A loader is also a part of this layer.
3. Resting on the support layer are the first of the basic language or programmer-communication components. An assembler is a primary example of this level of capability.
4. The next layer contains the actual procedure-oriented languages a system supplies. FORTRAN or ALGOL compilers would be components of this layer.

5.4 TRADE-OFFS

A trade-off represents the explicit recognition of some continuous reciprocal interaction among two or more system design objectives or constraints. We may have more of *this* at the expense of *that*; we select a design that represents some particular mixture of the elements of the trade-off.

All trade-offs are ultimately decided by economic considerations. Design trade-offs in the computer business lie primarily in the choice of whether to implement a desired capability in hardware or in software.

In our market-oriented economy, some trade-offs are based on predictions of salability. Other trade-offs are pushed onto the customer, who may purchase a hardware option to perform a function done by software in the basic sales package.

Job-oriented software

The software tools provided by the well-designed computer must be job-oriented. Rather than being so general that it satisfies many users slightly but none completely, the software must meet most of the needs of each specific type of user. For users interested in scientific and algebraic calculations, the software must include translators for languages such as FORTRAN and ALGOL. The business-oriented user will want COBOL. For string manipulation SNOBOL will be required; for list structures perhaps L^6 or LISP. Almost everyone will use an editor and file system. A library of frequently used subroutines for such applications as statistical analysis will be necessary. A set of problem-oriented languages (e.g., circuit analysis) is also desirable. And so it goes. A software hierarchy must be designed to provide the user with as much power and flexibility as he desires without unduly burdening him with details and tricks of accommodating himself to the software.

Job-oriented hardware

Similarly, the computer must offer job-oriented hardware. There should be specialized hardware features available that simplify the job of the scientific user, the business user, or the real-time user. These hardware features can be built in or optional, but they must permit the user to tailor his system to meet his needs.

Microprogramming and firmware

Microprogramming is based on the observation that the execution of a single instruction basically involves a sequence of information transfers from one register to another, either directly or through an adder or other logical circuitry. The execution of these instruction steps in a machine instruction is like the execution of the individual steps in a program. Thus arises the term *microprogramming*. Each elemental step of the instruction is called for by a *microinstruction*, which refers to one or more *microoperations* (the most elementary operations). The set of microinstructions that implement an instruction constitutes a *microprogram*. The microprograms are usually stored in a small, very fast memory (called the *control store*), which is usually *read-only* (there is no way to change its contents under program control). This memory controls the detailed sequence of logical operations which effect each programmed computer instruction.

Although not changeable during normal program execution, the contents of the control store can be altered by either hardware or software, depending on the implementation technology. The designer therefore has the option of providing some functions as microprograms. These *microprograms*, sometimes referred to as *firmware*, represent a gray area between hardware and software. They constitute another option for the systems architect.

Explicit hardware/software/firmware trade-offs

The architect, given every process or boundary condition, must make a fundamental decision concerning the way to implement it. What are the basic advantages and faults of each implementation?

A hardware approach to a process implies first and foremost that the process is compact and well defined and that it must be repeated frequently in the job to be done. Fixed-point integer division is an excellent example of such a process (if the job requires a lot of it). By way of contrast, a compiler could conceivably be implemented entirely in hardware, even though compilers are not compact and are periodically redefined or modified. Compilers are therefore poor candidates for hardware implementation.

Consider the question of floating-point arithmetic. In many small machines, the hardware to perform such operations is an extra-cost option; if the option is not taken, the operations are performed by software. Hardware is much faster; software is much cheaper. The purchaser has his choice, based—one hopes—on his utilization.

Hardware has certain clear advantages. It is fast, it uses no system resources such as memory for its retention or operation, it is simple to use (a single machine instruction per function), and it concomitantly produces simpler user code. Hardware's main disadvantage is its cost. A program supplied with a machine may be expensive to write, but further copies of it are essentially free. Each new copy of a hardware function, however, must still include the cost of the circuits. The presence of circuits for a function also necessarily implies that the circuits will occasionally break—a further expense.

Functions implemented in software are usually supplied free with the computer, and therefore their cost cannot be explicitly added to the purchase price, as the cost of hardware options may. They add a burden to system resources by taking memory space for their storage and by needing more running time for execution. As functions become more involved, less well defined, and less frequently required, they become more appropriate for software implementation, whether because the disadvantages are less applicable or because the advantages of software outweigh them.

Implementation in firmware may represent a good compromise for a large class of functions. When cost is a severe limitation, as it is in the extremely competitive small-computer market, a firmware organization for a machine can provide many hardware-like advantages without a large cost per advantage. The biggest problems are that firmware is not so fast or efficient as circuitry, and a computer organized around a read-only memory will have all or most pseudo-hardware functions handled by the microprogram. It is therefore usually necessary to have a total commitment to a firmware approach, and the natural continuity of the hardware/software trade-off boundary is disturbed. Trade-offs that would be simple between hardware and software are not so easily made between hardware and firmware or software and firmware.

5.5 TRADITIONAL COMPUTER DESIGN

Except for a rare innovation, computers are traditionally designed by iteration. Defects discovered in one product cycle are usually corrected in a subsequent cycle. Modifications occur as the result of information fed back to the designers from the users. This feedback may be looked at as a loop with a considerable amount of delay around it.

With notable exceptions, hardware design has preceded any software design and development, although the ultimate concern has been to design computers that matched the needs of most users—or for which such claims could be made as a basis for sales. The computer produced possibly differed in capability from that needed by the potential users. Some functions provided in the hardware saw little use or their use did not finally justify the cost of inclusion in the hardware. On the other hand, some processes omitted from the hardware had to be performed, perhaps inefficiently or with difficulty, so many times that, in retrospect, they should have been implemented in hardware. The nature and extent of the unsuitability of some of the hardware features must eventually become known to the hardware engineers if any better design is to emerge.

A traditional modification in computer equipment is the addition of more instructions to the current repertoire. Certain established criteria apply to changes brought about by the feedback from users to designers: the problem must be (1) obvious, (2) easily interpreted in software terms and by software people, (3) of sufficiently high expense, and (4) repetitive. On the basis of these factors, we can account for most changes. Operations which can be considered to be "naturally" hardware in behavior and are often simulated in software should yield new trade-offs; for example, floating-point hardware is the result of such a situation.

Generalizations of available instruction groups are obvious candidates for new hardware features. If you can "transfer if *AC* less than," why not "if greater than"? If these, then why not "if less than or equal" and "if greater than or equal"? Sometimes specific requirements of the software are reflected in new hardware instructions. For example, introduction of a subroutine jump instruction may be necessitated by a new software organization.

Variations on traditional hardware design

The long time delay in the feedback loop described above has forced serious thinking about how computer hardware design should proceed. We shall describe two schools of thought, each of which has proponents among commercial computer manufacturers. The philosophies have implications for machine organization as well as instruction sets. We can expect that changes in instruction sets will occur with much more ease and more often than changes in organization or other aspects of architecture.

The first school of thought holds that the first customer the hardware engineer should satisfy is the software engineer. In the design of all but the smallest computers there is sufficient work for separate hardware and software design teams. By forcing the hardware designers to meet the requirements of their own software design teams, proponents of this viewpoint intend that a workable cooperation will result. The supplier (the hardware engineer) must meet as many of the "customer's" demands as he can afford. When the customer (the software designer) makes unreasonable demands, he must be educated to recognize the difficulties his requests create. The marketplace question governing the relationship is "Is this feature worth the cost?"

The other school of thought holds that the software engineer cannot possibly know to what use the computer system will be put; therefore he cannot make intelligent specifications at design time which will apply for all situations. On the basis of history, the hardware and software designers should attempt to provide all the services and power possible. They should also allow for satisfaction of service requests that have not occurred to them. After the computer has been installed in the field, the software engineer must have a way of adding a new machine function or modifying an existing one. Such field modifications have always been possible. An engineer designs and constructs special circuits to effect the change. Such hardware modifications, however, are undesirable because they are equivalent to remanufacture under uncontrolled conditions.

The presence of a microprogram control store makes the problem much simpler. If the machine supports microprogramming, the software engineer may introduce very low-level machine-function modifications by creating new microprograms.

5.6 HARDWARE DESIGN ALTERNATIVES

Once the purposes of the system and the general description of the functions needed to accomplish those purposes have been defined, a large number of decisions about the detailed nature of the operation of the machine need to be made. These can be both hardware and software decisions, and in fact they are interrelated. The decision as to what that interrelation is and where the dividing line is to be drawn between hardware and software has presumably been made, and now it is necessary to do the detailed design of the hardware. What decisions are to be made?

Organization of the CPU

Perhaps the first decision to be made is the number of bits to be included in a computer word. It is indicated by the precision required of the data representation and by the speed with which arithmetic must be done. A machine with only 16-bit words, by means of multiple-precision arithmetic, may do the same calculations as a 60-bit processor, but it will take longer. This decision, of course, has a large effect on organization and design of all other parts of the system, such as memory and input/output. Another important parameter to be considered when choosing a word size is the design of an instruction format and addressing scheme. The bits in a word must be assigned according to the way they will be interpreted. If there are not enough bits in an instruction word to fully address all of memory, some scheme must be chosen to partition the memory into manageable pieces. If there are many bits available for an instruction, it may be advantageous to have an instruction contain more than one address.

How many operand address fields?

In Chapter 3 we introduced and discussed the operand address field and its employment during the execute portion of the major cycle. However, we did not discuss the number of address fields in the machine word. A typical arithmetic instruction should have four addresses associated with it: two operands, a result, and the location of the next instruction. In many cases some or all of these are assumed. Machines have been and are being built with one, two, or three address fields per machine instruction or with none. Furthermore, the choice is not always constant for a given machine; some instructions may devote an operand address field to other purposes in some instructions.

Let us examine the assignment of functions to the various numbers of address fields, as well as their implications for design and use.

The Three-Address-Field Instruction

Three address fields permit specification of two operands and the results of the operation. Using the general operator symbol, $\circ$,we can specify in one instruction

$$(C) \leftarrow (B) \circ (A)$$

which means that the contents of C are replaced by the results of the operation B $\circ$ A. The operands A and B and the result C are specified in the operand address fields, and $\circ$ is specified in the operation code field. Although the performance of the operation will undoubtedly involve a register, none is specified in the instruction. With a three-address operation it is unnecessary to know anything about registers.

In some systems with three-address machine organization, there is no program counter. Instead, one of the address fields contains the memory address of the next instruction to be executed. The remaining address fields are used as in two-address instructions. This organization is appropriate where main memory is not random access, and much time would be wasted fetching the next instruction if it were not next in numerical order.

The Two-Address-Field Instruction

Two address fields permit specification of two memory locations; they allow instructions of the form

$$(B) \leftarrow (B) \circ (A)$$

Like the three-address instruction, such an instruction does not require specification of a register.

Another type of instruction for which two addresses are an advantage is the comparison operation, which might be designed as

IF B ∘ A = 0, SKIP THE NEXT INSTRUCTION

or perhaps

IF B = 0, TAKE NEXT INSTRUCTION FROM A

The One-Address-Field Instruction

This instruction takes the familiar form

(AC)←(AC) ∘ (operand address)

where (AC) is some accumulator, perhaps specified in a register field, and it is always implied by the instruction. The next instruction is taken from the incremented contents of the program counter unless the instruction involved is a jump or skip instruction.

In small machines, the operand address is sometimes found in another computer word, either by indirect addressing or simply by using the word following the word containing the op code.

The No-Address-Field Instruction

In some machines, data and intermediate results are stored on *stacks.* In this context, a stack is a sequence of memory locations, and the address of the next datum or storage location is the final responsibility of the machine, not of the programmer. The instruction specifies no operand address, only the name of the stack where the result is to be placed or where the input data may be found.

Instructions with Register Designations

Any of the preceding address formats may be augmented by specification of a register that participates in the operation as the source and/or destination of an operand. When an instruction provides for such register designation, we append the phrase "plus one" to its name (e.g., one-plus-one-address-field instruction).

The matter of naming becomes more confused when register designations are the only operands permitted in an instruction. We then consider the register designations equivalent to regular operand addresses and use the names specifying one, two, or three addresses, as appropriate.

The instruction set

As already noted, the decision concerning which instructions to include as hardware functions is fundamental. Another necessary decision concerns whether certain functions, such as manipulations of bits within a word, should be included at all.

The instruction set defined in Chapter 7 for a general-purpose machine is a sample of the wide range of instructions that could be included. It is also necessary to decide on a method of internal representation of negative numbers. Hardware arithmetic functions must be chosen.

It is also desirable to choose an instruction set that is similar to the sets used with other machines, up to and including total program compatibility. A group of computers from one manufacturer which offers different capabilities based on the same instruction set is called a *family*.

Input/output

Many different schemes are available to the designer for implementation of I/O.

Word-by-Word I/O (Multiplexor channel)

The program running in the CPU tests whether the I/O device is ready for a data transfer and, if so, causes the transfer to occur. If the device is not ready, the program can wait or can do something else and check again later. This method is inexpensive as far as hardware goes, but it wastes computer time checking and waiting for devices. It also requires somewhat involved programs for device control.

Interrupt I/O

A provision is made for the device, when ready, to interrupt the CPU (at the end of the execute phase of the current instruction) and transfer a word into or out of memory, returning to the program after the transfer. The CPU no longer waits for the device to be ready, but software is still required to acknowledge the interrupt and effect the transfer of each word.

Channel I/O (cycle-stealing)

A separate group of logic circuits takes responsibility for data transfers. The CPU sends three parameters to the channel: initial address of block to be transferred, number of words to be transferred, and device name. The channel then acccsses memory by "stealing" control over it from the CPU whenever a data transfer is required by the device. The channel interrupts the CPU program at the end of the block transfer.

Memory hierarchy and organization

Most computers are equipped with auxiliary storage for data and programs in the form of bulk core, disk, drum, or tape. A decision must be made as to which of these media (or which combination of them) is appropriate for the problem at hand; speed of data transfer, speed of access, size of storage available, and cost must all be considered. This subject will be reconsidered in Chapter 12.

5.7 SOFTWARE DESIGN ALTERNATIVES

Software design, like hardware design, must be properly managed. The following are criteria which may be applied to proposed and ongoing software development.

1. What are the objectives of this piece of software? Is it necessary to provide system software to perform this function?
2. Is the function performed efficiently and effectively? Does the software put a minimal load on system resources?
3. Is the software easy to use?
4. Is the development worth the effort? Will the existence of the system software produce a sufficient return to warrant its creation? Will it sell?

   That these criteria are interrelated will become apparent in the following discussion.

Languages

As mentioned previously, the user deals with the computer through language interfaces. The convenience of using a computer is very dependent on the variety and quality of language translators available. In addition to translators for well-known, extensively used procedure- and problem-oriented languages, the manufacturer usually provides an assembly language translator and a system control language translator.

User convenience is an important criterion in translator design, but it is especially so with respect to the system control language. Users react adversely when they have difficulty telling the supervisor program what to do; their reaction is particularly strong when they cannot get the supervisor to perform a task that the hardware is capable of executing. It is therefore desirable that simple and frequently performed tasks correspond to simple statements in the system control language.

If the user community is heterogeneous, providing multiple translators for a single language may be worthwhile. One translator may feature very fast translation time but slow object code execution. Another may specialize in clear, tutorial, diagnostic messages. Still another may produce an object program that makes minimum use of system resources, such as time and memory, but the translator itself requires a lot of memory and/or runs slowly. Some of these attributes might be combined in one translator; perhaps some could be made available as translation-time options.

User services

It is not enough to provide the user with a tool; it is also necessary to explain how to use and, if necessary, to modify or fix that tool. System software must be documented! For a user community holding varying degrees of interest in the workings of the computer system, it is necessary to provide various levels of documentation. There are the novice programmers, the advanced applications-oriented programmers, the teachers, and the software systems programmers. Each requires a different level of detail in documentation.

The manufacturer assumes an obligation when he markets a piece of software—the obligation to make it work. Users will put software through gyrations the manufacturer never dreamed of; users will also attempt to perform very simple straightforward operations. In either situation, the software may fail to perform as expected. It is the manufacturer's responsibility to repair the errors. Modification of documentation may be one type of repair, but correction of programming errors in the system software is more common. Large user installations often employ systems programmers who can make some corrections locally. Many problems will be referred back to the manufacturer, who retains responsibility for repair. Keeping this responsibility in mind, the manufacturer is well advised to maintain extensive documentation on system programs. Such documentation will be of great assistance to customers or to the manufacturer's own systems programmers. To maximize user satisfaction, the manufacturer is advised to make his software "idiot proof," so that a minor error or a deliberate sabotage attempt results only in a no-operation or a warning message instead of a failure.

5.8 WHAT IS ARCHITECTURE?

The following definition of computer architecture has been proposed to the IEEE Technical Committee on Computer Architecture: "Computer architecture is the study and design of algorithms and logical control for the management of the physical resources of a computer system." The job of the computer systems architect is to develop an overall concept of a machine—what it can do and how that solves the problems for which the machine is intended. Just as an architect who designs houses must consider utility, appearance, and compatibility with the neighborhood, so must the computer designer balance requirements, user interface, and costs to make a viable design. The price of a large commercial computer will fall between two and three million dollars, and a hundred or more of them may be sold. The architect, to some extent, takes the final responsibility for the way in which this money is spent, and his systems concept can make the difference between a mediocre machine and an immensely successful one. Many of the necessary decisions can be made logically, but the overall problem is complex enough to require an artistic solution. An architect will be best prepared for this responsibility if his expertise extends over both hardware and software.

REFERENCES

C. Foster, *Computer Architecture*, Van Nostrand Reinhold, 1970.

H. S. Sobel, *Introduction to Digital Computer Design*, Addison-Wesley, 1970.

C. W. Gear, *Computer Organization and Programming*, McGraw-Hill, 1969.

I. Flores, *Computer Organization*, Prentice-Hall, 1969.

H. Hellerman, *Digital Computer System Principles*, McGraw-Hill, 1967.

C. G. Bell and A. Newell, *Computer Structures: Readings and Examples*, McGraw-Hill, 1970.

R. S. Ledley, *Digital Computer and Control Engineering*, McGraw-Hill, 1960.

QUESTIONS

1. What is a trade-off? Give several examples.
2. What constitutes the computer system as seen by the user?
3. What exactly is the interface between the computer and the user? Interpret the term "interface" both narrowly and broadly.
4. Do hardware and software affect each other? If so, how?
5. Is it possible to have a "digital computer" without hardware? Without software? Without firmware?
6. Draw a tree structure relating the hardware and software seen by a variety of computer users.
7. What are the criteria that determine a trade-off?
8. How does the marketplace affect the trade-offs in computer system design?
9. How should a well-designed computer system provide for its users?
10. How is firmware implemented and controlled?
11. Who programs the control store?
12. Explain in general how microprogramming works.
13. Compare the attributes of hardware, software, and firmware implementation of computer system capabilities.
14. Draw a flowchart describing the traditional computer design process.
15. Compare the two philosophies presented for improved hardware design. Do you have a better plan of your own?
16. Is there an optimum size for a computer word? Explain.
17. Compare the implications on word size, execution speed, and programmer convenience of instructions with one, two, and three address fields and with no address field.
18. Discuss the trade-offs you could make when providing a stack in hardware or software.
19. What are the speed, convenience, hardware, and software implications and requirements of multiplexor, interrupt, and cycle-stealing I/O?
20. What is your attitude toward computer makers who sell software and hardware separately?
21. Why are there so many computer languages? What functions do they serve?
22. What services does a computer manufacturer provide to the customer? Are they sufficient? Can you think of additional services?
23. Expand on the similarity of the computer system designer and the architect.

SECTION 2
HOW THE COMPUTER COMPUTES

6
THE HARDWARE IMPLEMENTATION OF THE INSTRUCTION SET

The entire work of the CPU consists of transferring binary numbers (data or instructions) among the various registers of the machine. This process follows the sequences ordered by the commands fetched from memory. The present chapter and the next one give specific details of how these sequences are carried out and, in turn, how this sequence of operations causes computing.

Figure 6.1 represents the typical computer we shall employ as the basis of discussion. Its operation is governed by a *fetch-execute* cycle, which was introduced in Section 2.5. An instruction, stored in main memory as a binary word, is obtained during the fetch phase. During the execute phase the meaning of the instruction is decoded, and the desired operation(s) are carried out.

6.1 INSTRUCTION DECODING

When an instruction is fetched into the *instruction register* (IR), it is interpreted to indicate to the machine what to do during the execute phase. The bits, in groups in the instruction, are each assigned specific functions. Figure 3.1 showed how the actual hardware bits could be grouped in a 36-bit machine; it is reproduced here as Fig. 6.2. Wires from the IR field decoding circuits are connected to other circuits which carry out the functions requested by the contents of that group of bits in the IR. This is not to be construed to mean that the assignment of bits in fields is done "any old way." An appreciable portion of a machine designer's effort goes into making the bit patterns clean, elegant, and therefore easy to implement. This task is all the more important in a microprogrammed machine, where the cost of implementing field organization may be quite high.

The operation field

The operation field contains the op code, which indicates the instruction to be performed. If the field is 8 bits long, as in our example, 256 possible instructions may be specified. A numerical code for the operation is assigned by the machine designer, and it is interpreted, either in hardware or in internal microprogramming, by the *op code decoder*. In this scheme, 33 might represent ADD, 45 LOAD or 28 DIVIDE. The effect of the operation decoder is to cause data to be copied from

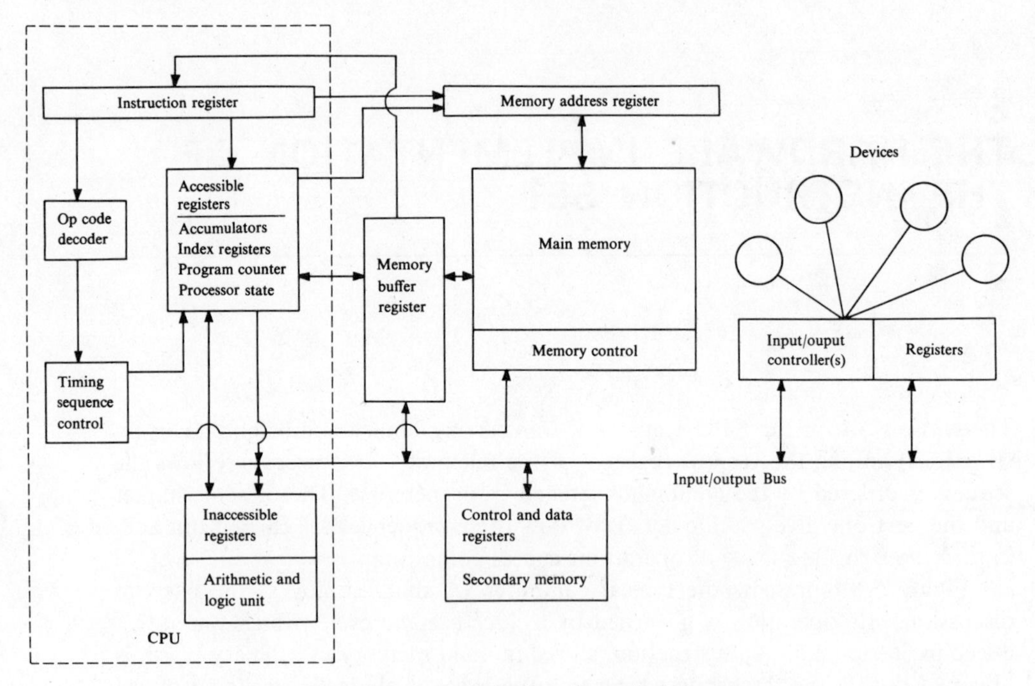

Fig. 6.1. Organization of a typical computer

one location to another, or to cause data to be subjected to a specified manipulation. The desired results are achieved by energizing certain specified circuits and de-energizing others in a well-determined sequence, so that the data are subjected to the transfer and/or operation specified. It is the op code that specifies and the operation decoder that carries out the movement of data among the registers, through the arithmetic unit, between the registers and main memory, and between the registers and the world.

The register field

If the machine contains more than one accumulator or general-purpose register, the register field indicates which register is to be used for the operation called for by the op code. A four-bit field could specify 16 registers. Depending on the op code, the register could be the source and/or destination for the data manipulated by the operation.

6.2 MEMORY REFERENCE INSTRUCTIONS

Whenever information is to be transferred between main memory and the registers of the CPU, some operation is executed which references memory locations. The set of these instructions comprises the *memory reference instructions*. (Nonmemory reference instructions are discussed in the next section.)

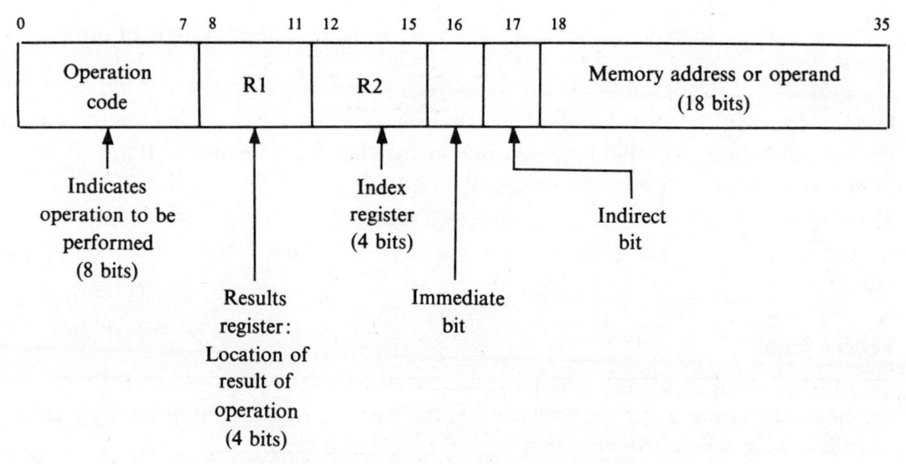

Fig. 6.2. Fields in an instruction for a typical 36-bit machine

The operand address field

A memory reference instruction calls for one word to be transferred from memory to a register, or vice versa. Sometimes another operation (such as subtraction) is done at the same time. The operand address field indicates the memory location from which this word is to be taken or in which the piece of data is to be stored. A typical memory reference instruction,

```
48 7 0 00 03336
```

would mean

```
LOAD ACCUMULATOR (48)
NUMBER 7 (7)
WITH THE CONTENTS OF MEMORY LOCATION 03336 (03336).
```

Figure 6.3 explains the assignment of fields in the description above. Complete

48	7	0	00	03336
Op code (2 hex digits) 8 bits	Register (1 hex digit) 4 bits	Index register (1 hex digit) 4 bits	Immediate (1 bit) Indirect (1 bit)	Operand address (18 bits)

Fig. 6.3. Bit assignments in 36-bit instruction

execution of this instruction requires many steps for the internal logic of the control unit.

Fetch phase—In this portion of the instruction cycle, the machine fetches from main memory the binary number representing the instruction to be done. It knows where to go to get this number because there is a register, the *program counter* (PC), whose purpose is to store the memory address of the next instruction to be done. When the instruction has been retrieved, it is placed in the instruction register. The actual sequence:

Fetch Phase

---

1. Send the contents of the program counter to the memory address register, indicating where the next instruction is to be found.
2. Cause the memory to do a read operation, which gets the actual instruction from main memory into the memory buffer register.
3. Transfer the contents of the memory buffer register into the instruction register, so that the instruction is now in place for execution.

Execute phase—In this portion of the instruction cycle, the machine decodes the binary number representing the desired operation. This procedure causes electrical signals to be generated on certain wires and inhibited on certain others in such a fashion as to actually carry out the programmer's order.

In a memory reference instruction, the operand address field specifies from where in memory the binary data word (involved in the operation) should be fetched (or where it should be stored). The execute phase of a LOAD REGISTER instruction:

Execute Phase

---

4. Decode instruction; find that it is a memory reference instruction.
5. Send operand address portion of instruction from instruction register to memory address register for retrieval of operand.
6. Cause the memory to do a read operation, loading the memory buffer with the contents of the address specified by the memory address register.
7. Transfer the operand from the memory buffer to the desired register.
8. Increment the program counter to prepare for the next instruction.

Note that the PC has been incremented to point to the next sequential instruction.

Alternatively, the PC could have been incremented at the end of the fetch phase instead of the end of the execute phase. As a matter of fact, exactly when the PC is incremented is a question of the detailed design of each machine, and the sequence given here is only an example.

For other memory reference instructions with arithmetic or other operations, the transfer at step 7 would include other circuitry to cause the operation. In general, the data would be routed through the arithmetic and logical unit (ALU) in going from memory to register. The operation decoder would have activated the appropriate circuits in the ALU to perform the operation specified. If there were no arithmetic operation, the ALU would simply pass the data through unmodified.

The index register field

Sometimes it is desirable to not use the operand address directly. If we wanted to do the same computational procedure to each member of a list of numbers, for example, we could use the same set of instructions for each. We would assign a pointer (a register, used as an index) to indicate which member of the list we were working on, as in Sections 3.5 and 3.6. The op code and register fields of the instruction would remain the same. The actual memory location used (as in the example above) would have its address computed as the sum of the operand address field in the instruction register and the contents of the indexing register. The operand address field would then specify the address of the first member of the list to be loaded, and the index register would point in turn to each succeeding member of the list as its contents were incremented, 0, 1, 2, ... A four-bit index register field could specify 15 index registers; a zero in the field, by convention, means no indexing.

The above example, rewritten for indexing, would read

```
48  7  3  00  03336
```

and would mean

```
ADD 3336 TO THE CONTENTS OF REGISTER 3
USE THAT AS A MEMORY ADDRESS, AND LOAD
THE CONTENTS OF THAT ADDRESS INTO REGISTER 7.
```

If REGISTER 3 contained 11, the actual address from which data were taken would be 3347. This address, the one actually arrived at after the completion of all address arithmetic, is called the *effective address*. Since the effective address is the arithmetic sum of the operand address and the index, it is proper (and indeed common) to have an operand address of zero or some other constant and to make the index point directly to the address desired.

The internal sequence of the execute phase is slightly modified when index register address modification is employed.

Execute phase with indexing

---

4. Decode instruction; find that it is a memory reference instruction.
5. a) Send operand address portion of instruction from instruction register to adder.
   b) Send contents of specified index register to adder.
   c) Perform summation, yielding effective address.
   d) Transfer effective address to memory address register for retrieval of operand.
6. Cause the memory to do a read operation, loading the memory buffer with the contents of the address specified by the memory address register.
7. Transfer the operand from the memory buffer to the desired register.
8. Increment the program counter to prepare for the next instruction.

Note that the discussion above assumed that the set of accumulators and the set of index registers are the same. This is often so, but it need not be.

Indirect field

Computation of the effective address may involve not just arithmetic but other memory references. Some machines can look at another location in memory for the effective address. Then the operand is *the address, in memory, of the effective address*. This procedure, called *indirect addressing*, is described in Section 3.5 and Fig. 3.3. In the same example

48 7 0 01 03336

the 01 would indicate indirect addressing. Since indirect addressing is a simple yes or no proposition, the indirect addressing field consists of a single bit. The number 03336 would be read from the instruction register and, instead of being loaded into REGISTER 7 would be loaded back into MAR to be used as the actual data-accessing location. If location 03336 contained 1023, REGISTER 7 would be loaded with the contents of memory location 1023. In some machines, this chain of indirectness may be carried beyond the first indirect reference and may even extend without limit until it reaches a command which does not specify indirect addressing.

The sequence of step 5 in the execute phase is again modified when indirect addressing is specified.

Execute phase with indirect addressing

---

4. Decode instruction; find that it is a memory reference instruction.

5. a) Send operand address portion of instruction register to memory address register to retrieve indirect address.
   b) Cause the memory to do a read operation, obtaining the address to be used as the effective address in the memory buffer register.
   c) Send operand address portion of memory buffer register to the memory address register.
6. Cause the memory to do a read operation, loading the memory buffer with the contents of the address specified by the memory address register.
7. Transfer the operand from the memory buffer to the desired register.
8. Increment the program counter to prepare for the next instruction.

It is further possible to specify both index register address modification and indirect addressing. When both are specified, index register address modification may occur before or after indirect addressing. We will discuss the case where indexing occurs before indirect addressing, called *pre-indexing*. The execute phase now becomes:

Execute phase with pre-indexing and indirect addressing

---

4. Decode instruction; find that it is a memory reference instruction.
5. a) Send operand address portion of instruction from instruction register to adder.
   b) Send contents of specified index register to adder.
   c) Perform summation, yielding effective address.
   d) Send effective address calculated by index register address modification to memory address register to retrieve indirect address.
   e) Cause the memory to do a read operation, obtaining the address to be used as the effective address in the memory buffer register. Send operand address portion of memory buffer register to memory address register for retrieval of operand.
6. Cause the memory to do a read operation, loading the memory buffer with the contents of the address specified by the memory address register.
7. Transfer the operand from the memory buffer to the desired register.
8. Increment the program counter to prepare for the next instruction.

If the contents of the memory word read in step 5(e) specifies indirect addressing (and if the machine organization is constructed to permit it), steps 5(d) and 5(e) may be repeated as many times as specified until a nonindirect address is reached.

6.3 NONMEMORY REFERENCE INSTRUCTIONS

Many instructions do not reference memory. They are used to manipulate or modify registers, store constants, test the results of arithmetic operations, and change the flow of control of the program.

Register-to-register instructions

Some instructions transfer data between registers and in addition may, at the same time, perform such operations as addition, clearing portions, shifting, etc. They may also be used to manipulate just one register. The whole machine word may be broken into portions—halves, quarters, sixths, or whatever—which can be treated separately. One register-to-register instruction might clear one half of a word, complement the other half, and move the entire result into another register. In some machines, instruction storage space is saved. Since this type of instruction does not require an operand address field, it might be possible to store two such instructions in one 36-bit word.

An 18-bit register-to-register instruction format is shown in Fig. 6.4. The 18-bit short instruction may, in fact, be used for instructions in other than a register-to-register form. A typical example would be an instruction to increment the contents of a register.

Immediates

Many of the programming functions carried out require the use of constants. They may be used to increment a register, mask a portion of some field, or indicate the number of an input/output device. Many early computers require that such constants be stored separately and that their memory addresses be referenced in the instruction stream. In an *immediate* operation (see Section 3.4), the constant is stored in the operand address field. It is necessary to specify that the contents of the operand address field is the immediate constant. Another field might be devoted to specification of an immediate, as shown below. Our previous example now reads

48 7 0 10 03337

The number 03337 itself is loaded into REGISTER 7. Sometimes a different op code is used to indicate an immediate rather than a separate field.

The instruction sequence for the execute phase would now read:

Execute phase; immediate addressing

---

4. Decode instruction; find that it is an immediate.
5. Load contents of operand address field into specified register.
6. Increment the program counter.

It is most important to note that immediate operands come directly from the in-

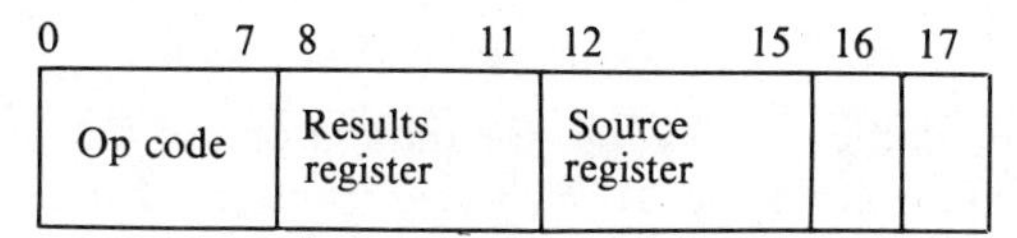

Fig. 6.4 18-Bit short instruction

struction; because of the increased convenience, that fact is the main reason such instructions are included in typical instruction sets.

Jumps

In order to make it possible to transfer control of the program to an instruction other than the one next in memory sequences, we have *jump* instructions. The operand address field contains the memory location from which the next instruction is to be taken. This number is transferred by the control unit from the IR to the PC. The normal incrementation of the PC at the end of the execute phase is suppressed, or the transfer occurs after the incrementation. The beginning of the next fetch phase finds the jump address in the PC, as required, and the next instruction is taken from there. The contents of the operand address field may be modified in the usual way in indexing and indirect addressing, so that tables of possible jump addresses may be constructed and used. It is also possible to have the jump done or not done, depending on the result of some previous operation, such as an arithmetic comparison. In this way, we may test the progress of our program periodically and change the flow or terminate it, depending on the results of the computations.

The sequence of operations is therefore:

Execute phase; jump instructor

---

4. Decode instruction; find that it is a jump.
5. Send operand address portion of instruction from instruction register to memory address register for retrieval of jump address. (Indexing and indirect addressing are also possible here.)
6. Cause the memory to do a read operation, loading the memory buffer register with the contents of memory at the effective address.
7. Transfer the jump address from the memory buffer to the program counter (perhaps conditional on some other fact).
8. Do not increment the program counter if it was just loaded.

As mentioned in Section 5.6, when there is no program counter, a field may be devoted to specifying the address of the next instruction to be executed.

6.4 INPUT/OUTPUT

Transfer of data to and from devices external to the CPU involves the programmer deeply in the internal structure of the particular I/O device used. However, there are two basic types of I/O instruction sets.

Single-word transfers

Instructions are available which cause the data in one of the general-purpose registers, or in one main memory location, to be transferred to an external device for output. Similar instructions retrieve data from the outside world and store them in memory. These instructions must be executed *by the CPU* for every single word to be transferred. In some machines, the external device for input/output is also specified in the same instruction.

Control and status

Before a single-word or other data transfer takes place, the device (which has its own cycle time, asynchronous with the computer) must be tested to see if it is ready to accept or transmit the next word. With the same instruction, other tests, such as for broken tape or an empty card hopper, may be made. It is also necessary to send control information to the device. These commands might cause a tape to rewind, a disk to search for a given track, or a printer to advance to the next page. *Status and control instructions* are available for such functions. In machines which do not have device selection fields in the data transfer instruction, control instructions are used to select and activate a device before the transfer takes place.

Block transfers

It is very inefficient for the CPU to fetch every word that is needed for transmission to an I/O device. The CPU must first test the device for ready and then fetch the word and transmit it. If many consecutive words must be transferred, as is usually true, the CPU can be relieved of this low-level task. Special hardware is available to perform this function with minimal interference to the CPU.

A block transfer device, usually called a "channel," is given an initial address (the first address of the block to be transferred), and a word count (the number of words in the block). It then carries out the transfer, including the fetching of each word in the block, without interfering with normal computation. The CPU is interrupted only after the entire transfer is finished. Of course, the channel must do memory read or store operations in the process of transferring the data. It will interrupt CPU memory fetches whenever it needs access; this process is called *cycle-stealing*. There is a "reason" for cycle-stealing. Since the I/O devices must be used almost continuously to be effective at all, it makes little sense to interrupt them when a very small drop in CPU performance (the lost cycle) won't hurt it quite so much. Some older reasons (historically) are that rotating devices and moving devices in general have inertia and thus tend to resist being stopped and started very abruptly; in other words, once you get the tape drive going, you had better keep it going. Priority of access is controlled by the designer or, in some machines, by the console operator. Normally, there is enough time between CPU memory fetches to allow cycle-stealing with very little interference with the main program. In any case, the conflicts are resolved by the hardware and are not apparent to the programmer.

REFERENCES

C. W. Gear, *Computer Organization and Programming*, McGraw-Hill, 1969.

I. Flores, *Computer Organization*, Prentice-Hall, 1969.

H. Hellerman, *Digital Computer System Principles*, McGraw-Hill, 1967.

Burroughs Corporation, *Digital Computer Principles*, McGraw-Hill, 1969.

Y. Chu, *Digital Computer Design Fundamentals*, McGraw-Hill, 1962.

P. Wegner, *Programming Languages, Information Structures, and Machine Organization*, McGraw-Hill, 1968.

H. S. Sobel, *Introduction to Digital Computer Design*, Addison-Wesley, 1970.

C. G. Bell and A. Newell, *Computer Structures: Readings and Examples*, McGraw-Hill, 1970.

J. K. Iliffe, *Basic Machine Principles*, American Elsevier, 1968.

QUESTIONS

1. What is the advantage of thinking of machine instructions as binary numbers? What other bases are employed for human convenience?
2. The fetch-execute cycle has been described as a two-phase cycle. In an attempt to speed up execution, some computers will overlap the fetch of one cycle with the execute of the previous cycle. What complications or additional features does this procedure introduce in the hardware and in the internal steps of the cycle?
3. Why is instruction decoding necessary?
4. How does the size of the instruction register relate to the word size in the machine?
5. What are the relative advantages of incrementing the program counter before or after the execute phase. What are the implications for jump and subroutine jump instructions?
6. As mentioned, it is possible to have indexing follow indirect addressing. This is called *post-indexing*. Write out the sequencing steps for the execute phase with post-indexing.
7. Explain the effect of an indexed indirect jump by listing the sequence followed during the execute phase.
8. List the sequence followed during the execute phase of a STORE REGISTER instruction.
9. In the process of cycle-stealing by an input/output channel, where must the channel put the memory address of the data it wishes to get?
10. A main memory has 8192 words of storage (sometimes called 8K), each word being 36 bits long.
    a) How many bits does the entire memory hold?
    b) How many bits of *register* storage are required for the MAR?
    c) For the MBR?

11. The IBM 650, one of the first commercial computers, had a drum memory, which is not a random-access device. Each word is written on a rotating drum, and the access time is the time required for the drum to get around to the read station. The 650 did not have a program counter. Each instruction contained the explicit address of the next instruction. Why does this organization produce much faster-running programs for a drum machine?

7
ASSEMBLY LANGUAGE

7.1 GENERALIZED ASSEMBLY LANGUAGE—INTRODUCTION

Since assembly language is intimately connected with the hardware organization of the machine, one should not expect the same assembly language commands to exist on different machines. However, there is a large set of commands that is common to many machines. It is with this set of common assembly language commands that we will concern ourselves. The specific additional instructions for any particular machine should be easily mastered once the basics have been learned.

One of the features of any practical assembly language is its use of abbreviations or mnemonics for the operations to be performed. Since the memorization of the mnemonic codes for any given machine requires a considerable amount of effort, we will use no abbreviations at all for the operations we will discuss. Each operation will be spelled out in complete detail. By avoiding the use of mnemonic abbreviations, we hope to make it possible for the student to easily understand the assembly language operations used in this text, as well as to make it easy for the student to transfer this knowledge to the assembly language for the machine he will have to program.

7.2 NOTATION USED TO DESCRIBE THE EFFECT OF AN ASSEMBLY COMMAND

Although there is not much specialized notation for describing the operation of assembly language, what does exist serves a very useful purpose. A location is *symbolically referenced* by the use of the name of that location or by relative addressing with that name. The *contents of a memory location*, however, are denoted by enclosing the symbolic name in parentheses. For instance, let us assume that SAM is the symbolic name referring to address 3176. The contents of memory at location 3176 are denoted by (SAM). If we wish to denote only some bits of the memory location's contents, we may attach a numerical subscript on the parentheses which will consist of one number, a dash, and another number. This will indicate the bits of the memory location that are to be used. For instance, $(\text{SAM})_{0-5}$ means bits 0 through 5 inclusive of the contents of the location symbolically referenced as SAM. A subscript such as $_{addr}$ means the entire operand address field, regardless of which bits it encompasses. An arrow is used as a *replacement operator* to indicate the direction of data transfer. For instance,

$$A + B \rightarrow (C)$$

means: add A to B, and store the sum in the location symbolically referenced as C. If we write

$$(A) + (B) \rightarrow (C)$$

it means: add the contents of the locations symbolically referenced as A and B, and store the sum in the location symbolically referenced as C.

Arithmetic operations will be indicated by the symbols $+$, $-$, $\cdot$, $\div$. The *logical operations* **AND, OR, EXCLUSIVE OR,** and **EQUIVALENCE** will be designated in boldface. The equals sign will be used only in a *test for equality*. Vertical bars denote *absolute values*. Finally, if a subscript symbol is used on a symbolic memory location reference, the subscript will describe an index register being used for *address modification*, as discussed in Chapter 3.

It is frequently necessary to differentiate between the *symbolic operand address* specified in the operand address field and the *effective address* formed by indexing or indirect addressing. The term addr will be used for the operand address as written in the instruction, and ADDR will be used for the result of the address arithmetic. For example, with indexing,

$$\text{addr} + (\text{index register}) = \text{ADDR}.$$

With indirect addressing,

$$(\text{addr}) = \text{ADDR}.$$

With pre-indexed indirect addressing,

$$\big(\text{addr} + (\text{index register})\big) = \text{ADDR}.$$

With post-indexed indirect addressing,

$$(\text{addr}) + (\text{index register}) = \text{ADDR}.$$

The operand address field may optionally contain a value to be used as an immediate or a literal. If an operand address is specified, that address may be used as a direct or indirect address. Direct addressing is the normal mode; when we mean something else we must specify our intentions. We may append one of the qualifiers IMMEDIATE, INDIRECT, or LITERAL to any instruction to specify the addressing mode.

Several examples and comments are in order. Note that the address pointed to by the replacement operator (arrow) is always enclosed in parentheses, indicating that the replacement is to the contents of the specified location. It would be quite meaningless to change the address. To slightly reduce the number of characters to be written, we could adopt a convention omitting these parentheses. We do not! To maximize clarity, we will always use enclosing parentheses to mean "contents of." For example,

VALUE

symbolically refers to an address using the name VALUE, whereas

(VALUE)

means the contents of the location whose address is symbolically referred to as VALUE. The form

((VALUE))

specifies one level of indirect addressing; that is to say, the contents of the location specified by the contents of the location symbolically referred to as VALUE. If VALUE corresponded to memory location 100, then

500 → (VALUE)

We would store 500 in location 100. Note that 500 is a literal (or immediate), not an address. If location 500 contained 631 and we executed

((VALUE)) → (VALUE)

we would store 631 in location 100.

We will use the notation XR for index register. Now consider the following illustrative sequence.

Notation	Meaning
5 → (XR1)	Replace the contents of XR1 with the number 5.
500 + (XR1) → (LOC)	Replace the contents of the location symbolically referred to as LOC by 500 + the contents of XR1 (which is 5). The result is 505 stored in LOC.
(500 + (XR1)) → (LOC)	Replace the contents of LOC by the contents of location 505.
((500 + (XR1))) → (LOC)	Replace the contents of LOC by the contents of the location specified in location 505.

If we consider RAY to be the symbolic name associated with the beginning (zeroth element) of an array, then we can use an index register to relatively address any element in the array. For example,

Notation	Meaning
(N) → (XR2)	Replace the contents of XR2 with the number stored in the location referred to as N.
(RAY + (XR2)) → (REGISTER1)	Replace the contents of REGISTER1 by the contents of the Nth location following RAY.

You will note from this example that in assembly language it is convenient to index the elements of an array beginning with zero.

7.3 REGISTER REFERENCE INSTRUCTIONS

The most basic set of assembly language instructions is that which directly references registers. These instructions serve to transfer information between register and primary storage and among the registers themselves. By definition, the *register reference instructions* deal only with the addressable registers; the registers which have designated hardware functions are not available to the programmer. The inaccessible registers include the memory buffer register, the memory address register, the instruction register, etc. The available commands follow.

Operation code	Description of operation
LOAD REGISTER	(ADDR) → (REGISTER)
LOAD NEGATIVE REGISTER	– (ADDR) → (REGISTER)
LOAD MAGNITUDE REGISTER	| (ADDR) | → (REGISTER)
STORE REGISTER	(REGISTER) → (ADDR)
STORE ADDRESS	$(\text{REGISTER})_{\text{addr}} \rightarrow (\text{ADDR})_{\text{addr}}$

Since these are the first instructions we have introduced, we cannot present examples that form a coherent program. However, the following examples show disconnected uses of register reference instructions.

Label field	Operation field	Register designation field	Operand address field	Index register designation	Comments
	LOAD REGISTER	REGISTER0	ALPHA		(ALPHA) → (REGISTER0)
	LOAD MAGNITUDE REGISTER	REGISTER1	MARY		| (MARY) | → (REGISTER1)
	LOAD NEGATIVE REGISTER	REGISTER2	X	XR1	– (X + (XR1)) → (REGISTER2)
	STORE REGISTER	REGISTER1	Y		(REGISTER1) → (Y)
	STORE REGISTER	REGISTER0	BETA	XR2	(REGISTER 0) → (BETA + (XR2))
	LOAD REGISTER	XR3	GAMMA		(GAMMA) → (XR3)

Label field	Operation field	Register designation field	Operand address field	Index register designation	Comments
	LOAD REGISTER	REGISTER3	MOE	XR3	(MOE + (XR3)) → (REGISTER3)
	STORE ADDRESS	REGISTER3	Z		$(\text{REGISTER3})_{\text{addr}} \rightarrow (\text{Z})_{\text{addr}}$
	LOAD REGISTER, INDIRECT	REGISTER0	ALPHA		((ALPHA)) → (REGISTER0)

7.4 INTEGER ARITHMETIC INSTRUCTIONS

The simplest *arithmetic operations* are those involving integers. These operations take place in the registers.

Operation code	Description of operation
INTEGER ADD TO REGISTER	(REGISTER) + (ADDR) → (REGISTER)
INTEGER SUBTRACT FROM REGISTER	(REGISTER) − (ADDR) → (REGISTER)
INTEGER MULTIPLY IN REGISTER	(REGISTER) · (ADDR) → (REGISTER)
INTEGER DIVIDE REGISTER	(REGISTER) ÷ (ADDR) → (REGISTER)

Combining integer arithmetic instructions and register reference instructions in our example, we can produce a segment of assembly language coding that possesses a minimal meaning. It must be assumed that the association of symbolic labels with memory locations has been provided elsewhere.

Label field	Operation field	Register designation field	Operand address field	Index register designation	Comments
	LOAD REGISTER	REGISTER1	X		(X) → (REGISTER1)
	LOAD REGISTER	XR1	REGISTER1		(REGISTER1) → (XR1)
	INTEGER ADD TO REGISTER	XR1	Y		(XR1) + (Y) → (XR1)
	INTEGER MULTIPLY IN REGISTER	REGISTER1	Z		(REGISTER1) · (Z) → (REGISTER1)

Label field	Operation field	Register designation field	Operand address field	Index register designation	Comments
	INTEGER DIVIDE REGISTER	REGISTER1	W		(REGISTER1) ÷ (W) → (REGISTER1)
	STORE REGISTER	REGISTER1	A	XR1	(REGISTER1) → (A + (XR1))
	INTEGER ADD TO REGISTER, IMMEDIATE	REGISTER1	5		(REGISTER1) + 5 → (REGISTER1)
	STORE REGISTER	REGISTER1	X		(REGISTER1) → (X)

7.5 INTEGER DOUBLE-PRECISION ARITHMETIC

In the examples above for multiplication and division, the product and the dividend were both small enough to fit into a single register. In general, this will not be the case. If only a single register were employed, the result would be truncated. Using decimal notation to illustrate, we have

$$\begin{array}{r} 546 \\ \times\ 772 \\ \hline 421512 \end{array} \quad \text{and} \quad 772\overline{)421515}\ \ \text{546 Remainder 3}$$

We need twice as many digits for the product and the dividend as for any of the other operands. In division, we need a place for both the quotient and the remainder. In either case, we can use two registers to hold the double-sized number. To meet this need, we have available *double-precision instructions*. In our example, register i holds the more significant portion (left-hand half), and register $i + 1$ holds the less significant (right-hand half). Only REGISTER or ADDR are named in the appropriate field of the instruction.

Operation code	Description of operation
DOUBLE MULTIPLY IN REGISTER	(REGISTER) · (ADDR) → (REGISTER i, REGISTER i + 1)
DOUBLE DIVIDE REGISTER	(REGISTER i, REGISTER i + 1) ÷ (ADDR) → (REGISTER i) [QUOTIENT] (REGISTER i + 1) [REMAINDER]

We can also add and subtract these numbers directly.

Operation code	Description of operation
DOUBLE LOAD REGISTER	(ADDR, ADDR + 1) → (REGISTER i, REGISTER i + 1)
DOUBLE ADD TO REGISTER	(REGISTER i, REGISTER i + 1) + (ADDR, ADDR + 1) → (REGISTER i, REGISTER i + 1)
DOUBLE SUBTRACT FROM REGISTER	(REGISTER i, REGISTER i + 1) − (ADDR, ADDR + 1) → (REGISTER i, REGISTER i + 1)
DOUBLE STORE REGISTER	(REGISTER i, REGISTER i + 1) → (ADDR, ADDR + 1)

The following example illustrates the use of double-precision integer arithmetic.

Label field	Operation field	Register designation field	Operand address field	Index register designation	Comments
	LOAD REGISTER	REGISTER1	X		(X) → (REGISTER1)
	DOUBLE MULTIPLY IN REGISTER	REGISTER1	Y		(REGISTER1) · (Y) → (REGISTER1, REGISTER2)
	DOUBLE ADD TO REGISTER	REGISTER1	Z		(REGISTER1, REGISTER2) + (Z, Z+1) → (REGISTER1, REGISTER2)
	DOUBLE STORE REGISTER	REGISTER1	W		(REGISTER1) → (W) (REGISTER2) → (W + 1)

7.6 FLOATING-POINT ARITHMETIC INSTRUCTIONS

To be useful for scientific and technical calculations, a computer must have a set of commands for performing arithmetic on floating-point numbers. The fastest way is to have these commands executed by hardware. On many small to medium-sized machines, floating-point hardware is an extra-cost option. When the hardware is not purchased, the programmer still has the *floating-point operations* available to him, but they are performed by software using integer arithmetic and shift instructions. Corresponding to the integer arithmetic instructions, floating-point arithmetic instructions follow.

Operation code	Description of operation
FLOATING ADD TO REGISTER	(REGISTER) + (ADDR) → (REGISTER)
FLOATING SUBTRACT FROM REGISTER	(REGISTER) − (ADDR) → (REGISTER)
FLOATING MULTIPLY IN REGISTER	(REGISTER) · (ADDR) → (REGISTER)
FLOATING DIVIDE REGISTER	(REGISTER) ÷ (ADDR) → (REGISTER)

You will note that in the description of operation there is no distinction between integer and floating-point arithmetic instructions, but the results they produce are quite different and are dependent on the representation of integer and floating-point numbers within the computer.

The following example determines the arithmetic mean of three numbers stored in locations symbolically referenced as X, Y, and Z. The mean is stored in the location as MEAN.

Label field	Operation field	Register designation field	Operand address field	Index register designation	Comments
	LOAD† REGISTER	REGISTER1	X		(X) → (REGISTER1)
	FLOATING ADD TO REGISTER	REGISTER1	Y		(REGISTER1) + (Y) → (REGISTER1)
	FLOATING ADD TO REGISTER	REGISTER1	Z		(REGISTER1) + (Z) → (REGISTER1)
					REGISTER1 contains X + Y + Z
	FLOATING DIVIDE REGISTER, LITERAL	REGISTER1	3.		(REGISTER1) ÷ 3. → (REGISTER1)
	STORE† REGISTER	REGISTER1	MEAN		(REGISTER1) → (MEAN)

7.7 SIMPLE INSTRUCTIONS AFFECTING THE PROGRAM COUNTER

As discussed in Chapter 4, the program counter is a nonaddressable register‡ which contains the location (address) of the next instruction to be executed. The program

† Double load and store instructions may be needed if the floating-point format uses two words and two registers.

‡ Nonaddressable means only that it may not be specified in the register designation field.

normally executes instructions stored in consecutive locations. When you are writing programs, however, it is frequently convenient and sometimes necessary to execute nonconsecutive instructions.

Directing the computer to execute an instruction other than the next sequential one is accomplished by changing the contents of the program counter, thereby specifying the next instruction to be executed. Conceptually, we may consider that control has been *transferred* to another location, that execution has *branched* to another address, or that flow has *jumped* to a specified instruction. The terms "transfer," "branch," and "jump" are all used to designate instructions which affect the program counter.

The unconditional jump

The simplest jump, known as unconditional, causes the effective address to be loaded into the program counter. At the beginning of the next fetch phase, the program counter is read and its contents are used as the address from which to fetch the next instruction.

Operation code	Description of operation
JUMP TO ADDR	ADDR → (PC)
JUMP TO ADDR, INDIRECT†	(ADDR) → (PC)

The conditional jump

Program flexibility is achieved by using the *conditional jump*, in which a specific comparison test is made. If the designated condition exists, the jump takes place; if it does not exist, the next sequential instruction is executed.

Operation code	Description of operation
JUMP IF REGISTER ZERO	IF (REGISTER) = 0, ADDR → (PC) ‡
JUMP IF REGISTER NONZERO	IF (REGISTER) ≠ 0, ADDR → (PC) ‡
JUMP IF REGISTER POSITIVE	IF [sign bit] = 0, ADDR → (PC) ‡
JUMP IF REGISTER NEGATIVE	IF [sign bit] = 1, ADDR → (PC) ‡

Note that the tests for positive and negative are concerned only with the most significant bit, the sign bit. While completely logical and internally consistent, a test of the sign bit treats positive zero as a positive number and negative zero as a negative number. Positive zero is a word of all zero bits. When sign-magnitude is employed to represent negative numbers, negative zero is a word wherein the most significant bit is 1 and all other bits are 0. When 1's complement is employed,

† In some small computers, where main memory is organized in sectors or areas, this instruction is also used to change to a different sector.

‡ Otherwise (PC + 1) → (PC) in all cases.

negative zero is a word of all 1's. In 2's complement there is no negative zero. You must take these specifications into account when programming a sign test.

A very simple example of the use of a conditional jump is the following sequence, which converts a negative zero to a positive zero. This program will work for all three methods of representing negative numbers.

Label field	Operation field	Register designation field	Operand address field	Index register designation	Comments
	LOAD REGISTER	REGISTER3	NUMB		(NUMB) → (REGISTER3)
	JUMP IF REGISTER NONZERO	REGISTER3	Present Location +4		No need to change sign
	JUMP IF REGISTER POSITIVE	REGISTER3	Present Location +3		No need to change sign
	INTEGER MULTIPLY IN REGISTER, LITERAL	REGISTER3	−1		(REGISTER3) · −1 → (REGISTER3)
	STORE REGISTER	REGISTER3	NUMB		(REGISTER3) → (NUMB)

Loops using the conditional jump

With the conditional jump instructions it is possible to build *loops*, which will repetitively execute a series of instructions, using a register to count the number of executions. The register used for counting loop executions may also be used as an index register for address modification. The direct enumeration employed above in the averaging of three numbers is inefficient when the number of memory locations to be addressed from within the loop is large. Instead, the data are stored in consecutive locations and treated as an array. An index register may then be employed to step through the array.

We will illustrate the use of looping and index register address modification to evaluate the following expression.

$$\text{AVG} = \frac{1}{N} \sum_{K=0}^{N-1} \text{ITEM}_K$$

We will assume that the zeroth element of data is stored in the location whose symbolic address is ITEM, the first element of data is stored in the location which may be symbolically addressed as ITEM + 1, etc. In assembly language, it is most common to write counting loops which run from N − 1 to 0, as in the following example. If you were to count from 1 to N, as in FORTRAN or most other higher-level languages, the instruction to test for completion of the loop would have to

Label field	Operation field	Register designation field	Operand address field	Index register designation	Comments
	LOAD REGISTER	XR1	N		Initialize XR1
	LOAD REGISTER, IMMEDIATE	REGISTER1	0		Set sum to zero
	INTEGER SUBTRACT FROM REGISTER, IMMEDIATE	XR1	1		(XR1) − 1 → (XR1)
	FLOATING ADD TO REGISTER	REGISTER1	ITEM	XR1	Sum + $ITEM_K$ → Sum
	JUMP IF REGISTER POSITIVE	XR1	Present location −2		Continue looping ?
	FLOATING DIVIDE REGISTER	REGISTER1	N		$\frac{1}{N}\sum_{K=0}^{N-1} ITEM_K$
	STORE REGISTER	REGISTER1	AVG		

test for N. Such a test might take two or three operations. A test for zero takes only one. This segment of assembly language code is represented by the flowchart of Fig. 7.1.

Looping instructions

Looping instructions are a subset of those instructions which affect the program counter. The looping capability permits the performance of a frequently used operation in a minimum number of instructions. Our general language has two looping instructions. One requires that the register be preset to a positive value, and it decrements the contents of the register to zero. The other requires that an upper limit be loaded into a register, which is used for comparison; the loop index then counts up to this limit. In the second instruction, both a general-purpose register and an index register are specified as part of the test. Therefore, only the contents of the operand address field are available to contain the symbolic location to which control is to be passed if the test is successful. The general form of the loop is shown in Fig. 7.2.

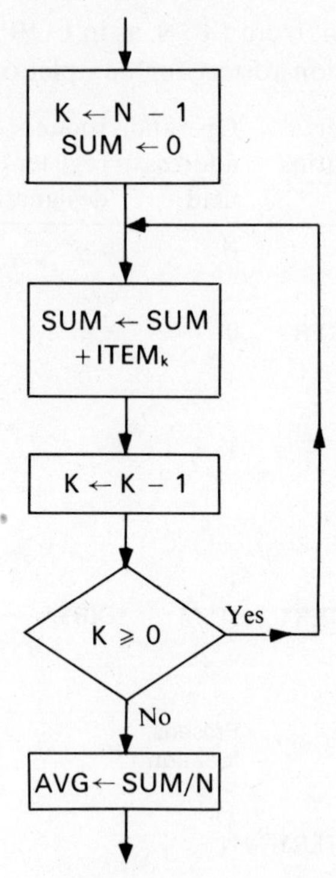

Fig. 7.1. Flowchart of $\text{AVG} = \frac{1}{N}\sum_{K=0}^{N-1} \text{ITEM}_K$

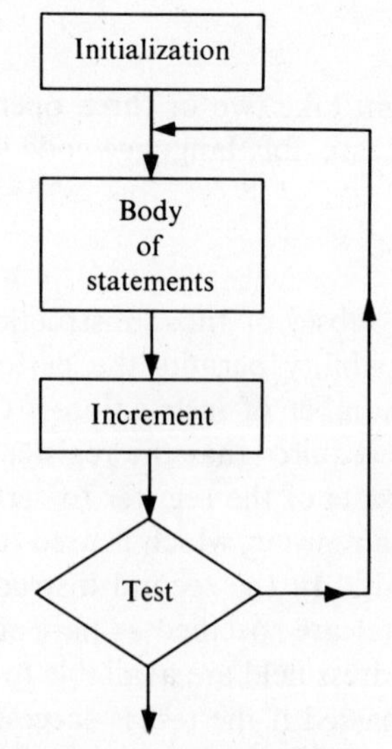

Fig. 7.2. A programmed loop

Operation code	Description of Operation
DECREMENT REGISTER JUMP IF NONNEGATIVE	(REGISTER) − 1 → (REGISTER) IF [sign bit] = 0, ADDR → (PC) †
INCREMENT INDEX REGISTER JUMP IF LESS THAN REGISTER	(XR) + 1 → (XR). IF (XR) < (REGISTER), addr → (PC) †

The previous example to evaluate

$$\text{AVG} = \frac{1}{\text{N}} \sum_{\text{K}=0}^{\text{N}-1} \text{ITEM}_{\text{K}}$$

may be recoded to use the looping instruction.

Label field	Operation field	Register designation field	Operand address field	Index register designation	Comments
	LOAD REGISTER	XR1	N		
	INTEGER SUBTRACT FROM REGISTER, IMMEDIATE	XR1	1		
	LOAD REGISTER, IMMEDIATE	REGISTER1	0		
	FLOATING ADD TO REGISTER	REGISTER1	ITEM	XR1	
	DECREMENT REGISTER JUMP IF NON-NEGATIVE	XR1	Present location −1		
	FLOATING DIVIDE REGISTER	REGISTER1	N		
	STORE REGISTER	REGISTER1	AVG		

As an example of the counting-up loop, let us search an ordered list, LIST, of N elements for the first entry greater than 999 and store this value in the location symbolically addressed as VALUE. If there is no such entry, the value −1 is to be stored in VALUE.

† Otherwise (PC) + 1 → (PC) in both cases.

Label field	Operation field	Register designation field	Operand address field	Index register designation	Comments
	LOAD REGISTER	REGISTER2	N		
	INTEGER SUBTRACT FROM REGISTER, IMMEDIATE	REGISTER2	1		Upper limit
	LOAD REGISTER, IMMEDIATE	XR1	0		
LOOP	LOAD REGISTER	REGISTER1	LIST	XR1	Get LIST entry (LIST + (XR1)) → (REGISTER1)
	INTEGER SUBTRACT FROM REGISTER, IMMEDIATE	REGISTER1	999		Compare with 999 by subtracting
	JUMP IF REGISTER ZERO	REGISTER1	Present location +2		If not positive (If zero or negative) Do LOOP instruction
	JUMP IF REGISTER POSITIVE	REGISTER1	GREAT		Test successful; number greater than 999. Jump out of loop
	INCREMENT INDEX REGISTER JUMP IF LESS THAN REGISTER	REGISTER2	LOOP	XR1	
	LOAD REGISTER, LITERAL	REGISTER1	−1.		No such entry Put −1. in (VALUE)
	JUMP TO ADDR		Present location +2		Jump around success
GREAT	LOAD REGISTER	REGISTER1	LIST	XR1	(LIST + (XR 1)) → (VALUE)
	STORE REGISTER	REGISTER1	VALUE		

7.8 SUBROUTINES

Another instruction that changes the contents of the program counter is the *subroutine jump*, which allows the use of *closed subroutines*. With a closed subroutine, a single block of code used for some specific, well-defined job (for example, a square root) can be jumped to from many places. Any program can be jumped to, but when the jump is to a closed subroutine, a path back to the *calling program* (the one jumped from) is saved. After the instructions constituting the subroutine are executed, control returns to the calling program at the instruction immediately following the subroutine jump. This means that the same subroutine may be used by many parts of the same calling program without duplicating the code in memory and without confusing the flow of control.

In addition to a convention for saving a return path, a way is needed to pass data or other variable information between the calling program and the subroutine. These data are called *arguments* or *parameters*.

There are three common techniques used for saving the return path.

1. Return address is in a register.
2. Return address is in the called subroutine.
3. Return address is in a stack.

It is possible, of course, to implement any of these in software. The frequency with which the function is used, however, strongly favors having hardware to do the job.

In all three techniques, the return address stored is actually† (PC) + 1, so that the return path pointer points to the instruction after the subroutine jump.

Return address in a register

A register is specified in the instruction showing where to save the return. Control then passes to the first instruction of the subroutine, at the address given in the jump instruction. Control is returned to the calling program by executing a JUMP INDIRECT, REGISTER instruction, which loads the PC with the contents of the specified register. A programmer may specify any register when he writes both the calling and called routines. When he uses other routines, such as ones from the library, some convention must be established indicating where the return is to be stored. We shall use XR7 in the examples shown here. The applicable instruction is

Operation code	Description of operation
STORE PC AND JUMP	(PC) + 1 → (REGISTER), ADDR → (PC)

Return address in the called subroutine

The first location of the subroutine text itself is reserved for storage of the return. The programmer is usually responsible for leaving a space. Execution of the sub-

† Whether (PC) or (PC) + 1 is stored depends on when PC is incremented.

routine begins with the second location. No convention for registers, etc., is needed since each subroutine has a separate place for storing the path back to the calling program. There are three alternatives for hardware implementation of this type of subroutine linkage. (1) The address of the next instruction of the calling program is placed into this location. Return is accomplished by an indirect jump via this location. (2) The hardware places the binary code corresponding to an unconditional jump to the next instruction of the calling program. Return is accomplished by a direct jump to this location. (3) The programmer is responsible for coding an unconditional jump instruction in this location; the hardware provides the address. Return is accomplished by a direct jump to this location. The instruction for all three types would be

Operation code	Description of operation
STORE RETURN AND JUMP	$(PC) + 1 \rightarrow (ADDR)_{addr}$ $ADDR + 1 \rightarrow (PC)$

A subroutine with return saved as in (1) might then be coded as follows:

Label field	Operation field	Register designation field	Operand address field	Index register designation	Comments
ROOT	JUMP		0		The starting address—leave blank for saving returns.
	LOAD REGISTER	REGISTER0	ARG1		The first line of executable code—what it is is irrelevant to this example.
⋮	⋮	⋮	⋮		⋮
	JUMP INDIRECT		ROOT		This instruction returns control to the calling program because the address of the next instruction in that calling program was stored at ROOT.

Return address in a stack

A stack is an ordered list whose contents are managed differently from those of an array. Items are added to and removed from the top of the stack on a last-in-first-out basis. It is as if you were to place the data (or return addresses) in turn on top of a pile and then *push* the pile down. You remove addresses or data from the top and then *pop* the stack. This data structure is sometimes called a *push-down stack*. The list may be managed either in hardware or in software. To call a subroutine in this way, you cause the return address to be placed on the stack and the stack to be pushed down. When the routine is ready to return, the stack is popped up and the address recovered. The applicable instructions are:

Operation code	Description of operation
PUSH RETURN AND JUMP	(PC) + 1 → (STACK), (ADDR) → (PC) (stack then pushed down)
POP RETURN AND JUMP	(stack popped up), then (STACK) → (PC)

Since many stacks may exist, the particular one involved must usually be specified.

Trade-offs

Which of these three methods is best? It depends on the application. Most very small computers save the return address in the subroutine. These machines have few, if any, index registers, and they have no need to save many returns, as they might do with a stack. Saving the return address in a subroutine is somewhat inefficient in its use of storage space, since every subroutine must have a separate reserved location. This scheme is especially awkward in a *multiprogramming* environment, in which more than one program may be sharing the machine at one time. Since each subroutine has only one return-saving location, each multiprogram user must have his own copy of it.

A machine with many general-purpose registers can handle multiple users by saving and reloading the contents of each user's registers each time his program is swapped out or brought back in. Under this arrangement, the register used for the subroutine return is saved along with the rest of the user's status. The same subroutine may then be used for many users, each in a different phase of execution. Whether they are subroutines or not, programs written so that they can be used in this way are called *re-entrant*. Library subroutines, system input-output handlers, commonly used systems programs, and language processors such as compilers and interpreters, are most efficient in a multiple-user environment when they are re-entrant.

Push-down stacks can also be used to write re-entrant code, but their most important application is in systems that may call a subroutine more than once from different places, even when that subroutine itself is running. The subroutine can even call itself, and when it does, it is said to be *recursive*. Each time the subroutine is called from a new place, the new return is placed on the stack. The various calls are then unraveled in a very natural fashion; each time the subroutine finishes with one call, it pops a return, taking it back to where it was called from, even from within itself.

This capability is most useful in *real-time* systems, where requests for service come from the outside world, and a request for a subroutine might come while that subroutine is running, working on someone else's problems.

As with most other computer trade-offs, any of these schemes can be used in any environment; there are no absolute circumstances where an approach *cannot* be used. The architect may choose to implement the most appropriate one in hardware. If a programmer feels that another method is needed, he can always program it.

Where the parameters are passed

The calling program often needs to pass one or more arguments to the called subroutine, and it in turn may pass arguments back. The square root program, for example, requires the surd to be passed in, and it passes back the root. Parameters are usually passed in one of three ways.

1. Parameters follow the subroutine jump instruction in the code sequence of the calling program.
2. Parameters are stored in defined registers.
3. Parameters are stored on a stack.

As specified above, each technique involves the passing of the parameters themselves. It is equally common to pass only one number, which is a pointer to the first parameter. The subroutine then indirectly fetches the parameters. In discussing the three techniques, we will describe passing the parameters themselves, but we could equally well be passing their addresses or a pointer to their addresses.

If the parameters follow the STORE PC AND JUMP instruction in the calling program, the subroutine must first get the return address from the location, register, or stack in which it was saved. This address is used as an indirect address. Since it contains (PC) + 1 at the time of the call to the subroutine, it points to the first parameter, *not* the next executable instruction of the main program. This pointer is then incremented each time another parameter is fetched. Before the subroutine exits, the pointer must again be incremented, at which time it *will* point to the return path to the calling program. The subroutine must then follow that path back.

If the parameters are located in registers and the return path is in another register, there is no interference. Parameter fetching is easy. Care must be taken in a multiple-user environment to see that a subroutine is not interrupted during parameter fetches lest a new set of register contents be swapped in. If all registers are saved and restored during a swap, this is not a problem; but if some subset is not saved, parameters may be lost.

If the parameters are saved on a stack, they may be saved either on the same stack as the addresses or on another stack—usually the latter. For either method, two more instructions are needed, since the previously defined stack instructions

(PUSH-JUMP and POP-JUMP) both transfer control, and here we need to transfer only data. The instructions are

Operation code	Description of operation
PUSH PARAMETER	(ADDR) → (STACK), (stack then pushed down)
POP PARAMETER TO REGISTER	(stack popped up), then (STACK) → (REGISTER)

Again, many stacks may be implemented, so one is specified with the instruction.

How the parameters are passed

Not only are there different ways of passing parameters; there are also different ways of using them. The two techniques we will discuss are *call by name* and *call by value*.

In a call by name the parameter passed to the subroutine is the address where the argument may be found. Since locations are symbolically addressed, the name of the argument appears in the parameter list. The occurrence of the argument name in the calling sequence gives rise to the term "call by name." The current value of the argument may be retrieved by reading the contents of the named location. The value of the argument may also be changed by storing into the location. It is by use of the call by name that subroutines may change the value of any of their arguments, thus passing information back to the calling program.

In a call by value the parameter passed to the subroutine is the value of the argument. That value appears in the calling list or is loaded into a register, whatever the subroutine linkage convention requires. Since only the value of the argument is passed to the subroutine, the subroutine cannot return information via the parameter list. In a call by value the calling program is protected from being modified by the called subroutine.

Analogously, subroutines may treat parameters as *use by name* and *use by value*. Parameters called by value must be used by value. In use by value the parameter is stored in a known internal location in the subroutine, used directly from a register, or used directly from the calling parameter list. The subroutine is written so that it knows where the argument value is to be found. If necessary, the entry part, or *preamble* of the subroutine stores the value into this location.

A parameter called by name may be used by value. This is accomplished in the preamble. The value of the argument is obtained from the location provided in the call, and it is stored in the internal location. Thereafter it is used by value.

Use by name does not copy the argument into the subroutine but uses it from its location in the calling program. Since the location of the actual argument is unknown to the subroutine until the call, the subroutine must do some work to retrieve the address and value of the actual argument which were passed as part of the subroutine call. Individual arguments (often called *scalars* to distinguish them from the elements of arrays or *vectors*) may be indirectly addressed, if indirect addressing is available. Otherwise, it is necessary to *build* the instructions referring to the argument. An instruction is built in the preamble by storing the address of the actual argument (as passed in the call by name) in every instruction which refers to that argument.

When the actual argument is an array, the array element may be referenced by a combination of indirect addressing, address-building, or index register address modification. In addition, one or more *base registers* may be supplied and used somewhat like pre-indexing index registers for address modification.† Further indexing may be used only if post-indexing is provided by the hardware.

Examples

We establish the convention that a subroutine may destroy the contents of any of the registers. Therefore we should not expect any register to have the same contents after a return from a subroutine that it had when the subroutine was called. Anything of value or importance should be copied from a register into a memory location before a subroutine is called.

Another necessary convention concerns the association of the actual arguments presented at the time of the subroutine call with the dummy arguments (formal parameters) defined in the subroutine. Your experience with compiler languages has taught you that this association is made by the order—the first actual argument associated with the first dummy argument, the second actual argument with the second dummy argument, etc. For compatibility, the same associations are used in assembly language. In fact, one of the major reasons for the establishment of conventions is that compilers need to be able to generate code to perform subroutine linkage.

We must also have a convention, when there is a function-type subroutine which calculates and returns a value, to establish where this result is to be stored by the subroutine before control is returned to the calling program. REGISTER 1 will be used in this book for this purpose. Another possibility is to leave it within the subroutine body and leave a pointer (possibly indirect) to its location; this has the advantage of being easily extended to provide the means to return more than one value since the pointer could be to a variable-length "record" of results.

The first location after the STORE PC AND JUMP will contain the (decimal) count of the number of arguments in the call on the subroutine (parameters in calling program text). This count is certainly not a universal feature in assembly language subroutine linkage, but it is useful for checking that the number of arguments presented is correct, as well as for determining the location to which control is to be returned when the subroutine has completed its task. In each of the loca-

---

† A base register is somewhat less flexible than an index register in that it must be separately loaded and usually has limited arithmetic capability. An effective address may be determined by adding the contents of the base register, index register, and address field.

tions following the count of parameters are the addresses of the locations of the actual arguments (or the arguments themselves if these are addresses); they may be used to access individual (scalar) arguments by indirect addressing, or to access an array by index register address modification. When the actual argument is an array, the address in the calling sequence will be the address of the base element of the array (the element which begins the array).

Sometimes alternative locations are provided for return of control from the subroutine. Control passes to them if, for example, the subroutine encounters special conditions, such as an overflow or other error. The subroutine must have been constructed with the knowledge that alternative return locations were provided as arguments. In the calling sequence, the alternative return path will be coded as a JUMP TO ADDR instruction. Control will be returned from the subroutine via a jump to the location of the argument, which then will cause a jump to the alternative return address.

In the subroutine linkage it is necessary to have parameters stored in the address field of an instruction word without having anything stored in the other fields. One scheme would be simply to leave these unused fields blank in our assembly language code, implying that the contents of these fields were unspecified.

Instead, we shall provide a NO OPERATION code that causes no operation to take place but does advance the program counter.

Operation code	Description of operation
NO OPERATION	(PC) + 1 → (PC)

The existence of NO OPERATION allows us to indicate explicitly that the contents of the operation field are not to be used to effect machine operation.† We shall adopt here a further convention sometimes used with the NO OPERATION, namely, that any numerals appearing in the operand address field are decimal immediates. (There will be further discussion of this sort of use in a later section on assembler directives.) We also assume that the numeric equivalent of NO OPERATION is zero.

As our first example, let us consider a subroutine, MOD, for calculating M modulus N. As noted in Chapter 4, the calculation consists of dividing M by N, discarding the quotient, and keeping the remainder as the desired result. In a higher-level language we might write a call of the form

X = MOD(M,N)

which would be compiled into machine language having the following assembly

† One very interesting use of the NO OPERATION in subroutine linkage is the stratagem of returning control to the first location after the STORE PC AND JUMP instruction. A series of NO OPERATIONS are then (not) executed until the first executable instruction is reached. The arguments or argument addresses can then be stored in the operand address fields of these NO OPERATIONS.

language equivalent (note that the return path is in a register, the parameter addresses are in the calling program after the jump, and parameters are called by name). The calling program follows.

Label field	Operation field	Register designation field	Operand address field	Index register designation	Comments
	STORE PC AND JUMP	XR7	MOD		Jump to subroutine
	NO OPERATION		2		Two arguments
	NO OPERATION		M		First argument
	NO OPERATION		N		Second argument
	STORE REGISTER	REGISTER1	X		Value returned by MOD in Register 1 stored in X

The subroutine MOD follows.

Label field	Operation field	Register designation field	Operand address field	Index register designation	Comments
MOD	LOAD REGISTER	REGISTER0	0	XR7	Number arguments
	INTEGER SUBTRACT FROM REGISTER, IMMEDIATE	REGISTER0	2		Equal two?
	JUMP IF REGISTER NONZERO	REGISTER0	BAD		No!
	LOAD REGISTER, IMMEDIATE	REGISTER0	0		
	LOAD REGISTER, INDIRECT	REGISTER1	1	XR7	First argument
	DOUBLE DIVIDE REGISTER, INDIRECT	REGISTER1	2	XR7	Divided by second argument, remainder in register
	JUMP TO ADDR		3	XR7	Return

Label field	Operation field	Register designation field	Operand address field	Index register designation	Comments
BAD	LOAD NEGATIVE REGISTER, IMMEDIATE	REGISTER1	1		Error flag = −1
	JUMP TO ADDR		3	XR7	Return

Our second example will introduce a software push-down stack. Compare the cumbersomeness of this subroutine with the ease of using hardware instructions for handling stacks, as introduced earlier. We will allow any list to be used as a push-down stack. The first location in the list will be used to count the number of items presently stored in the list; all other locations will provide storage for these items.

Three entry points to the subroutine are required. The first will name the array to be utilized as a pushdown stack. In a higher-level language this entry point would be accessed by one of the following subroutine calls.

```
CALL LIST(NAME)
CALL LIST(NAME,LENGTH)
```

The second call contains the optional argument stating the length of the list. The assembly language calling sequences for these two subroutines follow.

Label field	Operation field	Register designation field	Operand address field	Index register designation	Comments
	STORE PC AND JUMP	XR7	LIST		
	NO OPERATION		1		One argument
	NO OPERATION		NAME		

or

Label field	Operation field	Register designation field	Operand address field	Index register designation	Comments
	STORE PC AND JUMP	XR7	LIST		
	NO OPERATION		2		Two arguments
	NO OPERATION		NAME		
	NO OPERATION		LENGTH		

To save or add an item to the list, the higher-level subroutine call would be

CALL PUSH(ITEM)

To retrieve, return, or extract the last item stored,

CALL POP(ITEM)

It is also desirable to use the stack for jump addresses. In this case, you POP the data item into a register and jump to it indirectly. The corresponding subroutine linkages in assembly language follow.

Label field	Operation field	Register designation field	Operand address field	Index register designation	Comments
	STORE PC AND JUMP	XR7	PUSH		
	NO OPERATION		1		
	NO OPERATION		ITEM		

Label field	Operation field	Register designation field	Operand address field	Index register designation	Comments
	STORE PC AND JUMP	XR7	POP		
	NO OPERATION		1		
	NO OPERATION		ITEM		

The flowchart of this subroutine is shown in Fig. 7.3.

Two internal locations, called WHICH and MANY, hold the base address of the array and its declared length, respectively. If the length of the array is not declared, it is stored as -1. If the declared length of the array is exceeded, the subroutine will cause the computer to cease processing the program by execution of the operation HALT.

Label field	Operation field	Register designation field	Operand address field	Index register designation	Comments
LIST	LOAD REGISTER	REGISTER0	1	XR7	NAME of array
	STORE REGISTER	REGISTER0	WHICH		

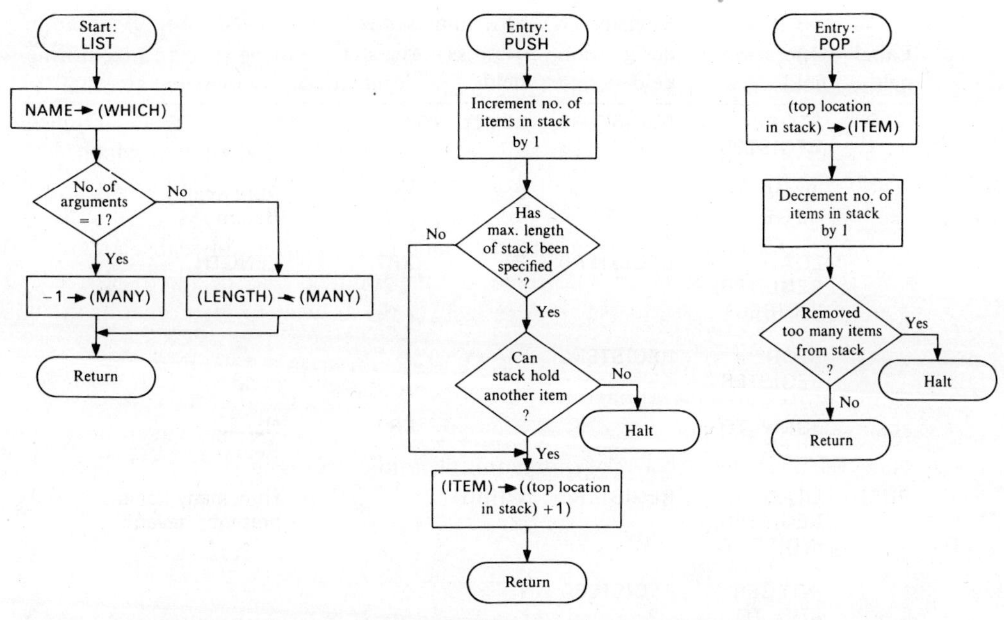

Fig. 7.3. Flowchart for software push-down stack. WHICH **and** MANY **are symbolic operands used by the subroutine.** (WHICH) **is the address of the first element of the array** NAME, **which is used as a stack.** ((WHICH)) **is the number of items in the stack; we can assume that the stack is empty initially.** (MANY) **is the length of the array used as a stack; if no length is specified,** (MANY) **is arbitrarily set to —1.**

Label field	Operation field	Register designation field	Operand address field	Index register designation	Comments
	LOAD REGISTER	REGISTER0	0	XR7	How many arguments?
	INTEGER SUBTRACT FROM REGISTER, IMMEDIATE	REGISTER0	1		
	JUMP IF REGISTER ZERO	REGISTER0	Present location +2		1 argument
	JUMP IF REGISTER POSITIVE	REGISTER0	Present location +4		> 1 argument
	LOAD NEGATIVE REGISTER, IMMEDIATE	REGISTER0	1		Flag of −1

Label field	Operation field	Register designation field	Operand address field	Index register designation	Comments
	STORE REGISTER	REGISTER0	MANY		
	JUMP TO ADDR		2	XR7	Abnormal Return
	LOAD REGISTER, INDIRECT	REGISTER0	2	XR7	LENGTH
	STORE REGISTER	REGISTER0	MANY		
	JUMP TO ADDR		3	XR7	Normal Return
PUSH	LOAD REGISTER, INDIRECT	REGISTER0	WHICH		How many items presently saved?
	INTEGER ADD TO REGISTER, IMMEDIATE	REGISTER0	1		
	STORE REGISTER, INDIRECT	REGISTER0	WHICH		((WHICH)) + 1 → ((WHICH))
	LOAD REGISTER	REGISTER1	MANY		
	JUMP IF REGISTER NEGATIVE	REGISTER1	STORE		LENGTH not stated
	INTEGER SUBTRACT FROM REGISTER, INDIRECT	REGISTER1	WHICH		Too many?
	JUMP IF REGISTER POSITIVE	REGISTER1	STORE		ok
	HALT				Too many!
STORE	LOAD REGISTER, INDIRECT	REGISTER0	1	XR7	ITEM
	LOAD REGISTER	REGISTER1	WHICH		LIST base address

Label field	Operation field	Register designation field	Operand address field	Index register designation	Comments
	INTEGER ADD TO REGISTER, INDIRECT	REGISTER1	WHICH		Storage address WHICH + (WHICH)
	STORE REGISTER INDIRECT	REGISTER0	REGISTER1		(ITEM) → $(\text{NAME})_{\text{NAME}}$
	JUMP TO ADDR		2	XR7	Return
POP	LOAD REGISTER	REGISTER0	WHICH		
	INTEGER ADD TO REGISTER, INDIRECT	REGISTER0	WHICH		LIST element last in
	LOAD REGISTER, INDIRECT	REGISTER1	REGISTER0		
	STORE REGISTER, INDIRECT	REGISTER1	1	XR7	$(\text{NAME})_{\text{NAME}}$ → (ITEM)
	LOAD REGISTER, INDIRECT	REGISTER0	WHICH		Decrement count
	INTEGER SUBTRACT FROM REGISTER, IMMEDIATE	REGISTER0	1		
	STORE REGISTER, INDIRECT	REGISTER0	WHICH		((WHICH)) − 1 → ((WHICH))
	JUMP IF REGISTER POSITIVE	REGISTER0	2	XR7	Return
	HALT				Removed too many from NAME
MANY	NO OPERATION				
WHICH	NO OPERATION				

The following example shows a subroutine linkage using hardware stacks. It uses the natural ability of the stack implementation to recursively call itself. The subroutine calculates N factorial (N!).

$$N! = N(N-1)! \qquad (N > 1)$$
$$1! = 0! = 1$$

The call would consist of the following statement pair embedded in a larger program.

Label field	Operation field	Register designation field	Operand address field	Index register designation	Comments
	PUSH PARAMETER		N		
	PUSH RETURN AND JUMP		NFACT		Call subroutine

The subroutine would be:

Label field	Operation field	Register designation field	Operand address field	Index register designation	Comments
NFACT	POP PARAMETER TO REGISTER	REGISTER1			
	PUSH PARAMETER		REGISTER1		Put it back on
	INTEGER SUBTRACT FROM REGISTER, IMMEDIATE	REGISTER1	1		
	JUMP IF REGISTER NONZERO		CMPT		Not 1
	POP RETURN AND JUMP				1's left on stack
CMPT	PUSH PARAMETER		REGISTER1		
	PUSH RETURN AND JUMP		NFACT		call (N − 1)!

Label field	Operation field	Register designation field	Operand address field	Index register designation	Comments
	POP PARAMETER TO REGISTER	REGISTER1			(N − 1)!
	POP PARAMETER TO REGISTER	REGISTER0			
	INTEGER MULTIPLY IN REGISTER	REGISTER1	REGISTER0		N(N − 1)!
	PUSH PARAMETER		REGISTER1		
	POP RETURN AND JUMP				

This routine was conceived as a tour de force for the stack organization; the example is illustrative of the power of the technique.

7.9 INPUT/OUTPUT INSTRUCTIONS

Input and *output* are perhaps the most diverse of the operations in a computer system. Data transfers occur between the CPU and main memory, between main and auxiliary memory, and between memory and input/output devices. The control and data transfer instructions are wholly dependent on the particular devices.

In discussing *data transfer*, we shall be as general as possible. We will not explicitly specify any device involved; rather, all devices will be uniquely numbered, and therefore we will need to specify only the device number. Some provision will have to be made to indicate when the transfer has been completed. This indication will sometimes allow the program to continue calculation (when possible) concurrent with data transfer. The fact that computation and data transfer can proceed concurrently implies independent operation of the CPU and the data transfer mechanism.

The most general example of data transfer is that from or to a contiguous sequence of memory locations, called a *block*. A single-word transfer is merely a special case. To accomplish a *block transfer*, it is necessary to specify the address of the first word in the block and the number of words to be transferred.

The information required for specification of an input/output operation is more than can be included in a single-word instruction. We will adopt the convention that the data transfer instruction forms the beginning of a fixed-length block or vector employed for communication of the required specifications. The vector is not executed; the first location after the command vector contains the next executable instruction. If the instruction contains a pointer to the command vector, the next instruction is executable.

We will employ the following two commands for data transfer.

Operation code	Description of operation
TRANSMIT DATA TO DEVICE	Initiate data transfer from main memory to device specified in addr
RECEIVE DATA FROM DEVICE	Initiate data transfer from device specified in addr to main memory

Three words following each of the commands are employed for the necessary command vector. The first word contains the address of the first data word in the block in the ADDR (effective address) field. The second word contains the (integer immediate) count of the number of words to be transferred. The third word contains a flag to indicate completion of the transfer. When the operation is completed, this flag, or status word, will have been set to some value indicating the status of the transmission. For example, a value of +1 would mean successful completion of the entire transfer; +2 would indicate a transmission error; +3 would show that the device was empty (such as a card reader with an empty hopper).

The following vector transfers 10 words from main memory, beginning at the location symbolically addressed as BLOCK, to external device number 4.

Label field	Operation field	Register designation field	Operand address field	Index register designation	Comments
	TRANSMIT TO DEVICE		4		
	NO OPERATION		BLOCK		Transmit starting at location BLOCK
	NO OPERATION		10		Transmit 10 words
	NO OPERATION		0		Completion flag

The following vector transfers 22 words from external device number 11 to main memory, beginning at the location symbolically referenced as ARRAY.

Label field	Operation field	Register designation field	Operand address field	Index register designation	Comments
	RECEIVE FROM DEVICE		11		
	NO OPERATION		ARRAY		
	NO OPERATION		22		
	NO OPERATION		0		

There are two other functions needed for an input/output device. We may wish to sense the status of the equipment to determine whether it is ready or busy. In this mode of operation, one word of status information is read from the device into the designated register. Since the assignment of the bits within the word is usually device dependent, it cannot be generalized here. An assembly language sequence to write a block on magnetic tape may ask first if the tape controller is busy and then either wait or continue calculations if it is not available. As soon as it becomes available, the transfer instruction can be executed.

We can also execute an instruction that sends one word of information to the device. Since these bits are also interpreted differently in different devices, they will not be discussed here. A typical command function to a magnetic tape would be to rewind the tape.

Operation code	Description of operation
SENSE STATUS OF DEVICE	Load status word of device specified in addr into register specified
OUTPUT COMMAND TO DEVICE	Send command word in register specified to device specified in addr

Let us now present an example that combines input/output and a subroutine call. We shall read two binary integers from device 5, calculate the second modulo the first, and write the resultant binary integer onto device 6. We call subroutine MOD, previously presented, to perform the calculations. Note that in this example no checking is done to see if the data transmission was completed properly.

Label field	Operation field	Register designation field	Operand address field	Index register designation	Comments
	RECEIVE FROM DEVICE		5		
	NO OPERATION		A1		First integer
	NO OPERATION		1		
	NO OPERATION		0		Flag
	LOAD REGISTER	REGISTER0	Present location −1		Transmission complete?
	JUMP IF REGISTER ZERO	REGISTER0	Present location −1		No. Loop until complete
	RECEIVE FROM DEVICE		5		
	NO OPERATION		A2		Second integer
	NO OPERATION		1		
	NO OPERATION		0		Flag
	LOAD REGISTER	REGISTER0	Present location −1		Complete?
	JUMP IF REGISTER ZERO	REGISTER0	Present location −1		No
	STORE PC AND JUMP	XR7	MOD		Jump to subroutine
	NO OPERATION		2		
	NO OPERATION		A2		
	NO OPERATION		A1		

Label field	Operation field	Register designation field	Operand address field	Index register designation	Comments
	TRANSMIT TO DEVICE		6		
	NO OPERATION		REGISTER1		
	NO OPERATION		1		
	NO OPERATION		0		

7.10 LOGICAL INSTRUCTIONS

We are concerned here with the logical operations: **AND**, **OR**, and **EXCLUSIVE OR**. These operations between two computer words are applied on a bit-by-bit basis. That is to say, the least significant bit of each of the two operands is treated as input to the truth table, and the least significant bit of the result is the output of the truth table. The process is carried out for each bit position in the computer word with no carries or other interactions between positions. The following truth tables define the operations for two one-bit operands and a one-bit result.

AND

Inputs		Result
X	Y	R
0	0	0
0	1	0
1	0	0
1	1	1

OR

Inputs		Result
X	Y	R
0	0	0
0	1	1
1	0	1
1	1	1

EXCLUSIVE OR

Inputs		Result
X	Y	R
0	0	0
0	1	1
1	0	1
1	1	0

The three assembly language instructions which perform the logical operations are:

Operation code	Description of operation
LOGICAL AND TO REGISTER	(REGISTER) **AND** (ADDR) → (REGISTER)
LOGICAL OR TO REGISTER	(REGISTER) **OR** (ADDR) → (REGISTER)
LOGICAL EXCLUSIVE OR TO REGISTER	(REGISTER) **EXCLUSIVE OR** (ADDR) → (REGISTER)

7.11 SHIFT INSTRUCTIONS

When performing operations with characters, or when performing double-precision multiplication or division, you will find it quite useful to be able to shift the contents of a register. The effect of a shift instruction is to copy the contents of every bit position in the register into an adjacent bit position.

Shifts may be made to the right or left. A digit shifted beyond the limits of the register is lost; you might say it "falls off the end." Positions at the end of the word being shifted away from are filled with 0's. This is equivalent to shifting 0's into the end of the word.

To avoid losing information off the end of the word, we provide two additional shifts. The circular shift acts as if the register formed a circular ring. If the bit positions in the register are numbered from 0 to n, then 0 and n are assumed to be adjacent, as shown in Fig. 7.4(a). A double shift merely shifts the contents of two contiguous registers as if they were one, with bit n of the higher-numbered register contiguous to bit 0 of the lower-numbered register. This arrangement is shown in Fig. 7.4(b). In a double-length register, we may also provide a double circular shift.

To provide for right and left shifts, single, circular, and double, we require eight shift instructions, which are given below, using a slight modification of our notation. The brace enclosure, { }, means select one of the items enclosed.

Operation code	Description of operation
SHIFT REGISTER {LEFT / RIGHT} {SINGLE / DOUBLE} {ARITHMETIC / LOGICAL} {CYCLIC / (null)}	Shift (REGISTER) addr places as specified

In machines with 1's or 2's complement negative numbers, shifting a negative number right should include shifting 1's in at the left end (called *sign fill*) in order to preserve the correct arithmetic relationship (shifting right N places is equivalent to dividing by 2^N). In a sign-magnitude machine, 0's are filled in for right shifts, but only from the second bit position on, because the sign bit must be unchanged. These shift functions are referred to as ARITHMETIC shifts. ARITHMETIC left

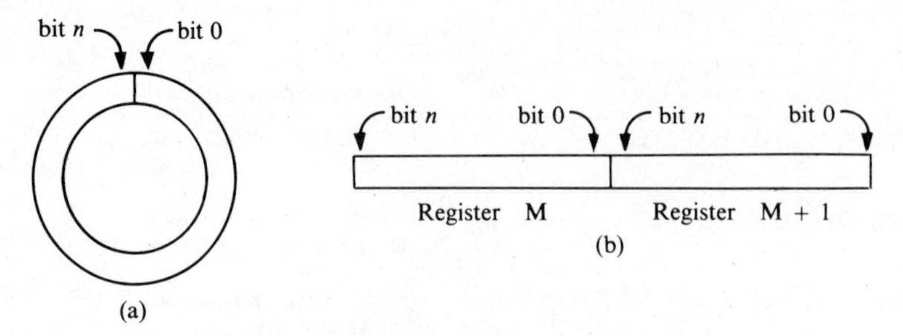

Fig. 7.4. (a) Circular register (b) double register

shifts always fill with 0's from the right, but the sign bit is not included in the shifting. ARITHMETIC double shifts usually treat the sign bit of the less significant word as another magnitude bit when the machine employs 1's or 2's complement representation of negative numbers and as a null (nonexistent) bit when sign-magnitude is employed.

The LOGICAL shift instructions treat all bits the same, shifting 0's into locations voided as a result of the shift.

7.12 PSEUDO-OPERATIONS

Each assembly language instruction we have discussed so far corresponds exactly to one machine language instruction. Considering the machine language fields, the use of assembly language has only the advantage of mnemonic operation codes instead of the numerical ones of machine language.

In this and the following sections we introduce operation mnemonics which do not have a one-to-one correspondence with machine language. Each of the *pseudo-operations* introduced in this section does actually cause an instruction word to be generated. But the use of the pseudo-operation is only a convenience; it is not the only way that the word could have been created.

One interpretation of a pseudo-operation is as a directive to the assembler program to generate a certain kind or format of machine word. The exact pseudo-operations available will of course depend on the assembler program being used. We will discuss a few typical examples.

In the generation of constants, it is usually most convenient to write decimal numbers and have the assembler translate them into binary internal representation. There are times, however, when it is more convenient to write numbers in binary, octal, or hexadecimal form and have the assembler generate the corresponding internal machine words.

By inspection of a group of numerals, it is not always obvious what the base is or what number they represent. To clarify and simplify the task of the assembler, pseudo-operations are provided to declare the radix of the numeral that appears in the address field.

Pseudo-operation code	Description of operation
BINARY	Exactly reproduce the binary number in the operand as a machine word
OCTAL	Transform the octal number in the operand into a machine word
DECIMAL	Transform the decimal number in the operand into a machine word
HEXADECIMAL	Transform the hexadecimal number in the operand into a machine word

We will assume that the binary, octal, and hexadecimal numbers are integers in their respective bases, having no fractional parts, and that they are to be translated into binary integers for internal representation. The decimal number, however, may be either integer or floating point and must be translated into a binary integer or a floating-point number, as appropriate.

A floating-point decimal is recognized by the presence of a decimal point among the characters encountered in the operand. In addition, the scale factor or power of ten multiplier may be included following the least significant digit of the fractional part. The power of ten is written as E ± dd, where dd are two decimal digits. Examples are:

$$.1795E - 03 = .1795 \times 10^{-3}$$
$$-.7763E + 22 = -.7763 \times 10^{22}$$

An additional advantage of these pseudo-operations is that an entire machine word may be employed for the internal representation. The word is not an instruction word but a data word. In the input/output instructions, instead of a NO OPERATION, we might have used one of these four numerical pseudo-operations. Note that interpretation of the pseudo-operation takes place when the program is being assembled, *not* when it is being run.

In another class of pseudo-operations, a short operation code is substituted for a logically equivalent but longer or more cumbersome one. Consider, for example, loading a register with zero. The pseudo-operation LOAD ZERO

Pseudo-operation Code	Description of operation
LOAD ZERO	0 → (REGISTER)

is exactly equivalent to the following LOAD REGISTER command, assembling exactly the same machine language instruction word.

Label field	Operation field	Register designation field	Operand address field	Index register designation	Comments
	LOAD REGISTER, IMMEDIATE	REGISTER i	0		

The only advantage of this pseudo-operation is simplicity.

Several of the instructions we have now introduced may be combined in an example. Let us assume that we have a subroutine, SUB, which was entered by a STORE PC AND JUMP type of call, and that the first parameter was either a NO OPERATION or a JUMP, the latter being provided in case an error condition should occur. Assume further that the subroutine has encountered an error. It must then sense whether the programmer has provided an explicit error return. Since a NO OPERATION assembles as zero, we must test for a non-zero operation code field if we are able to perform the error return jump. If the field is zero, we then do a STORE RETURN AND JUMP to a system library routine called ERROR. The calling program would be either

Label field	Operation field	Register designation field	Operand address field	Index register designation	Comments
	STORE PC AND JUMP	XR7	SUB		
	NO OPERATION		1		
	NO OPERATION		0		

or

Label field	Operation field	Register designation field	Operand address field	Index register designation	Comments
	STORE PC AND JUMP	XR7	SUB		
	NO OPERATION		1		
	JUMP		LOC		

The relevant positions of the subroutine are shown below. It is assumed that the op code is contained in bits 12_{10} through 17_{10} (numbered from 0 as the rightmost bit).

Label field	Operation field	Register designation field	Operand address field	Index register designation	Comments
MASK	OCTAL		770000		Corresponds to binary 111111000000000000
SUB	LOAD REGISTER	REGISTER0	1	XR7	
	LOGICAL AND TO REGISTER	REGISTER0	MASK		Is it a NO OPERATION?
	JUMP IF REGISTER NONZERO	REGISTER0	1	XR7	It's a JUMP
	STORE RETURN AND JUMP		ERROR		It was a NO OPERATION

7.13 ASSEMBLER DIRECTIVES

Sometimes it is necessary to convey to the assembler program information that does not cause any machine language instruction word to be generated. Instructions of this class, known as *assembler directives*, allow the programmer to specify the way the assembler should do its job.

One obvious assembler directive informs the assembler that it has reached the end of the program unit. Without the END directive, the assembler might go on forever.

Assembler directive operation code	Description of operation
END	Marks end of program unit

Another extremely useful assembler directive advances the location counter being used to assign memory addresses to the machine words generated. No machine instructions are generated by the RESERVE directive, but locations are saved for the storage of scalar variables and arrays.

Operation code	Description of operation
RESERVE	(Location counter) + addr → (Location counter)

Note that the location counter is an assembler function, not a piece of hardware. By convention, if the addr field is blank, a value of 1 is assumed. Thus the RESERVE directive may be used to assign a symbolic variable name to a memory address.

7.14 SUMMARY AND FINAL EXAMPLE

Our final example will attempt to bring together many of the instructions and concepts we have introduced. The example is a mathematical library subroutine to calculate the arctangent of a number and express the answer in radians. The calculation is a Chebychev approximation to a Taylor series. Normal polynomial expansions converge slowly, and small adjustments of the individual coefficients make them converge to within acceptable limits after taking fewer terms in the polynomial. These "adjusted" coefficients are stored as constants in a table.

For arguments whose absolute value exceeds 1, the algorithm does not converge. In this case, we calculate the arctangent of the reciprocal of the argument and subtract the result from $\pi/2$. This produces the correct answer for the sectors where the angle is between $\pi/4$ and $3\pi/4$.

For small values of the argument, the small angle approximation is used, and the answer is set equal to the argument.

There is sufficient complexity in this subroutine to require a flowchart, shown in Fig. 7.5. The program itself is given below. The statement labels in the program are those used in the flowchart, making it possible to follow what the program does.

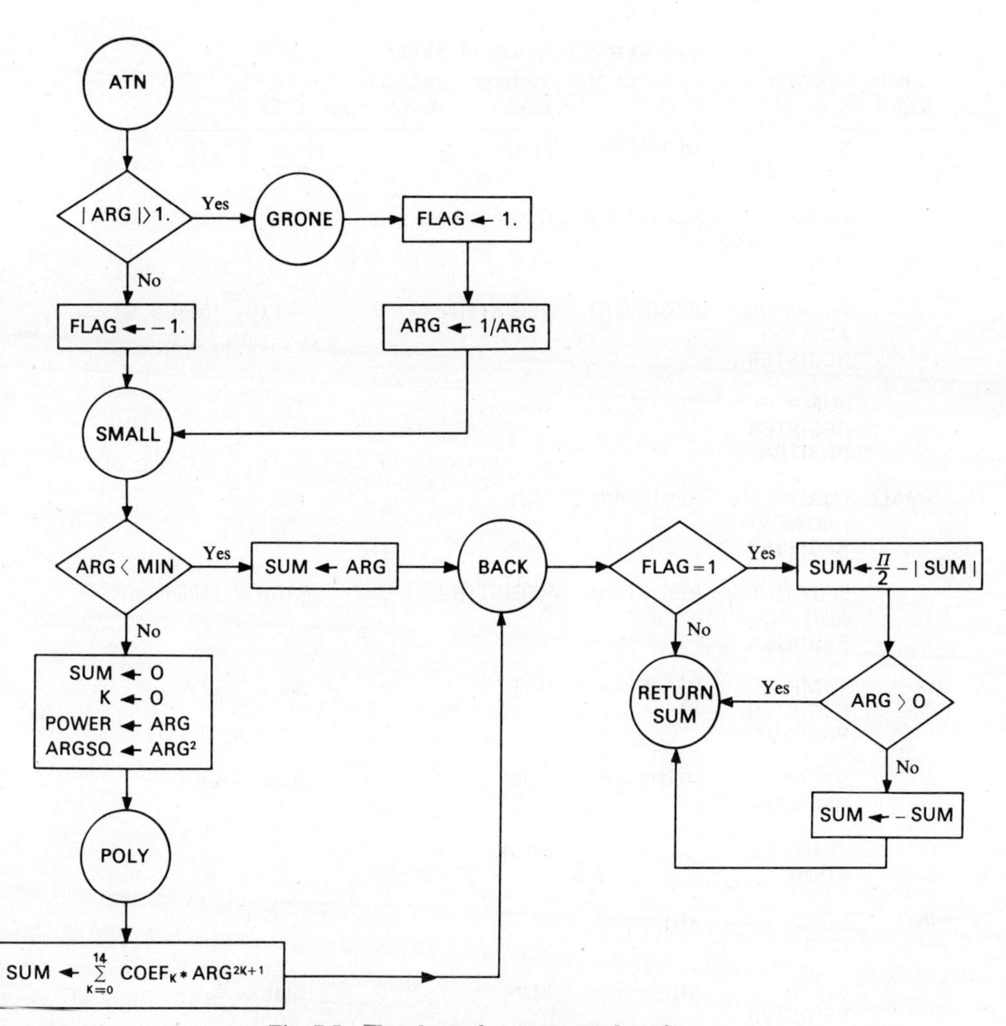

Fig. 7.5. Flowchart of arctangent subroutine

Label field	Operation field	Register designation field	Operand address field	Index register designation	Comments
ATN	LOAD REGISTER, INDIRECT	REGISTER0	1	XR7	Argument
	LOAD NEGATIVE REGISTER, LITERAL	REGISTER1	1.		Flag = −1.0

Label field	Operation field	Register designation field	Operand address field	Index register designation	Comments
	STORE REGISTER	REGISTER1	FLAG		
	LOAD MAGNITUDE REGISTER	REGISTER2	REGISTER0		|Arg|
	FLOATING ADD TO REGISTER	REGISTER1	REGISTER2		$-1.0 + \|Arg\| > 0$?
	JUMP IF REGISTER POSITIVE	REGISTER1	GRONE		Yes
SMALL	LOAD NEGATIVE REGISTER	REGISTER1	MIN		No
	FLOATING ADD TO REGISTER	REGISTER1	REGISTER2		|Arg| < Minimum?
	JUMP IF REGISTER POSITIVE	REGISTER1	INIT		No
	STORE REGISTER	REGISTER0	SUM		Sum ← Arg
	JUMP TO ADDR		BACK		
INIT	LOAD ZERO	XR1			K ← 0
	STORE REGISTER	XR1	SUM		Sum ← 0
	STORE REGISTER	REGISTER0	POWER		Power ← Arg
	LOAD REGISTER	REGISTER2	REGISTER0		
	FLOATING MULTIPLY REGISTER	REGISTER2	REGISTER0		
	STORE REGISTER	REGISTER2	ARGSQ		Arg sq ← Arg^2
	LOAD REGISTER, IMMEDIATE	REGISTER2	15		Upper limit on K is 14

Label field	Operation field	Register designation field	Operand address field	Index register designation	Comments
POLY	LOAD REGISTER	REGISTER1	POWER		
	FLOATING MULTIPLY IN REGISTER	REGISTER1	COEF	XR1	
	FLOATING ADD TO REGISTER	REGISTER1	SUM		
	STORE REGISTER	REGISTER1	SUM		Sum ← Sum + $Coef_K * Arg^{2K+1}$
	LOAD REGISTER	REGISTER1	POWER		
	FLOATING MULTIPLY REGISTER	REGISTER1	ARGSQ		
	STORE REGISTER	REGISTER1	POWER		Power ← Power * Arg^2
	INCREMENT INDEX REGISTER JUMP IF LESS THAN REGISTER	REGISTER2	POLY	XR1	Jump if K < 15
BACK	LOAD REGISTER	REGISTER1	FLAG		$\|Arg\| > 1$?
	JUMP IF REGISTER POSITIVE	REGISTER1	Present location +3		Yes
	LOAD REGISTER	REGISTER1	SUM		
	JUMP TO ADDR		2	XR7	Return
	LOAD MAGNITUDE REGISTER	REGISTER1	SUM		
	LOAD NEGATIVE REGISTER	REGISTER1	REGISTER1		
	FLOATING ADD TO REGISTER	REGISTER1	HALFPI		

Label field	Operation field	Register designation field	Operand address field	Index register designation	Comments
	STORE REGISTER	REGISTER1	SUM		Atn ← $\pi/2$ − |Sum|
	JUMP IF REGISTER POSITIVE	REGISTER0	BACK + 2		Return if arg > 0
	LOAD NEGATIVE REGISTER	REGISTER1	SUM		Atn ← |Sum| − $\pi/2$
	JUMP TO ADDR		2	XR7	Return
GRONE	LOAD REGISTER, LITERAL	REGISTER1	1.		
	STORE REGISTER	REGISTER1	FLAG		|Arg| > 1.
	FLOATING DIVIDE REGISTER	REGISTER1	REGISTER0		
	STORE REGISTER	REGISTER1	REGISTER0		Arg ← 1.0/Arg
	JUMP TO ADDR		SMALL		
ARGSQ	RESERVE				
POWER	RESERVE				
FLAG	RESERVE				
SUM	RESERVE				
MIN	DECIMAL		.03		
HALFPI	DECIMAL		1.5707963		
COEF	DECIMAL		.99999999		
	DECIMAL		−.33333333		
	DECIMAL		.19999998		
	DECIMAL		−.14285658		
	DECIMAL		.11110348		
	DECIMAL		−.90843663E−01		
	DECIMAL		.76541001E−01		

Label field	Operation field	Register designation field	Operand address field	Index register designation	Comments
	DECIMAL		−.65075421E−01		
	DECIMAL		.53939992E−01		
	DECIMAL		−.41298271E−01		
	DECIMAL		.27248480E−01		
	DECIMAL		−.14321572E−01		
	DECIMAL		.54663727E−02		
	DECIMAL		−.13230859E−02		
	DECIMAL		.15078680E−03		
	END				

Writing in assembly language is a potentially frustrating exercise in bookkeeping. Whenever possible, therefore, programming should be done in a higher-level language. If the higher-level languages available do not permit the desired operations, it may be necessary to employ assembly language.

There is a more subtle trade-off that can also lead to the use of assembly language. The crucial point here is run-time efficiency. Even the best optimizing routine cannot match the cunning of a skillful programmer. If code is to be a large part of a big system, it may pay to write in assembly language.

For reference purposes, the assembly language instructions described in this chapter are listed in Fig. 7.6. You are strongly advised to work the exercise in Question 10 at the end of the chapter to relate these instructions to those available on your computer.

Operation code	Description of operation
Load and store instructions	
LOAD REGISTER	(ADDR) → (REGISTER)
LOAD NEGATIVE REGISTER	− (ADDR) → (REGISTER)
LOAD MAGNITUDE REGISTER	|(ADDR)| → (REGISTER)
STORE REGISTER	(REGISTER) → (ADDR)
STORE ADDRESS	$(\text{REGISTER})_{\text{addr}} \rightarrow (\text{ADDR})_{\text{addr}}$
Integer arithmetic	
INTEGER ADD TO REGISTER	(REGISTER) + (ADDR) → (REGISTER)
INTEGER SUBTRACT FROM REGISTER	(REGISTER) − (ADDR) → (REGISTER)

(continued)

Fig. 7.6. (*continued*)

Operation code	Description of operation
INTEGER MULTIPLY IN REGISTER	(REGISTER) · (ADDR) → (REGISTER)
INTEGER DIVIDE REGISTER	(REGISTER) ÷ (ADDR) → (REGISTER)
Double-register instructions	
DOUBLE MULTIPLY IN REGISTER	(REGISTER) · (ADDR) → (REGISTER i, REGISTER i + 1)
DOUBLE DIVIDE REGISTER	(REGISTER i, REGISTER i + 1) ÷ (ADDR) → (REGISTER i) [QUOTIENT] (REGISTER i + 1) [REMAINDER]
DOUBLE LOAD REGISTER	(ADDR, ADDR + 1) → (REGISTER i, REGISTER i + 1)
DOUBLE ADD TO REGISTER	(REGISTER i, REGISTER i + 1) (ADDR, ADDR + 1) + → (REGISTER i, REGISTER i + 1)
DOUBLE SUBTRACT FROM REGISTER	(REGISTER i, REGISTER i + 1) − (ADDR, ADDR + 1) → (REGISTER i, REGISTER i + 1)
DOUBLE STORE REGISTER	(REGISTER i, REGISTER i + 1) → (ADDR, ADDR + 1)
Floating-point operations	
FLOATING ADD TO REGISTER	(REGISTER) + (ADDR) → (REGISTER)
FLOATING SUBTRACT FROM REGISTER	(REGISTER) − (ADDR) → (REGISTER)
FLOATING MULTIPLY IN REGISTER	(REGISTER) · (ADDR) → (REGISTER)
FLOATING DIVIDE REGISTER	(REGISTER) ÷ (ADDR) → (REGISTER)
Logical operations	
LOGICAL AND TO REGISTER	(REGISTER) **AND** (ADDR) → (REGISTER)
LOGICAL OR TO REGISTER	(REGISTER) **OR** (ADDR) → (REGISTER)
LOGICAL EXCLUSIVE OR TO REGISTER	(REGISTER) **EXCLUSIVE OR** (ADDR) → (REGISTER)
Transfer instructions	
JUMP TO ADDR	ADDR → (PC)
JUMP TO ADDR, INDIRECT	(ADDR) → (PC)

Operation code	Description of operation
JUMP IF REGISTER ZERO	IF (REGISTER) = 0, ADDR → (PC)†
JUMP IF REGISTER NONZERO	IF (REGISTER) ≠ 0, ADDR → (PC)†
JUMP IF REGISTER POSITIVE	IF [sign bit] = 0, ADDR → (PC)†
JUMP IF REGISTER NEGATIVE	IF [sign bit] = 1, ADDR → (PC)†
DECREMENT REGISTER JUMP IF NONNEGATIVE	(REGISTER) − 1 → (REGISTER) IF [sign bit] = 0, ADDR → (PC)†
INCREMENT INDEX REGISTER JUMP IF LESS THAN REGISTER	(XR) + 1 → (XR) IF (XR) < (REGISTER), addr → (PC)†
STORE PC AND JUMP	(PC) + 1 → (REGISTER), ADDR → (PC)
STORE RETURN AND JUMP	(PC) + 1 → $(ADDR)_{addr}$, ADDR + 1 → (PC)
PUSH RETURN AND JUMP	(PC) + 1 → STACK, (ADDR) → (PC) (stack pushed)
POP RETURN AND JUMP	(stack popped), then STACK → (PC)
PUSH PARAMETER	(ADDR) → (STACK), (stack then pushed down)
POP PARAMETER TO REGISTER	(stack popped up), then (STACK) → (REGISTER)
Input/output operations	
TRANSMIT DATA TO DEVICE	Initiate data transfer from main memory to device specified in addr
RECEIVE DATA FROM DEVICE	Initiate data transfer from device specified in addr to main memory
SENSE STATUS OF DEVICE	Load status word of device specified in addr into register specified
OUTPUT COMMAND TO DEVICE	Send command word in register specified to device specified in addr
Shift instructions	
SHIFT REGISTER {LEFT / RIGHT} {SINGLE / DOUBLE} {ARITHMETIC / LOGICAL} {CYCLIC / (null)}	Shift (REGISTER) addr places as specified

† Otherwise (PC) + 1 → (PC) in all cases

(continued)

Operation code	Description of operation
Pseudo-operations	
BINARY	Exactly reproduce the binary number in the operand as a machine word
OCTAL	Transform the octal number in the operand into a machine word
DECIMAL	Transform the decimal number in the operand into a machine word
HEXADECIMAL	Transform the hexadecimal number in the operand into a machine word
LOAD ZERO	0 → (REGISTER)
NO OPERATION	(PC) + 1 → (PC)
Assembler directives	
END	Marks end of program unit
RESERVE	(Location counter) + addr → (Location counter)

Fig. 7.6. Generalized assembly language instructions

REFERENCES

C. W. Gear, *Computer Organization and Programming*, McGraw-Hill, 1969.

W. D. Maurer, *Programming: An Introduction to Computer Languages and Techniques*, Holden-Day, 1969.

F. K. Walnut, *Introduction to Computer Programming and Coding*, Prentice-Hall, 1968.

J. K. Iliffe, *Basic Machine Principles*, American Elsevier, 1968.

QUESTIONS

1. Differentiate clearly between the symbolic operand address, the numerical operand address, and the effective address for a memory reference instruction.
2. In a register-to-register transfer instruction, is it possible to use indexing or indirect addressing?
3. What are the possible dangers and the possible benefits of mixing floating-point and integer arithmetic instructions operating on the same data in a program?
4. In some small machines, conditional jumps are not allowed; instead, a SKIP instruction is implemented. This instruction increments the contents of the program counter by 2, thereby skipping the next instruction. The SKIP may be done conditionally,

that is, testing for accumulator zero, etc. Write a short program using SKIP instructions that tests two input/output devices to see whether they are busy. Wait if they are busy; branch to service routines for them if they are not.

5. What are the advantages and disadvantages of two-way branches, in which control is transferred to one address location if a condition is satisfied and to another address location if it is not?
6. Referring to the example in Section 7.7 in which a negative zero is converted to positive zero, rewrite this program using the LOAD MAGNITUDE REGISTER instruction.
7. One hundred numbers are stored in a linear array. Write a program to calculate the sum of these numbers, their average, and the sum of their squares. Use looping and indexing techniques.
8. Using conditional JUMP and ADD instructions, write a small routine that simulates the behavior of the single instruction, INCREMENT INDEX REGISTER JUMP IF LESS THAN REGISTER.
9. Using conditional JUMP and arithmetic instructions, write a small program that simulates or is equivalent to the machine instruction, DECREMENT REGISTER JUMP IF NONNEGATIVE.
10. *Assembly language correlation exercise.* It is assumed that readers of this book have access to a computer. Using the reference manual for that computer in conjunction with Fig. 7.6, construct a table comparing the symbolic operation code and the actual operation of each instruction in your machine with one of the instruction types in the generalized assembly language. Make two additional lists, one of instructions available in generalized assembly language but not available in your machine, the other of instructions available in your machine but not in the generalized assembly language. For each instruction missing from one or the other language, indicate how you think it might be implemented by means of instructions available in that language. For example, if your machine does not have floating-point add instructions, block out a way to accomplish the task involved, using the instructions your machine does have.
11. What precautions must be taken with data or instructions stored in various machine registers before branching to a subroutine? Describe how the JUMP TO SUBROUTINE instruction is different from the unconditional JUMP instruction.
12. Is it possible or feasible to have a conditional JUMP TO SUBROUTINE instruction? What would be the advantages of such a code?
13. What precautions would you take if, in writing a closed subroutine, you had to call another subroutine from it?
14. Why is it not a good idea to use an unconditional JUMP to a given address in the main program as the main exit from a closed subroutine?
15. Why is it necessary to have an explicit computer instruction NO OPERATION?
16. Write at least three computer instructions that have the same actual effect as the hardware NO OPERATION instruction.
17. What would be the advantages, if any, in storing the return locations for subroutines

in a push-down stack? Would this procedure be more important if any given subroutine called another, which in turn called another?

18. Consider the following column of numbers:

 1
37
54
83
91
14
 0

If this column is a linear array, what is the fifth element entered in the list? If it is a push-down stack, with the most recent entry to the stack at the bottom of the column, what is the fifth element entered in the list? If the column is a linear array, what must be done to the index register pointing to it to access the $n + 1$ element? If it is a push-down stack, what must be done to access the $n + 1$ element?

19. Prepare a flowchart for and write in assembly language a subroutine to execute the PUSH JUMP function. You may call the PUSH subroutine given in the text.

20. Prepare a flowchart and then write a subroutine to execute the POP JUMP instruction. You may use the POP subroutine given in the text.

21. What are the possible advantages of using a push-down stack to store subroutine return addresses? Explain why this procedure is particularly useful when subroutines are nested several levels deep.

22. Design the logic for the main memory of a computer in such a way that data may be accessed from memory, either by the central processing unit for the purpose of computation or data formatting, or by an input/output device for the purpose of loading or unloading the contents of the memory. Indicate what protection devices are necessary to prevent interference between the two different ways of accessing the memory.

23. What parameters are necessary to completely specify a block transfer of data?

24. Given a computer without block transfer instructions, i.e., with instructions, READ DATA and WRITE DATA that can read or write only single characters of words, write a subroutine, using the SENSE STATUS, OUTPUT command, and READ DATA or WRITE DATA commands, which would effect a block transfer. Do so for both inputting blocks and outputting blocks.

25. A computer uses a 32-bit word in main memory. Alphanumeric data are stored in 8-bit bytes, packed 4 to a word. Show how the logical instructions can be used to unpack the word so that the 8-bit bytes are stored one to the word, left-justified.

26. Construct a truth table for binary addition, and compare it to the **AND**, **OR**, and **EXCLUSIVE OR** functions. Be sure to include both the sum and the carry to the next digit.

27. Using the TRANSMIT DATA or RECEIVE DATA instructions for single bytes, write an assembly language program to accept 8-bit bytes from a Teletype. Use the shift and logical instructions to pack the 8-bit bytes 4 to the word.

28. Using the TRANSMIT DATA instruction for single bytes, write a computer program to unpack 8-bit bytes that are packed 4 to the word, and write them on an 8-bit device, such as a Teletype.
29. What would be the effect of shifting any computer word four places to the left by means of a LOGICAL SHIFT and then four places to the right by means of an ARITHMETIC SHIFT?
30. Explain why it is necessary to shift 1's in at the left end of a register that is subject to an ARITHMETIC SHIFT RIGHT. Is it necessary for a sign-magnitude machine?
31. Explain why it might be useful to have the direction and magnitude parameters in a shift instruction indexable or indirectly addressable.
32. Distinguish clearly between a pseudo-operation and an actual operation in assembly language.
33. Give some examples in which binary, octal, or hexadecimal constants would be more convenient than the decimal constants normally generated by an assembler.
34. Explain how the RESERVE assembler directive can be used to give a symbolic name to the first element of an array. Using relative addressing, show that any member of the array may be similarly addressed.
35. *Computer concepts.* Define or explain briefly each of the following.
    a) accumulator
    b) index register
    c) double-precision arithmetic
    d) pointer
    e) indirect addressing
    f) push-down stack
    g) buffer
    h) calling sequence
    i) pseudo-operation
    j) immediate addressing
    k) effective addressing
    l) assembler language and machine language
    m) indexing
    n) subroutine and subroutine linkage
    o) relative addressing
36. Suppose that you are to write in assembler language a subprogram to be called as a function (as opposed to a subroutine) from a FORTRAN program. If your subprogram is to work correctly, what must you know about (a) the conventions used by FORTRAN and (b) the things "being done" in the calling program?
37. Suppose that a computer has all the features of the general machine of this chapter, but only one double-length accumulator and hardware for only the following two left-shift operations.

Operation code	Description of operation
LEFT SHIFT LOGICAL	Shift contents of accumulator left N places (as specified in address field). Zero-fill from right.
LEFT CIRCULAR SHIFT	Shift contents of accumulator left N places (as specified in address field). Shift contents of leftmost bit into rightmost bit.

Assuming a 16-bit word and 1's complement representation of negative numbers, write software subroutines to accomplish the following two right shift operations.

RIGHT SHIFT ARITHMETIC	Shift contents of accumulator right N places (as specified in address field). Sign fill from left.
RIGHT SHIFT CIRCULAR	Shift contents of accumulator right N bits (as specified in address field). Shift contents of rightmost bit into leftmost bit.

For the purpose of subroutine linkage, assume that the count N is stored in a location symbolically referenced as PLACES. You need to provide subroutine entry and return linkage.

38. We wish to write a piece of program in assembly language for the following machine. The machine uses 18-bit words (integers are 17 bits plus sign; the leading digit is 1 if the integer is negative and 0 if it is positive or zero). Overflow is not detected, but the sum of two positive numbers is negative if overflow occurs. The machine has eight general-purpose registers, each of which can be used as an accumulator.

    The assembly language instructions are of the form

    L OP β, α

    where L is a label, OP is the operation code, β is the register involved (0, . . ., 7), and α is the memory address. All instructions require two memory cycles.

    The available instructions follow.

STR β, α	$(\beta) \rightarrow (\alpha)$	that is, the contents of register β are stored in memory location α.
CLA β, α	$(\alpha) \rightarrow (\beta)$	
ADD β, α	$(\beta) + (\alpha) \rightarrow (\beta)$	
SUB β, α	$(\beta) - (\alpha) \rightarrow (\beta)$	
TRA β, α		Take next instruction from location α (β not used).
TRN β, α		If $(\beta) < 0$, take next instruction from location α; otherwise, continue with next instructions in order.
W 1		Defines the contents of location W as 1.

    Using only these instructions, write a segment of code to find the sum of the magnitudes of a set of n integers. Assume that the number n has been stored in location N and that the integers are in locations A, A + 1, . . ., A + N − 1. The result is to be left in register 0. If overflow occurs, register 0 is to contain −1.

    Hints: The location of A can be found with address modification; no indexing is available; you must assume that the address field is in the right part of word.

39. Compare the three techniques of subroutine linkage by postulating subroutines for which each linkage technique is in some sense optimal.

40. Examine the code produced by the compiler(s) you use to determine the existence and frequency of use of each of the subroutine linkage techniques.

SECTION 3
HARDWARE DESIGN

8
COMPUTER LOGIC

8.1 BOOLEAN ALGEBRA

We are now familiar with the CPU and the various subsystems that constitute a computer system. From our study of assembly language we know what functions these units perform and how to use them to achieve a desired result. We have not yet looked closely into the internal operation of these units to learn how they are designed or how they work. To study the design and construction of the electronic digital computer we must first have a working knowledge of *Boolean algebra.*

Propositional algebra

Boolean algebra, named after George Boole whose work [1847, 1854] defined the field, was designed as a propositional algebra and is so used by logicians today. A *proposition* is an assertion or statement which may be either correct or not correct. If a proposition is correct, we say it is "true" and symbolically note this state of being true by the letter T or the numeral 1. If a proposition is not true, we say it is "false" and symbolically note this state by the letter F or the numeral 0. Boolean algebra is a two-state algebra; there are no other possible states than true or false.

To clarify the concept of propositions, let us examine a few.

1. Alaska is the westernmost state in the United States.
2. Stop signs are triangular in shape.
3. The sun is shining.

From the facts of U.S. geography and traffic regulations in the 1970's, we know that proposition 1 is true and proposition 2 is false. To evaluate proposition 3 we have to look out the window. Propositions 1 and 2 are constants; proposition 3 is a variable.

Boolean algebra is concerned with defining relationships among propositions so as to make it possible to evaluate the truth or falsehood of a combination of propositions. Since it would be exceedingly cumbersome to write out completely each proposition in such a combination, we resort to symbols to represent propositions, just as in ordinary algebra. For example, let us say that a represents the proposition "Alaska is the westernmost state in the United States," and b represents the proposition "The sun is shining." The symbols a and b are thus Boolean symbols, representing propositions. We then may write $a = 1$, expressing the truth of proposition a, but we may not say anything about proposition b until we have looked out the window.

To clarify this representation of propositions, let us introduce one Boolean operator, called **AND**, to be formally defined later. The expression

$$f = a \textbf{ AND } b \tag{8.1}$$

is true if and only if a is true *and* b is true. Then f represents the combination of propositions "Alaska is the westernmost state in the United States **AND** the sun is shining." There is no requirement that propositions have any relationship to each other.

An algebra of classes

Another equally valid application of Boolean algebra is as an algebra of classes. A *class* is a set of objects or concepts which have a definable property in common. Sample classes are "men over six feet tall," which we may represent as class a, and "policemen," which we may represent as class b.

A class may be empty; that is, there may be no objects which meet the qualifications set forth in defining the class. Any empty class is an example of the *null class*. The symbol for the null class is the numeral 0.

In a discussion of classes, it is often convenient to use a graphical representation as an aid to comprehension. A simple Venn diagram for a single class, a, is shown as Fig. 8.1. The circle represents the bounds of class a. All objects belonging to class a are within the circle; all other objects are outside the circle.

By our definition of classes, either an object is a member of a class or it is not. There is no in-between state. All objects not belonging to class a are outside the circle of the Venn diagram. The rectangle in the diagram represents the boundary on the objects we are going to consider. This class of all objects under consideration, called the *universe of discourse*, is denoted by the numeral 1.

Consider a Venn diagram representing the two classes a and b introduced above and illustrated in Fig. 8.2. All "objects" satisfying the definition of class a, men over six feet tall, lie within circle a. All objects satisfying the definition of class b, policemen, lie within circle b. Note that there are some objects which

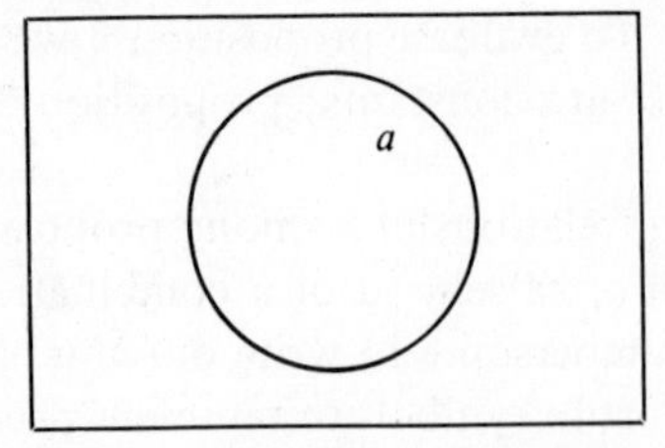

Fig. 8.1. Venn diagram of class *a*

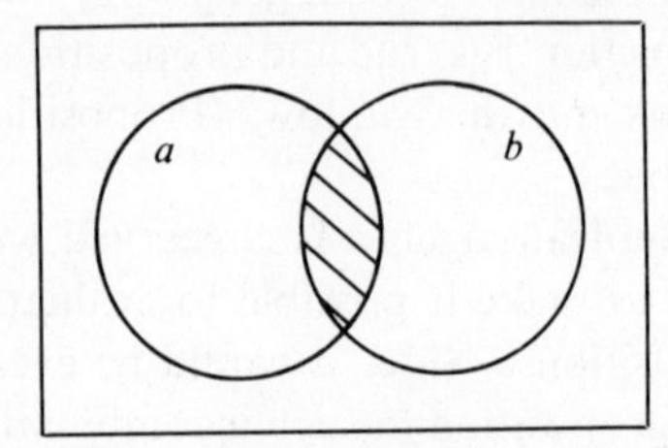

Fig. 8.2. Intersection of two classes

satisfy the definitions of *both* classes, namely, policemen over six feet tall. These objects lie in the shaded area of Fig. 8.2. This area represents a new class, called the *intersection* of classes a **AND** b.

Switching algebra

The application of Boolean algebra to a practical problem gives rise to a third interpretation. The problem is to describe the necessary conditions for continuity of transmission through a circuit. The circuit is composed of interconnected switches. When a switch is "closed" there is transmission through it, represented by the numeral 1; when "open" there is no transmission, represented by the numeral 0. The schematic representation of open and closed switches is shown in Fig. 8.3.

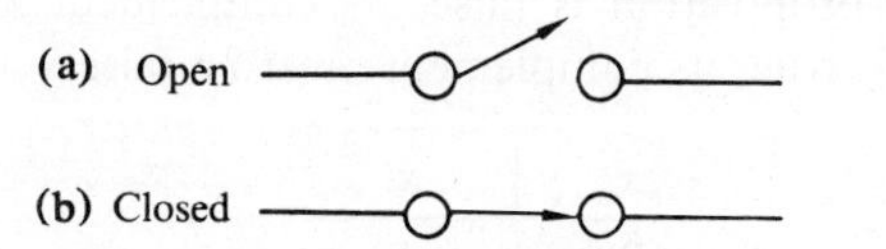

Fig. 8.3. A switch: (a) Open switch; (b) Closed switch

Fig. 8.4. Two switches in series

Very complicated switching circuits may be described by Boolean algebra. Such description is a very powerful tool for logic circuit designers. As an informal introduction to switching algebra involving two switches and thus two switching variables, consider the circuit of Fig. 8.4. Since this circuit has two switches in series there must be transmission through both switches in order that there be transmission through the circuit. If we denote transmission by the variable t, then we write

$$t = a \textbf{ AND } b \tag{8.2}$$

Note the similarity to Eq. (8.1).

8.2 TRUTH TABLES

In all fields of numerical mathematics, theorems may be proved and operations defined by the process of enumeration of all possible results. Enumeration is particularly well suited to Boolean algebra, which possesses only the two states of 0 and 1. Boolean algebra tables of enumeration are called *truth tables*.

A truth table has one column for each input operand and one column for each output result. When discussing operations, we will use the truth table to define the operation in question by specifying the results for all combinations of inputs. When discussing proofs of theorems, we will prepare truth tables for each expression of a pair that is to be proved equal. If the truth tables are identical, then the expressions they represent must also be equal. If the truth tables are different, then the expressions cannot be equal.

8.3 NEGATION

Negation is the only *unary* operation in the Boolean algebra (each operator is associated with *one* operand). If we let this single operand be x, then negation produces the *complement* of x, denoted $\bar{x}$, x', $\neg x$, or **NOT** (x). The notation for negation is largely a matter of individual preference and convenience. The bar and prime are sometimes intermixed. The four notations are vocalized as "x bar," "x prime," "not x," and "not x," respectively. This book will use "x prime" notation. The truth table for negation is extremely simple (Fig. 8.5). Its interpretation is that whenever a proposition is false, its complement must be true, and whenever a proposition is true, its complement must be false.

x	x'
0	1
1	0

Fig. 8.5. Truth table for negation

Applied to classes, negation states that those objects which do not satisfy the definition of class a do satisfy the definition of those objects not in class a, and vice versa. As shown in Fig. 8.6, class a lies within the circle of the Venn diagram, and class a' lies outside the circle.

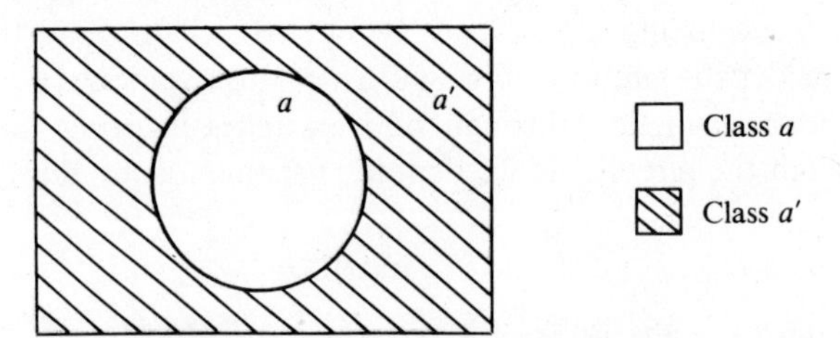

Fig. 8.6. Venn diagram illustrating negation

Transmission through a switch may have one of two possible relationships with the state of the variable represented by the switch. Let t be the transmission through the switch, and let x be the variable represented by the switch. The *normally open* switch, shown in Fig. 8.7(a), is described by the relationship

$$t = x \tag{8.3}$$

and the *normally closed* switch (Fig. 8.7b), is described by

$$t = x' \tag{8.4}$$

Note that the transmission through the normally open switch is 1 only when the

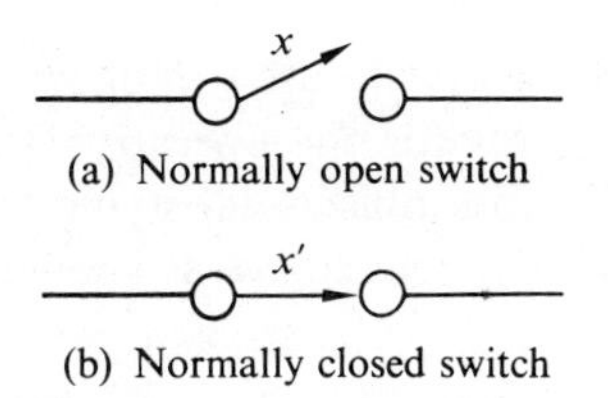

Fig. 8.7. Negation applied to switches: (a) Normally open switch; (b) Normally closed switch

variable represented is 1. Transmission through the normally closed switch is 1 only when the variable is 0.

8.4 OR

An expression formed with the **OR** operator (sometimes called **INCLUSIVE OR**) is true if either proposition is true or if both propositions are true. If the two propositions are denoted a and b, then the expression may be written a **OR** b, $a + b$, or $a \bigcup b$. The symbol + is properly called **OR**; $\bigcup$ is called "cup." The truth table is given in Fig. 8.8. The symbol $\bigcup$ stands for the operation known as *union*. The expression $a + b$ represents all those objects satisfying the definition of class a or of class b or of both. The union of two classes is illustrated in Fig. 8.9. Hereafter in this book, + will be used to represent the **OR** operation.

a	b	$a + b$
0	0	0
0	1	1
1	0	1
1	1	1

Fig. 8.8. Truth table for OR

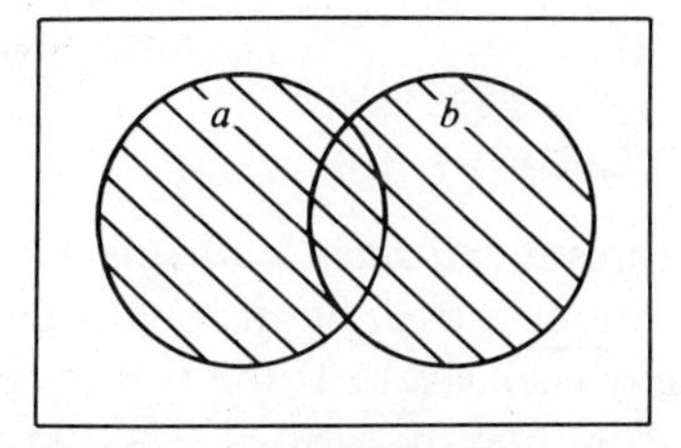

Fig. 8.9. Venn diagram of $a + b$

Applied to switches, the logical **OR** may be represented as parallel branches (Fig. 8.10). Transmission occurs when either switch is closed or when both are closed.

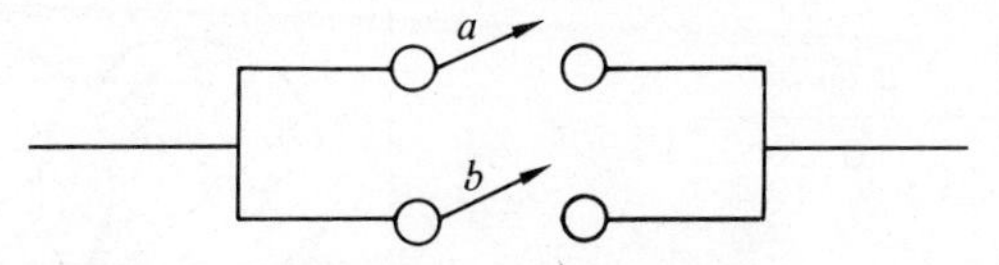

Fig. 8.10. OR of switches

8.5 AND

An expression formed with the **AND** is true if and only if both propositions are true. If the two propositions are represented as a and b, then the expression may be written a **AND** b, $a \cdot b$, ab, or $a \bigcap b$. The symbol $\cdot$ is pronounced "and"; the

$\cap$ is called "cap." In this book we will use $a \cdot b$ or ab. The truth table is given in Fig. 8.11. The symbol $\cap$ stands for the operation known as *intersection.* The expression $a \cdot b$ represents all those objects satisfying the definitions of both class a and class b. The intersection of two classes is shown in Fig. 8.12.

a	b	$a \cdot b$
0	0	0
0	1	0
1	0	0
1	1	1

Fig. 8.11. Truth table for AND

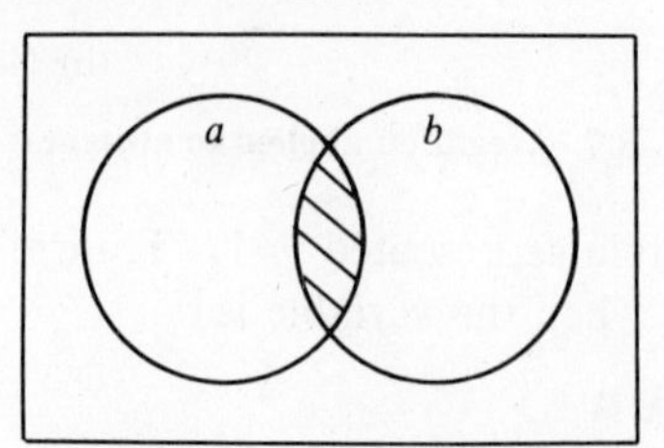

Fig. 8.12. Venn diagram of $a \cdot b$

Since the logical **AND** is a binary operator, it may be represented by two switches. From the truth table (Fig. 8.11) we observe that there is to be transmission only when switches a **AND** b are both closed. This is accomplished with two switches in series, as shown in Fig. 8.13.

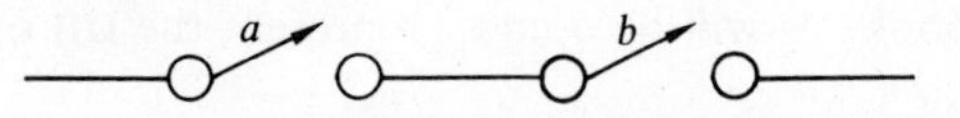

Fig. 8.13. Series switches forming AND

8.6 EXCLUSIVE OR

An expression formed with the **EXCLUSIVE OR** is true if either one of the two propositions is true, but it is not true if both propositions are true (that is, they are mutually exclusive). If the two propositions are represented as a and b, then the expression may be written a **EXCLUSIVE OR** b, or $a \oplus b$. The truth table for this operator is given in Fig. 8.14. In terms of classes, $a \oplus b$ represents all those objects satisfying the definition of class a or of class b but not of both. For the Venn diagram, see Fig. 8.15.

a	b	$a \oplus b$
0	0	0
0	1	1
1	0	1
1	1	0

Fig. 8.14. Truth table for EXCLUSIVE OR

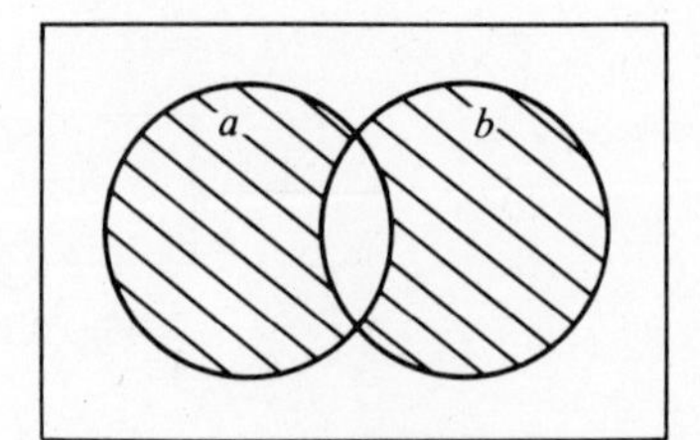

Fig. 8.15. Venn diagram of $a \oplus b$

When implemented with switches the **EXCLUSIVE OR** appears to require four switches, as drawn in Fig. 8.16. If a special type of switch is used, the **EXCLUSIVE OR** can be implemented with only two switches. The **EXCLUSIVE OR**

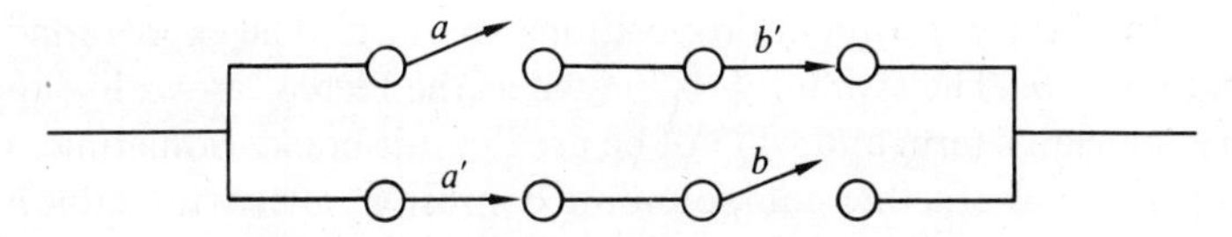

Fig. 8.16. Switches forming EXCLUSIVE OR

is extensively employed in logic circuits for the formation of binary sums, as we shall see later.

8.7 NAND

The **NAND** may be thought of as an **AND** followed by negation. An expression formed with the logical **NAND** is true if either one of the two propositions is false, or if both are false. The expression may be written as a **NAND** b, or $a \uparrow b$. The symbol $\uparrow$ is known as the Sheffer stroke; this name is employed primarily by logicians and will not be used in this book. The truth table is shown in Fig. 8.17.

a	b	a NAND b
0	0	1
0	1	1
1	0	1
1	1	0

Fig. 8.17. Truth table of NAND

Interpreted in terms of classes, the **NAND** operation defines all those objects which are not in both class a and class b; in other words, it defines the universe of discourse except for the intersection of a and b. The Venn diagram is presented as Fig. 8.18. The **NAND** operation is simply performed with two parallel switches, as shown in Fig. 8.19.

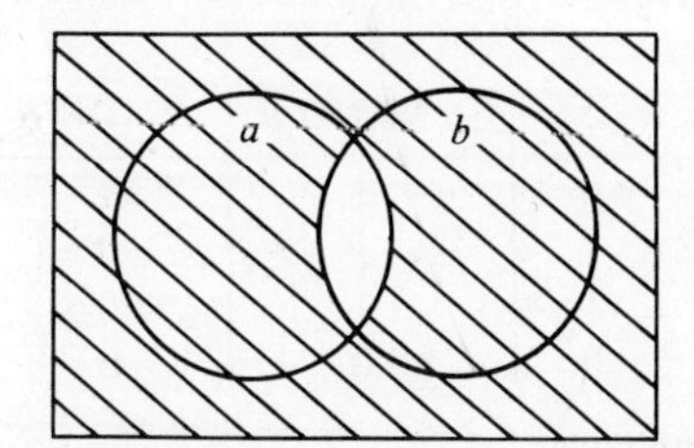

Fig. 8.18. Venn diagram of a NAND b

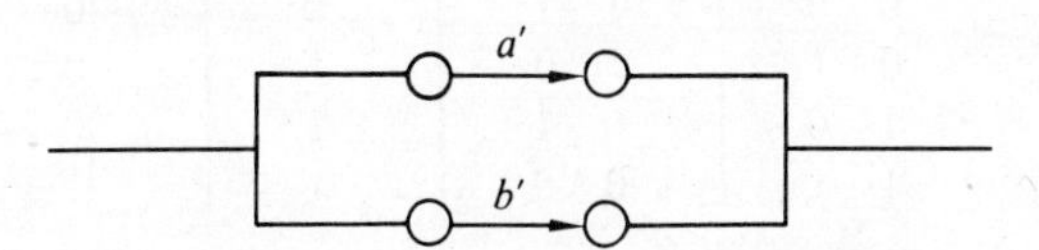

Fig. 8.19. Switches performing NAND

8.8 NOR

The final elementary binary operator of Boolean algebra we shall consider is **NOR**, which is an **OR** followed by negation. An expression formed with **NOR** is true if and only if both of the constituent propositions are false. The expression is written as *a* **NOR** *b*, or $a \downarrow b$. The symbol $\downarrow$ is known as the Pierce arrow; like the Sheffer stroke, it is a logician's term and will not be used in this book. Sometimes the slash, /, is used in place of $\downarrow$, further compounding confusion. The truth table is given in Fig. 8.20.

a	b	a NOR b
0	0	1
0	1	0
1	0	0
1	1	0

Fig. 8.20. Truth table of NOR

As a class operator, the **NOR** defines all those objects which are in neither class *a* nor class *b*, namely, the universe of discourse except for the union of *a* and *b*. The Venn diagram is presented as Fig. 8.21. Two switches in series serve to execute the **NOR** operation, as shown in Fig. 8.22.

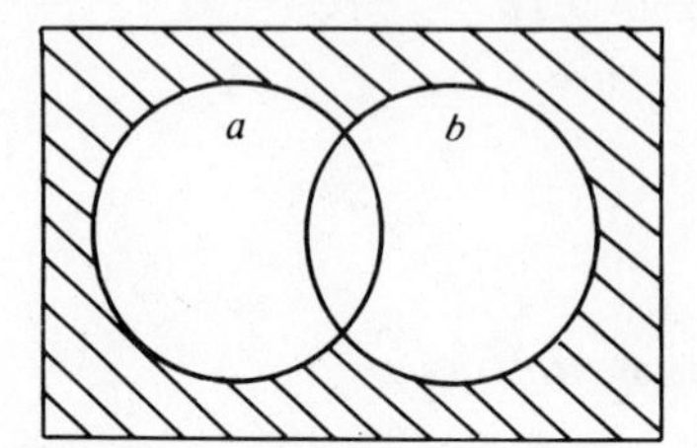

Fig. 8.21. Venn diagram of *a* NOR *b*

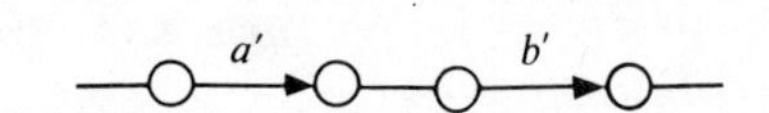

Fig. 8.22. Switches performing NOR

8.9 COMPOSITE TRUTH TABLE

To summarize the operations introduced above, we present all their truth tables together in Fig. 8.23.

a	b	a'	b'	$a + b$	$a \cdot b$	$a \oplus b$	a NAND b	a NOR b
0	0	1	1	0	0	0	1	1
0	1	1	0	1	0	1	1	0
1	0	0	1	1	0	1	1	0
1	1	0	0	1	1	0	0	0

Fig. 8.23. Composite truth table

8.10 BOOLEAN EXPRESSIONS

Using any of the binary Boolean operators just defined, we may write a simple Boolean expression in the form

$$p = a \langle \text{Boolean operator} \rangle b$$

where the truth of p is dependent on the truth of a and b and on the definition of the ⟨Boolean operator⟩, as expressed in its defining truth table. In general, we are interested in more complex expressions involving many variables and operators.

8.11 PRECEDENCE AND PARENTHESES

In writing more complex Boolean expressions, we must take pains to avoid ambiguities. Ambiguities occur whenever it is unclear which Boolean operator is to be applied first. We solve this problem by establishing the following order of precedence for the Boolean operators.

NOT
AND NAND
OR EXCLUSIVE OR NOR

The higher operation is always performed first, and operators on the same line have equal precedence.

Applying precedence relations to operators does not permit the identification of all possible and desirable expressions. Therefore parentheses are used to enclose Boolean subexpressions which are to be evaluated first. Thus the evaluation of a Boolean expression must proceed from inside to out, following precedence. Finally, evaluation is carried out from left to right if any ambiguities remain. The following expressions are examples of somewhat involved Boolean expressions. The development of truth tables is left as an exercise.

$$f = ((a + b)' + (c + d')' \oplus (b'd + a'b'c))' + (abc')' \tag{8.5}$$

$$g = (b'd' + c'd' + a'b)' \cdot (cd + ab + a'b') + (a'b'd' + b'c'd') \tag{8.6}$$

8.12 PROPERTIES OF BOOLEAN ALGEBRA

Our discussion of Boolean algebra would be incomplete, as would our ability to work with Boolean expressions, without discussion of the properties of this algebra.

Boolean algebra is different from the algebra of numbers. Since it deals directly with binary (true-false) states, it is especially appropriate for logic circuits. For purposes of mathematical elegance, it is customary to present a minimum number of postulates and to deduce theorems from them. The set of postulates and theorems constitute the properties of the Boolean algebra.

The set of postulates and theorems is not unique. It is possible to make a new selection among these properties, calling some postulates and the remainder theorems to be proved by application of the new postulates.

Postulates of Boolean algebra

The following set of postulates is known as Huntington's postulates [1904]:

Postulate 1 0 and 1 are unique identity elements such that

$$a + 0 = a \tag{8.7a}$$
$$a \cdot 1 = a \tag{8.7b}$$

Postulate 2 The Boolean algebra is *complete*

$$a + a' = 1 \tag{8.8a}$$
$$a \cdot a' = 0 \tag{8.8b}$$

Postulate 3 The operations **OR** and **AND** are *commutative*

$$a + b = b + a \tag{8.9a}$$
$$a \cdot b = b \cdot a \tag{8.9b}$$

Postulate 4 The operations **OR** and **AND** are *distributive*

$$a \cdot (b + c) = a \cdot b + a \cdot c \tag{8.10a}$$
$$a + (b \cdot c) = (a + b) \cdot (a + c) \tag{8.10b}$$

Theorems of Boolean Algebra

Theorem 1 The Boolean algebra is *idempotent.*

$$a + a = a \tag{8.11a}$$
$$a \cdot a = a \tag{8.11b}$$

The next five theorems are useful when manipulating Boolean expressions.

Theorem 2

$$a + 1 = 1 \tag{8.12a}$$
$$a \cdot 0 = 0 \tag{8.12b}$$

Theorem 3 The Boolean algebra is *absorptive.*

$$a + (a \cdot b) = a \tag{8.13a}$$
$$a \cdot (a + b) = a \tag{8.13b}$$

Theorem 4

$$a + (a' \cdot b) = a + b \tag{8.14}$$

Theorem 5

$$a \cdot b + a' \cdot c + bc = ab + a'c \tag{8.15}$$

Theorem 6

$$(a + b) \cdot (a + c) = a + bc \tag{8.16}$$

Theorem 7 The Boolean algebra is *associative.*

$$(ab)c = a(bc) \tag{8.17a}$$
$$(a + b) + c = a + (b + c) \tag{8.17b}$$

Using the property of associativity, one may employ the following notation to indicate that a function is formed by the **OR** of the variable a_1 through a_n:

$$f = \sum_{i=1}^{n} a_i = a_1 + a_2 + a_3 + \cdots + a_n \tag{8.18}$$

Similarly, the notation that a function is formed by the **AND** of the variable:

$$g = \prod_{i+1}^{n} a_i = a_1 \cdot a_2 \cdot a_3 \cdot \quad \cdots \quad \cdot a_n \tag{8.19}$$

Theorem 8 The Boolean algebra is *involuted.*

$$(a')' = a \tag{8.20}$$

The final two theorems, known as *DeMorgan's theorems*, deal with the process of negating **AND** and **OR** expressions of multiple terms. The proof of one will be given completely; the proof of the other is left as an exercise. The process we will use to prove the first DeMorgan theorem is a proof by induction. That is, given a measure of size n, we will prove the theorem true for a specified minimum value of n; we will also prove that *if* the theorem is true for $n = k$, it is also true for $n = k + 1$. Thus it is true for all values of n.

Theorem 9 $$(a_1 + a_2 + \cdots + a_n)' = a_1' \cdot a_2' \cdot a_3' \cdot \quad \cdots \quad \cdot a_n' \tag{8.21}$$

If we let $n = 2$, the form simplifies to

$$(a_1 + a_2)' = a_1' \cdot a_2' \tag{8.22}$$

which is easily proved by the truth table in Fig. 8.24. For the second half of the

a_1	a_2	$a_1 + a_2$	$(a_1 + a_2)'$	a_1'	a_2'	$a_1' \cdot a_2'$
0	0	0	1	1	1	1
0	1	1	0	1	0	0
1	0	1	0	0	1	0
1	1	1	0	0	0	0

Fig. 8.24. Truth table for proof of DeMorgan's theorem for $n = 2$

proofs we assume the theorem to be true for $n = k$. That is, we assume the following expressions to be true.

$$f_1 = a_1 + a_2 + \cdots + a_k \tag{8.23a}$$
$$f_1' = (a_1 + a_2 + \cdots + a_k)' \tag{8.23b}$$
$$= a_1' \cdot a_2' \cdot \quad \cdots \quad \cdot a_k' \tag{8.23c}$$

Now let us form a new Boolean function, f_2.

$$f_2 = f_1 + a_{k+1} \tag{8.24a}$$
$$= a_1 + a_2 + \cdots + a_k + a_{k+1} \tag{8.24b}$$

We may write the negative of f_2, using the truth table of Fig. 8.24 as justification; it is, after all, a function of two Boolean variables

$$f_2' = f_1' \cdot a'_{k+1} \tag{8.25}$$

We may expand Eq. (8.25) by inserting the definition of f_1' from Eq. (8.23c)

$$f_2' = a_1' \cdot a_2' \cdot \quad \cdots \quad \cdot a_k' \cdot a_{k+1}' \tag{8.26}$$

Combining Eqs. (8.24b) and (8.26), we obtain (8.21), which is the original statement of DeMorgan's theorem. Our proof is complete.

The second DeMorgan theorem takes the following form.

Theorem 10 $\quad (a_1 \cdot a_2 \cdot \quad \cdots \quad \cdot a_n)' = a_1' + a_2' + \cdots + a_n' \quad (8.27)$

The proof is left as an exercise.

8.13 DUALITY

The principle of duality for Boolean algebra may be employed to advantage. Duality provides a way of transforming one proposition or Boolean expression into another. If the original proposition is true, then so is its dual.

There are three steps in forming a *dual*.

1. Replace all 1's by 0's and 0's by 1's.
2. Replace all **AND**s by **OR**s and **OR**s by **AND**s.
3. Leave all **NOT**s alone.

The reader may review the properties of Boolean algebra presented in this chapter to discover all the duals so far presented. When employing duality to transform an expression, you may have to apply DeMorgan's theorems to remove parentheses.

Duality is a simple consequence of Huntington's postulates. The postulates themselves are duals, being symmetric with respect to **AND** and **OR**, 0 and 1. Since all proofs in Boolean algebra are based on these postulates, duality is a natural consequence.

8.14 TERMINOLOGY AND STANDARD FORMS

For ease in organizing Boolean expressions and for communicating with others, certain phrases have defined meanings which are applicable to the description of these Boolean expressions.

A *product term* is one or more Boolean variables connected by the **AND** operator. A *sum term* is one or more Boolean variables connected by the **OR** operator. Note that in the trivial case of one variable, the sum term and the product term are identical.

A *sum of products* is a Boolean expression composed of product terms connected by the **OR** operator. A *product of sums* is a Boolean expression composed of sum terms connected by the **AND** operator.

Example

$ab + c'd$ sum of products
$(a + b)(c' + d)$ product of sums

If a Boolean expression involves m Boolean variables, then a *standard product*, also known as a *minterm*, is a product term in which all m variables are explicit. A *standard sum*, also known as a *maxterm*, is a sum term in which all m variables are explicit. The eight minterms and eight maxterms of three variables are presented in Fig. 8.25. A *sum of standard products*, also known as a *canonical sum*,

All possible minterms of three variables	All possible maxterms of three variables
$a'b'c'$	$a' + b' + c'$
$a'b'c$	$a' + b' + c$
$a'bc'$	$a' + b + c'$
$a'bc$	$a' + b + c$
$ab'c'$	$a + b' + c'$
$ab'c$	$a + b' + c$
abc'	$a + b + c'$
abc	$a + b + c$

Fig. 8.25

or as *disjunctive normal form*, is, as the name implies, a sum of product terms, each of which is a standard product. Similarly, a *product of standard sums*, also known as a *canonical product*, or as *conjunctive normal form*, is a product of sum terms, each of which is a standard sum.

Frequently a switching function will be stated in nonstandard form. You may wish to put it into standard form. If the given function is a sum of products, you employ Eq. (8.7b) of Postulate 1 and Eq. (8.8a) of Postulate 2 to expand to standard sum of products form. Consider the switching function

$$f = ab + bc' + a'd \tag{8.28}$$

where f is a function of the four variables a, b, c, and d. We expand to standard form by **AND**ing to each product term 1 in the form $(a + a')$, as follows:

$$f = ab(c + c') + (a + a')bc' + a'd(c + c') \tag{8.29a}$$
$$= abc + abc' + abc' + a'bc' + a'cd + a'c'd \tag{8.29b}$$
$$= abc + abc' + abc' + a'cd + a'c'd \tag{8.29c}$$
$$= abc(d + d') + abc'(d + d') + abc'(d + d') + a'cd(b + b') + a'c'd(b + b') \tag{8.29d}$$
$$= abcd + abcd' + abc'd + abc'd' + abc'd + abc'd' + a'bcd + a'b'cd + a'bc'd + a'b'c'd \tag{8.29e}$$
$$= abcd + abcd' + abc'd + abc'd' + a'bcd + a'b'cd + a'bc'd + a'b'c'd \tag{8.29f}$$

Note the use of Theorem 1 (Eq. 8.11a) in going from (8.29b) to (8.29c) and from (8.29e) to (8.29f).

To expand a product of sums to standard form, the duals in Postulates 1 and 2 (8.7a) and (8.8b) are employed.

8.15 NUMBERING OF STANDARD PRODUCTS

A shorthand notation that assigns a (decimal) number to each standard product term makes it easy to write the canonical sum of standard products. The rules for assigning a number are simple.

1. Write the term in alphabetic order of the variables.
2. Replace all primed variables by 0 and all unprimed variables by 1.
3. Viewing the result as a binary number, express the value as a decimal number.

For example, consider the sum of standard products

$$f = abcd + a'bc'd + ab'cd' \tag{8.30a}$$

Converting to binary, we have

$$f = 1111 + 0101 + 1010 \tag{8.30b}$$

which in decimal is

$$f = 15 + 5 + 10 \tag{8.30c}$$

The shorthand notation, in the form of Eq. (8.18), is

$$f = \sum 5, 10, 15 \tag{8.30d}$$

which is certainly a convenience.

Although several notations have been proposed as standardized shorthand notation for the product of standard sums, none is universally accepted.

8.16 CIRCUIT IMPLEMENTATION OF BOOLEAN FUNCTIONS

The practical objective of our work with Boolean functions is their implementation in physical hardware as components of a computer system. To begin, let us consider representing the Boolean operators schematically.

In our schematic representation of the Boolean operators we shall show operator *gates* with certain numbers of input and output wires. Except for the single input on the **NOT** gate, the number of inputs shown should not be taken as an indication of what is physically realizable and available. Physical realizability and practical limitations will be discussed in Chapter 10. Some of the commonly employed logic circuit symbol sets used in schematic drawings are given in Fig. 8.26. The set in column (a) is standardized, as indicated.

The conventional way of drawing logic circuits is with inputs at the left and outputs at the right. All of the first-level gates are arranged in a column, with each input wire to each gate identified as to the variable it represents. The outputs of the

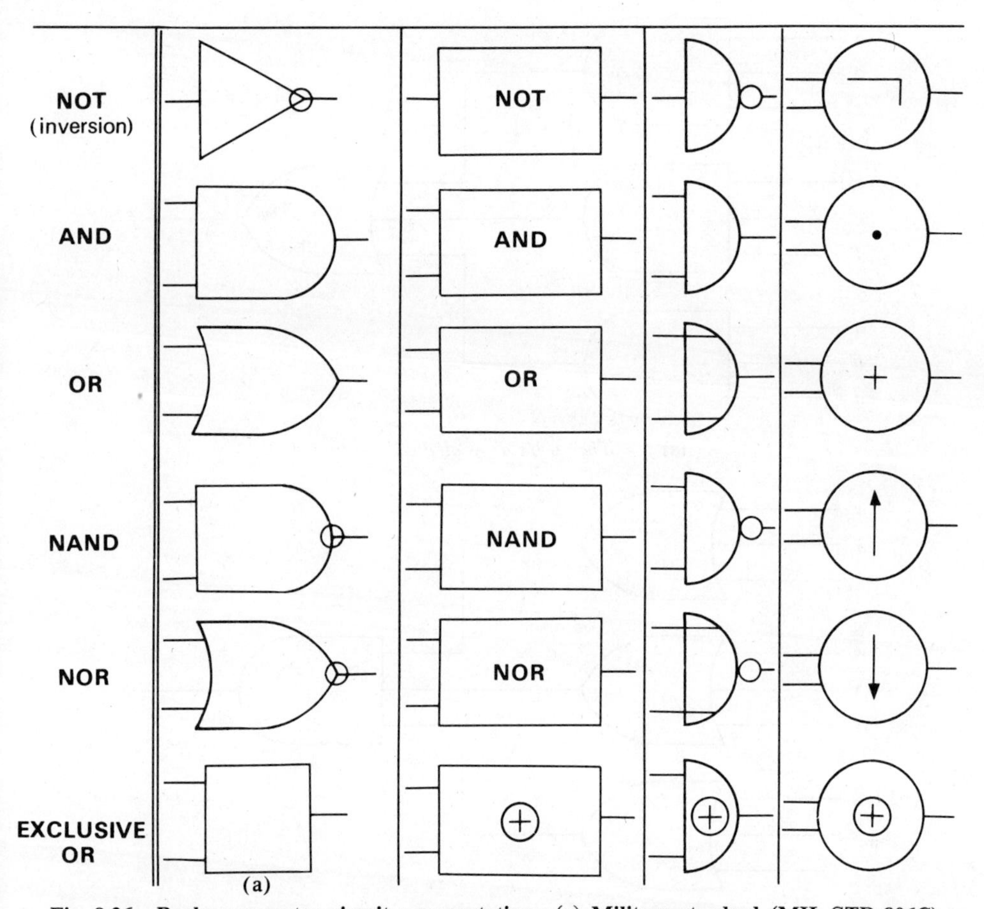

Fig. 8.26. Boolean operator circuit representations: (a) Military standard (MIL-STD-806C)

gates are then treated as inputs to succeeding gates (levels) until the desired function is formed.

When the Boolean function has been written in standard form, the logic diagram produced will be what is termed *two-level logic*. The first level, a column on the schematic diagram, will be all **AND**s for the sum-of-products form followed by a second level of **OR**s. For the product-of-sums form, the first level will be composed of **OR**s and the second level of **AND**s (see Fig. 8.27).

There was a time when passive devices were much less expensive than active devices (see Chapter 10 for elaboration), making it desirable to design two-level logic. Although this restriction no longer exists, it is still convenient and common to design logic circuits with even numbers of levels. An additional reason is presented in the following section.

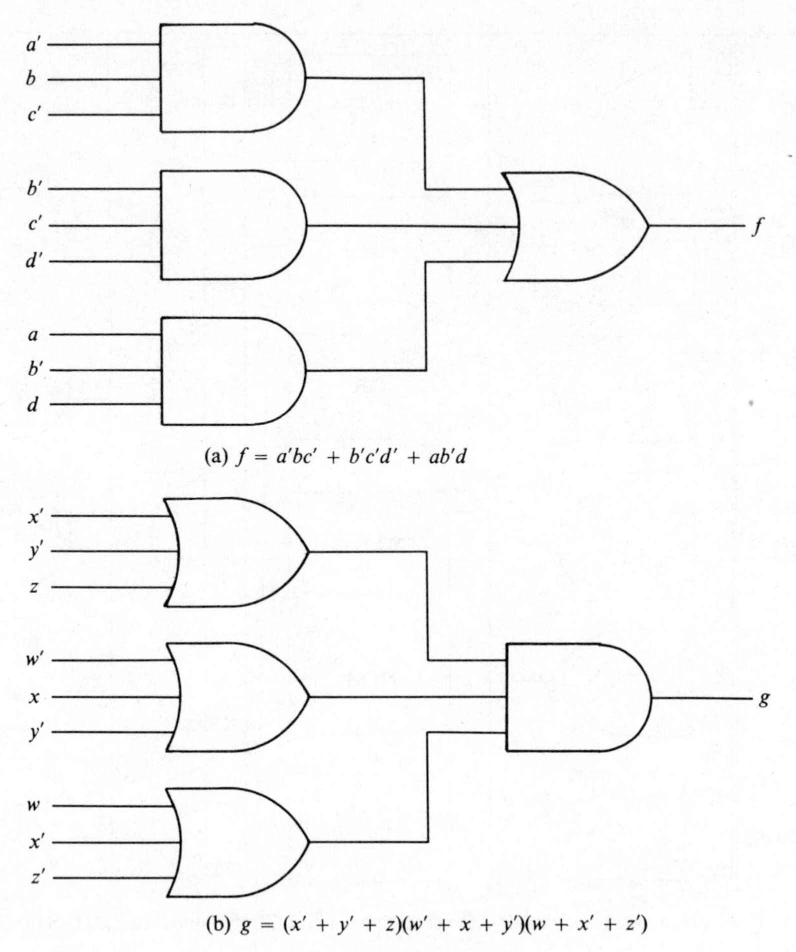

(a) $f = a'bc' + b'c'd' + ab'd$

(b) $g = (x' + y' + z)(w' + x + y')(w + x' + z')$

Fig. 8.27. Sample logic diagrams

8.17 COMPLETENESS OF NANDs AND NORs

It may appear that we have neglected the **NAND** and **NOR** operators in favor of **AND** and **OR**. Any Boolean expression can be implemented using only **NAND**s or only **NOR**s. In this sense, the **NAND** and **NOR** are each said to be *complete*.

The easiest function is inversion. We simply use one input of a **NAND** or **NOR** gate; the other input is forced to be 1 or 0, as appropriate. Using the two-input gates, we have, for example,

$$a \text{ NAND } 1 = (a \cdot 1)' = (a)' = a' \tag{8.31a}$$

$$a \text{ NOR } 0 = (a + 0)' = (a)' = a' \tag{8.31b}$$

In addition to being thought of as an **AND** followed by a **NOT**, the **NAND**

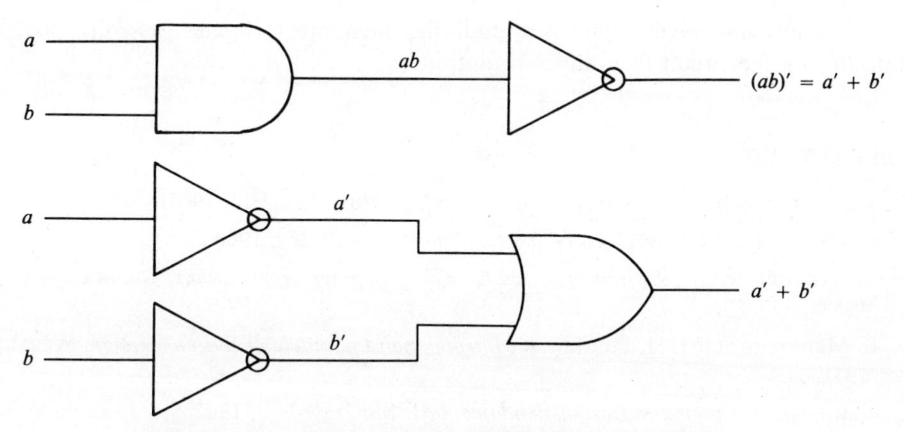

Fig. 8.28. The NAND as combinations of AND, OR, and NOT

may also be thought of as an **OR** with a **NOT** connected before each input, as shown in Fig. 8.28. Similarly, the **NOR** may be thought of as an **OR** followed by a **NOT** or an **AND** preceded by **NOT**s, as shown in Fig. 8.29. This structure follows from DeMorgan's theorems, Eqs. (8.21) and (8.27).

Circuit diagrams drawn for even-level **AND/OR** logic may be converted to all **NAND**s or all **NOR**s simply by direct replacement of all the **AND**s and **OR**s by either **NAND**s or **NOR**s, as appropriate. In a circuit designed as sums of products, all the **AND/OR** gates are replaced by **NAND** gates. In a circuit designed as products of sums, the **AND/OR** gates are replaced by **NOR** gates. A brief reflection on Figs. 8.28 and 8.29 shows that when even levels are used, the **NOT**s one may think of as included in the **NAND**s and **NOR**s will indeed cancel and be of no effect, because they come two in a row, according to Eq. (8.20).

If the **AND/OR** circuit design produces an odd number of levels of logic, it is necessary to add a **NOT** to restore evenness. The odd or even nature, if not evident

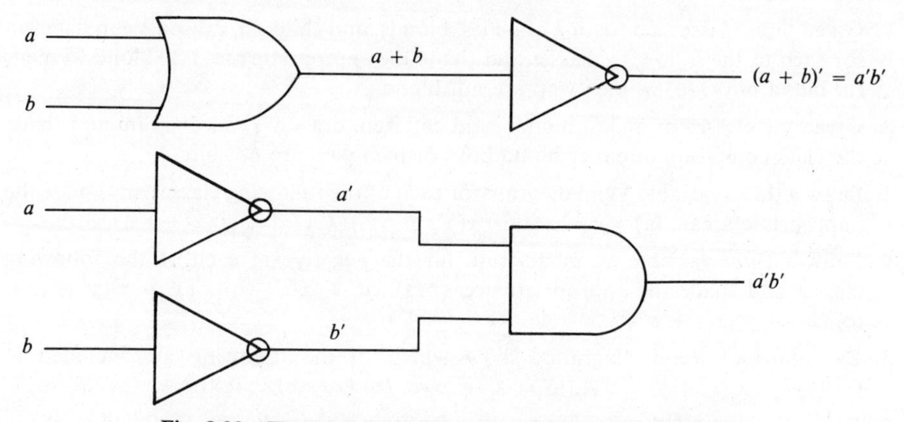

Fig. 8.29. The NOR as combinations of AND, OR, and NOT

by inspection, may be determined by counting the number of gates passed through in following the circuit from input to output.

REFERENCES

H. Hellerman, *Digital Computer System Principles*, McGraw-Hill, 1967.

S. H. Caldwell, *Switching Circuits and Logical Design*, Wiley, 1960.

W. S. Humphrey, Jr., *Switching Circuits with Computer Applications*, McGraw-Hill, 1958.

A. B. Marcovitz and J. H. Pugsley, *An Introduction to Switching System Design*, Wiley, 1971.

Y. Chu, *Digital Computer Design Fundamentals*, McGraw-Hill, 1962.

R. R. Korfhage, *Logic and Algorithms*, Wiley, 1966.

B. H. Arnold, *Logic and Boolean Algebra*, Prentice-Hall, 1962.

R. Serrell, "Elements of Boolean Algebra for the Study of Information-Handling Systems," *Proceedings of the I.R.E.*, October 1953, pp. 1366–1379.

G. Boole, *The Mathematical Analysis of Logic*, Cambridge, 1847; *An Investigation of the Laws of Thought on Which Are Founded the Mathematical Theories of Logic and Probabilities*, London, 1854.

C. Shannon, "A Symbolic Analysis of Relay and Switching Circuits," *Trans. AIEE*, **57**, 1938, pp. 713–723; "The Synthesis of Two-Terminal Switching Circuits," *Bell Syst. Tech. J.*, **28**, 1949, pp. 59–98.

E. V. Huntington, "Sets of Independent Postulates for the Algebra of Logic," *Trans. Amer. Math. Soc.*, **5**, 1904, pp. 288–309.

QUESTIONS

1. Given three classes, consisting of males, blonds, and children, draw a Venn diagram for each of the following classes and shade the appropriate area: (a) blond women; (b) blond boys; (c) women who are not blond.
2. Given the classes of males, blonds, and children, draw a Venn diagram and shade the class consisting of either blond boys or men who are not blond.
3. Draw a three-variable Venn diagram for each of the following classes and shade the appropriate areas: (a) $x'y + xz$; (b) $x(y'z + yz')$; (c) $(x' + y)(x + z')$.
4. Draw a three-variable Venn diagram for the negative of each of the following classes and shade the appropriate areas: (a) $xy' + x'z'$; (b) $x'(y + z)(y' + z')$; (c) $(xz' + y)(x'z + y')$.
5. By means of Venn diagrams, show which of the following are included in $(x + y' + z)(x' + y + z')$; (a) $xy'z' + x'yz$; (b) $(x + z)y'$; (c) $(x + y')z'$.
6. Prove Theorems 1 through 8 and 10. Identify each step by the postulate used.

7. Apply the appropriate DeMorgan theorem directly to the following expressions: (a) $(x + y')z$; (b) $(ab' + cd')e'$; (c) $wx' + y(z' + v')$.
8. Apply DeMorgan's theorems to the following expressions (leaving no primed parentheses): (a) $(a'b + c)'d + e'$; (b) $[(x + y'z)w + (a + b'c)'d + f'](g - h')$; (c) $\{z' + y[x + a'b(c' + d'e)']'\}'$.
9. Reduce the following expressions to simplest form (with no primed parentheses): (a) $a\{b[a'(db)']'\}'$; (b) $u' + \{vw + [v' + xy(u + v)']'\}'$; (c) $\{[(xy')'(x' + y + z')]' x(z' + y)'\} + y'$.
10. Construct the three-variable minterm canonical forms for (a) $(a' + b)(c' + a) + b'$; (b) $(x' + z)'(x + y' + z') + y$; (c) $(x + y')(x' + z)' + (x' + y)(y' + z)$.
11. Construct the three-variable maxterm canonical forms for (a) $xy + y'z + x'y'z'$; (b) $(a + b + c)' + abc' + b'c$; (c) $xyz' + x'y'z$.
12. Prove that the product of all 2^n maxterms of n variables is equal to zero.
13. Illustrate Theorems 1 through 7 by drawing a switching circuit representing the proof of the theorem.
14. Combine two binary numbers of arbitrary length, a and b, in some logical fashion, using **AND**, **OR** and other logical operators, to produce a result called R. There are, in fact, 16 different ways of combining a and b and their complements, a' and b'. List each combination in terms of the Boolean equation that defines it, and write a short sentence explaining the meaning of each.
15. Take each of the 16 Boolean functions you have just defined and rewrite them in the following way. Each Boolean function may be represented as the **OR** function of one or more **AND** functions of the two variables. The four **AND** functions of the variables are: a **AND** b, a' **AND** b, a **AND** b', a' **AND** b'. By **OR**ing together combinations of these four **AND** terms, show how to generate each of the defined 16 Boolean functions.
16. Prove that as an inherent result of the fact that these functions are generated as the **OR** of four **AND** terms, there are in fact exactly sixteen possible combinations.
17. Invent a computer instruction called, "generate Boolean," in which two computer words, one in a register and one from memory, are combined according to any one of the sixteen Boolean functions described earlier. The instructions should have a way of describing which function is to be used and the results left in a register.
18. Invent a computer instruction called, "compare Boolean," which has the same effect as a "generate Boolean" instruction, except that after the result is generated, it is compared with zero and the program branches if the result is zero.
19. Using only the **AND**, **OR**, and **EXCLUSIVE OR** functions, write a subroutine to execute the "generate Boolean" and "compare Boolean" instructions that you have just invented.

9 MINIMIZATION AND OPTIMIZATION

9.1 THE OBJECTIVE

Having mastered the techniques of Boolean algebra, you should now be in a position to specify the equations of a switching circuit to perform a given logic function. Your design may work, but it will probably not be the "best" circuit for the job. Our purpose here is to discuss the nature of this "best" circuit design.

Optimization is a regular engineering function. It is desirable to produce a final product that satisfies the customer for whom it was designed. In the process of producing a satisfactory design, the engineer should first define the problem. Each participant in the specification and solution of a problem brings with him a predisposition based on his environment and experience. When the predispositions of the participants differ, misunderstandings can result. It is far better to reconcile any differences during the design stages of a project than to discover that the end result does not conform to some unwritten specification.

The subsequent steps in solving an engineering design problem are plan of solution, execution of plan, and testing of results. An important first step is a verbal specification. The logic designer should realize, however, that descriptive specifications, oral or written, serve only to provide a "ball park" problem definition. There is too much ambiguity in natural language to provide a precise statement of a problem in logic design. The problem statement should therefore be reduced to an inherently logical form. The truth table is an ideal medium for such a precise problem definition. The functional relationship of input and output variable states is clearly defined. To obtain a truth table from verbal specifications may require that certain assumptions be made about the problem. Better yet, the engineer can confer with the person or persons who prepared the specifications to better determine their intent.

In addition to input/output relationships, there are constraints imposed by the problem specifications. These constraints come from the engineering environment in which the logic circuitry will be designed. The most common constraint is cost. For certain special applications, such as satellites or space vehicles, weight and volume may be very important. In these and other applications, reliability is extremely important. One way of treating all these criteria is to "price them out," that is, to determine how important these boundary conditions are to the customer by presenting alternatives—various levels of performance and the cost of each. Some compromise will always be reached.

Even after the boundary conditions have been assigned a cost, the engineer is not yet ready to get to work. If he is to minimize cost, he must know the ways in which current technology affect design; he must know how alternative implementations vary the cost of a design. The implementation of digital electronics is a rapidly changing field. As manufacturers improve their products, the design engineer needs to keep abreast of current components and costs so that he can produce a good design.

The sequence of steps for designing a digital logic network:

a) Translating the verbal problem specifications into a truth table.
b) Obtaining from the truth table the Boolean equations which describe the circuit operation.
c) Minimizing the equations.
d) Translating the equations into the form for the gates to be employed.
e) Implementing the circuit.

9.2 ALGEBRAIC MINIMIZATION TOOLS

The Boolean algebra can be manipulated to minimize a switching function by application of the properties of the algebra discussed in Chapter 8. There are a few properties which are especially useful in minimization. They are stated below. Postulate 1 deals with reduction to a single switch:

$$x + 0 = x \quad (9.1a)$$
$$x \cdot 1 = x \quad (9.1b)$$

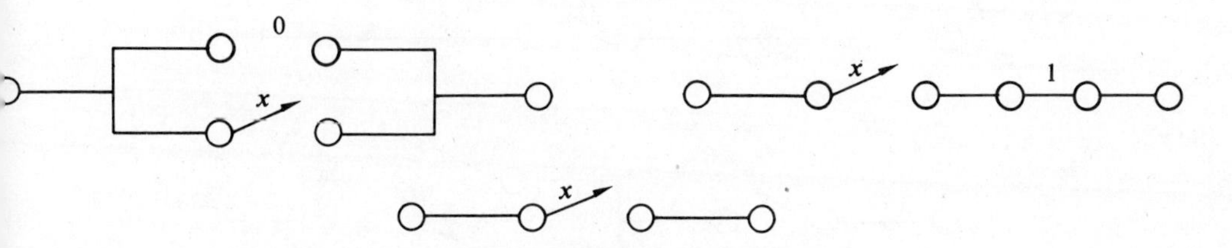

Fig. 9.1. Reduction to one switch: $x + 0 = x \cdot 1 = x$

as illustrated in switching circuits in Fig. 9.1. Theorem 2 deals with reduction to no switches at all:

$$x \cdot 0 = 0 \quad (9.2a)$$
$$x + 1 = 1 \quad (9.2b)$$

as shown in Fig. 9.2.

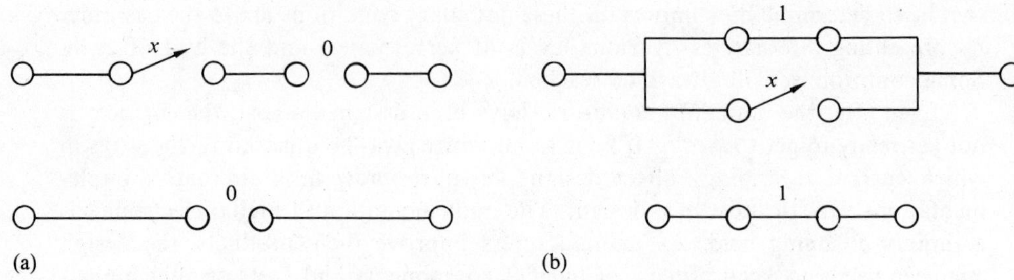

Fig. 9.2. Reduction to no switch: (a) $x \cdot 0 = 0$; (b) $x + 1 = 1$

Another property warrants special mention. Theorem 3 is used for elimination of a redundant term:

$$x + xy = x \tag{9.3a}$$

$$x(x + y) = x \tag{9.3b}$$

The switching circuits are shown in Fig. 9.3.

Redundancy can also be reduced using Theorem 4

$$x + x'y = x + y \tag{9.4}$$

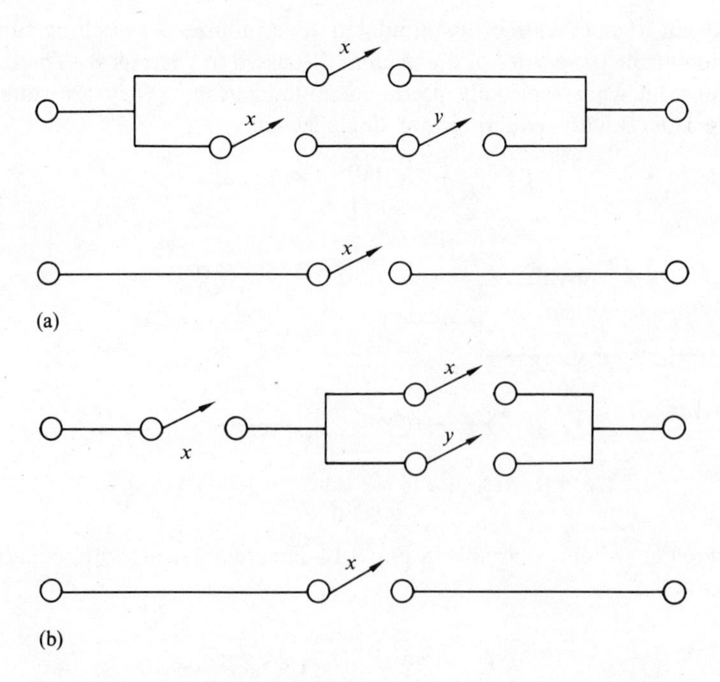

Fig. 9.3. Elimination of redundancies: (a) $x + xy = x$; (b) $x(x + y) = x$

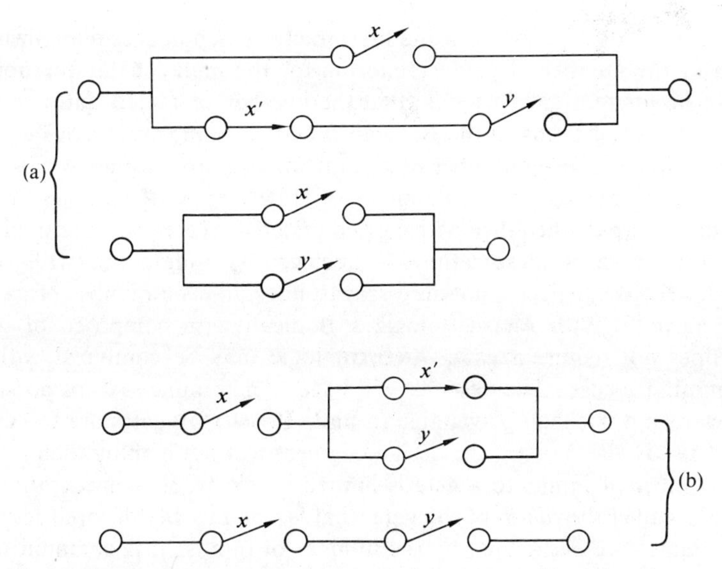

Fig. 9.4. Redundancy reduction: (a) $x + x'y = x + y$; (b) $x(x' + y) = xy$

and, as shown in Fig. 9.4.

$$x(x' + y) = xy \tag{9.5}$$

which is simply proved as follows:

$$x(x' + y) = xx' + xy = 0 + xy = xy$$

A corollary on the completeness property, Postulate 2, is expressible for two variables:

$$ab + a'b + ab' + a'b' = 1 \tag{9.6}$$

Note that all the possible combination states of the two variables have been expressed; therefore their union is 1.

9.3 MINIMIZATION DEPENDENCE ON IMPLEMENTATION

Our tools in hand, we next turn to formulating a solution within our boundary conditions. Regardless of how inexpensive an individual component may be, it is even less expensive not to have the component at all. The postulates and theorems given in Eqs. (9.1) through (9.6) may be applied to any Boolean expression to reduce the number of occurrences of variables and thereby reduce the number of switches represented by these variables. If we paid by the switch, and cost reduction was our objective, it would clearly be advantageous to reduce the number of switches.

But we are not planning to implement our Boolean expression with switches. We will use electronic gates, primarily **NAND** gates. If we pay by the gate, and if gates are our major expense, then the first practical step in cost minimization is a reduction of the number of gates. Ignoring for the moment the question of the number of inputs to a gate, we note that the number of **NAND** gates required is twice the number of terms in the Boolean function minus one; provided that the expression implies an even number of gates from input to output. As discussed in Chapter 8, there must be an even number of **NAND** or **NOR** gates in every path from input to output when disjunctive or conjunctive expressions are implemented with inverters. Savings can sometimes be achieved by using noninverting **AND/OR** logic; with **AND/OR** logic it is not necessary to have an even number of gates from input to output. With **AND/OR** logic a Boolean term composed of only one variable does not require a gate. **AND/OR** logic may be combined with **NAND** logic to minimize cost. The designer will have to determine savings possible with the hardware and packaging available to him. If **AND/OR** gates are too expensive compared to his alternatives, the design engineer will not employ them.

The number of inputs to a gate is limited by electronic considerations. This limitation is called the *fan-in* of the gate, that is, the number of input terminals of the gate. Gates are designated by the number of inputs; it is common to speak, for example, of a four-input **NAND**. What do you do if the number of variables in a term exceeds the number of inputs to the available gate? You implement the offending term with multiple gates in accordance with the associative property. If inverter logic is used, it may be necessary to insert an additional level of gating to restore evenness. When the lowest common denominator is the gate, the design engineer must manipulate his algebra to permit implementation at the lowest cost. Such cost is dependent on the number of gates of each type employed.

Technological progress has provided the designer with larger and larger basic building blocks. The individual components of the previous technology become internal parts of the new and larger whole. On one hand, the number and complexity of the electronic functions performed within a single package have grown. On the other hand, the designer specifies and purchases his components mounted on etched-circuit wiring boards called *cards* which usually plug into larger chassis. When a hardware malfunction occurs, cards are simply replaced; they are repaired or remanufactured elsewhere.

When the building block is the hardware logic component card, the task of the design engineer is in some senses harder. He must design his circuitry to use a minimum number of cards at the lowest cost possible. This may mean manipulating the logic expressions so that terms are grouped to use a maximum number of functions available on a given card, thus reducing the number of cards. Another factor is the cost and number of pins of the connector used to plug the card into the larger system. The connector will usually be a standard for the cards chosen, so the designer has no choice. Along with such standardization comes a limit on the number of connections which can be made to the card. The connector sets a fixed upper limit on the number of logic signals which can be communicated to and from the card. The designer must arrange his logic functions so that this restriction is not violated. When applying cost criteria, the designer may often find that the number of connections between the card and the larger system is to be minimized even at the expense of additional electronic components; the connector and wiring may far outweigh the cost of components.

9.4 OBTAINING THE SWITCHING FUNCTION

Within the boundary conditions of the technology to be employed, we now begin to execute the plan by minimizing the hardware requirements. The minimization process requires, as its starting point, a Boolean expression relating output to inputs. Consider how this expression is obtained: The truth table is an unambiguous statement of the input-output relationship. If the problem has first been specified verbally, it will be necessary to reduce such nonrigorous specifications to an unequivocal form.

As mentioned previously, the truth table is formed by listing every possible combination of the inputs and then tabulating the unique output state desired for each of these input combinations. Since the input states can take on only the truth values of 0 or 1, it is customary and convenient to tabulate the possible inputs by writing them as a binary number counting sequence. For only two inputs the sequence counts from 0 to 3; for three inputs, from 0 to 7; for four, from 0 to 15_{10}, etc. This part of the truth table is shown in Fig. 9.5.

Given the input side of the truth table as invariant, we proceed to complete the output side for the particular situation at hand. If the output is to be true when a particular combination of inputs occurs, we enter a 1 in the output column for

Input				Output
a	*b*	*c*	*d*	
0	0	0	0	
0	0	0	1	
0	0	1	0	
0	0	1	1	
0	1	0	0	
0	1	0	1	
0	1	1	0	
0	1	1	1	
1	0	0	0	
1	0	0	1	
1	0	1	0	
1	0	1	1	
1	1	0	0	
1	1	0	1	
1	1	1	0	
1	1	1	1	

Fig. 9.5. Four-input truth table as a counting sequence

the row of the term being considered. Conversely, if a given combination of inputs is to result in a false output, we enter a 0 in the output column for that combination of inputs. In enumerating the output for combinations of the inputs, we may discover that certain combinations of inputs cannot occur. That is to say, the propositions which the input variables represent are logically incompatible and/or physically nonrealizable. This situation has an appropriate name; it is called a *don't care condition.* The entry in the truth table for a don't care condition is a superimposed 0 and 1. This symbol is often written ϕ and referred to as a *phi condition.* In writing and reducing the Boolean expression, you may treat a don't care as a 0 or a 1, whichever is convenient and will lead to best minimization. When writing the terms of a Boolean function, you segregate the don't care terms.

After the truth table has been completed, the Boolean output function may be written as a sum of standard products or a product of standard sums. The sum of standard products is obtained by directly writing the sum of those standard products for which the output is 1, plus those for which the output is ϕ. If the state of the input variable is 1, it appears in the product term uncomplemented; if the state of the input variable is 0, it appears complemented in the product term. The product of standard sums is obtained by employing DeMorgan's theorem. The sum of products form is obtained for the complement and then negated to obtain the desired output. When ϕ conditions are present, it is necessary to attempt minimization on both minterm and maxterm forms independently, accepting the better result for implementation. The minterm minimization and the maxterm minimization may achieve different results because different ϕ terms were included in each.

Input				Output
a	*b*	*c*	*d*	*f*
0	0	0	0	0
0	0	0	1	0
0	0	1	0	0
0	0	1	1	0
0	1	0	0	1
0	1	0	1	1
0	1	1	0	0
0	1	1	1	1
1	0	0	0	1
1	0	0	1	1
1	0	1	0	1
1	0	1	1	1
1	1	0	0	1
1	1	0	1	1
1	1	1	0	0
1	1	1	1	1

Fig. 9.6. Truth table for the automobile manufacturers

As an example of entering the specifications into a truth table and obtaining Boolean equations from it, consider the automobile manufacturer who implements a warning buzzer system to remind the motorist of things he should not have forgotten. The buzzer is to sound if the headlights are on and the ignition is off; if the ignition is on and the seat belts are not fastened, or if the ignition is on and a door is open. We recognize four Boolean variables, which we represent as follows: let a represent the proposition that the headlights are on; let b represent the proposition that the ignition is on; let c represent the proposition that the seat belts are fastened; and let d represent the proposition that a door is open. Let f represent the output function (the buzzer sounding). The truth table for f is shown in Fig. 9.6. The sum of products and product of sums functions are given in Eqs. (9.7) and (9.8), respectively.

$$\begin{aligned} f &= a'bc'd' + a'bcd' + a'bcd + ab'c'd' + ab'c'd + ab'cd' + ab'cd + abc'd' + \\ &\quad abc'd + abcd \\ f &= \textstyle\sum 4, 5, 7, 8, 9, 10, 11, 12, 13, 15 \end{aligned} \tag{9.7}$$

$$\begin{aligned} f' &= a'b'c'd' + a'b'c'd + a'b'cd' + a'b'cd + a'bcd' + abcd' \\ f &= (a + b + c + d)(a + b + c + d')(a + b + c' + d)(a + b + c' + d') \\ &\quad \cdot(a + b' + c' + d)(a' + b' + c' + d) \end{aligned} \tag{9.8}$$

For an example involving don't care conditions, consider a simplified traffic control signal. The light is to change if a car has been waiting for three minutes or more. Suppose that one street runs east-west, the other north-south, and that we have a timer which indicates the passage of a three-minute interval. Let a represent the proposition that the light controlling the north-south street is red; let b represent the proposition that three minutes have passed since the light changed; let c represent the proposition that a car is waiting northbound or southbound; and let d represent the proposition that a car is waiting eastbound or westbound. Let f be the function that the light should change. We do not concern ourselves with the light-changing mechanism; we simply assume that one direction is red when the other is green. The truth table is given in Fig. 9.7.

$$\begin{aligned} f &= \textstyle\sum 5, 7, 14, 15 + \sum_\phi 2, 4, 6, 9, 12, 13 \\ &= a'bc'd + a'bcd + abcd' + abcd \\ &\quad + [a'b'cd' + a'bc'd' + a'bcd' + ab'c'd + abc'd' + abc'd]_\phi \end{aligned} \tag{9.9}$$

$$\begin{aligned} f' &= \textstyle\sum 0, 1, 3, 8, 10, 11 + \sum_\phi 2, 4, 6, 9, 12, 13 \\ &= a'b'c'd' + a'b'c'd + a'b'cd + ab'c'd' + ab'cd' + ab'cd \\ &\quad + [a'b'cd' + a'bc'd' + a'bcd' + ab'c'd + abc'd' + abc'd]_\phi \end{aligned} \tag{9.10a}$$

$$\begin{aligned} f' &= (a + b + c + d)(a + b + c + d')(a + b + c' + d')(a' + b + c + d) \\ &\quad \cdot(a' + b + c' + d)(a' + b + c' + d)[(a + b + c' + d) \\ &\quad \cdot(a + b' + c + d)(a + b' + c' + d)(a' + b' + c' + d)(a' + b' + c + d) \\ &\quad \cdot(a' + b' + c + d')]_\phi \end{aligned} \tag{9.10b}$$

a	b	c	d	f
0	0	0	0	0
0	0	0	1	0
0	0	1	0	ϕ
0	0	1	1	0
0	1	0	0	ϕ
0	1	0	1	1
0	1	1	0	ϕ
0	1	1	1	1
1	0	0	0	0
1	0	0	1	ϕ
1	0	1	0	0
1	0	1	1	0
1	1	0	0	ϕ
1	1	0	1	ϕ
1	1	1	0	1
1	1	1	1	1

Fig. 9.7. Truth table for the traffic signal

The don't care conditions arise in two ways, based on our interpretation of the rule for changing the signal. The first is that it is impossible for a car to be waiting when the light is green. The second is that we don't care if the light changes every three minutes if no car is waiting. The sum of products and product of sums forms are given in Eqs. (9.9) and (9.10), respectively. Try to state verbally, for practice, the meaning of each combination in Fig. 9.7, and verify the statement of the desired output.

9.5 ALGEBRAIC MINIMIZATION PROCEDURES

Once minimization criteria and objectives have been set, the logic designer may proceed to optimize his circuit implementation of a Boolean expression.

Algebraic simplification is a sort of pattern recognition. Like many cognitive processes, it is actually easier to perform than to describe. Practice with the exercises or with real-life problems will bring the necessary familiarity with the technique.

The first step is to remove the obviously unnecessary terms by applying the properties shown in Figs. 9.1, 9.2 and 9.3. Application of even these basic tools requires a pairwise comparison of all the terms in the expression. The beginning designer and the computer proceed algorithmically, making a comparison of terms in the order in which they were written. With experience, this first step of simplification can be done "by inspection." The minimization process continues iteratively until further progress is no longer possible.

What is our criterion for combining terms and utilizing don't care conditions? Do it only when it results in a simplification! We simplify by reducing the number of terms and the number of variables in each term. This is equivalent to reducing the number of gates and the number of inputs to each gate. Unless some other minimization criterion conflicts, this is a good procedure to follow.

Let us now apply these minimization techniques to the traffic light problem (Eq. 9.9).

$$\begin{aligned} f &= a'bc'd + a'bcd + abcd' + abcd \\ &\quad + [a'b'cd' + a'bc'd' + a'bcd' + ab'c'd + abc'd' + abc'd]_\phi \\ &= a'bd + bcd + abc \\ &\quad + (a'bc' + bc'd + a'bc + bcd' + abd + a'cd' + a'bd' + bc'd' + ac'd + abc')_\phi \\ &= bd + a'b + bc + ab + (bc' + bc' + bd')_\phi \\ &= b \end{aligned} \tag{9.11}$$

Note especially the use of Theorems 3 and 4. You should verify verbally that Eq. (9.11) is indeed a solution to the problem.

Our attempt at simplification is not complete. We must also try to minimize the product of sums form; then we can select the better one for our purposes. The product of sums form is minimized by working on its complement, Eq. (9.10a), and then complementing the resultant.

Of course the simplicity of Eq. (9.11) makes it difficult to expect a better result. Nevertheless, we shall try another example of algebraic techniques.

$$\begin{aligned} f' &= a'b'c'd' + a'b'c'd + a'b'cd + ab'c'd' + ab'cd' + ab'cd \\ &\quad + [a'b'cd' + a'bc'd' + a'bcd + ab'c'd + abc'd' + abc'd]_\phi \\ &= a'b'c' + b'c'd' + a'b'd + b'cd + ab'd' + ab'c' \\ &\quad + [a'b'd' + a'c'd' + b'c'd + a'b'c + ab'c' + ac'd' + b'cd' + ab'd]_\phi \\ &= a'b' + b'c' + b'd' + b'd + b'c \\ &= b' \\ f &= b \end{aligned} \tag{9.12}$$

This result may also be obtained from the product of sums form directly, starting with the form shown in Eq. (9.10b).

In following these examples and in doing your own problems, you can find it exceedingly difficult to remember which terms have been employed in some reduction combination. One suggested technique is to write a check mark above the terms thus included. A check does not prohibit using that term again, as indeed has been done above. A term checked and included in a reduction combination is said to be *covered* by that combination. The fewer the variables appearing in the covering term, the better the cover.

9.6 PARTITIONING SUBOPTIMIZATION

With newer technology it is almost as simple to make large arrays of interconnected gates as to make single ones. Minimizing gate count is less important. Designers who use these *large scale integration* or LSI devices must be concerned with partitioning the desired function to make best use of available LSI modules.

Designers of the internal logic within one LSI module are more concerned with minimizing propagation time (signal delay through the gate) and external pin connections than with reducing gate count.

The logic card designer tries to put a general-purpose assortment of gates on a card so that a large design, needing some of each type, can use most of the gates on each card. The designer of LSI modules, however, tries to produce entire self-contained functions, such as a four-bit accumulator or an analog-to-digital converter.

An engineer faced with the task of building a large system from LSI chips, which he may specify, must optimize his design to best partition his logic functions among many building blocks. This is still an art, and no algebraic minimization is possible. In fact, Boolean manipulation may give wrong "best" solutions, simply because gate count is no longer of prime importance.

9.7 THE KARNAUGH MAP

Algebraic equations do not readily lend themselves to manual minimization. Although the process is algorithmic, the abundance of terms in the Boolean expressions makes it likely that an error will occur unless careful precautions are taken. Being algorithmic, such procedures are better performed by computer than by human being. In fact, there are two well-known algorithms, the *Quine-McCluskey method* and the *method of iterated consensus*, which are well suited to machine implementation. Discussions of these algorithms may be found in texts devoted to switching algebra.

Another scheme, one that is better suited to manual manipulation, utilizes the *unit distance* representation for minimization by the completeness property, namely,

$$a + a' = 1$$

This method, called a *Karnaugh map*, provides a graphical representation such that minterms appear physically adjacent to all other minterms algebraically a unit distance removed. Unfortunately, it is necessary to draw a Karnaugh map in two dimensions. This physical boundary condition makes the map method too cumbersome for problems involving more than six variables; it is best with four variables, as we shall show.

The three-variable map

The three-variable map is illustrated in Fig. 9.8. Two variables are displayed along the top of the map, one along the side. Since $2^3 = 8$, there are eight squares on the map. In the map of Fig. 9.8(a), the variables a, b, c are shown assigned to the squares. The shorthand notation, in which the true state of a variable is represented by 1 and the complemented state by 0, is employed as shown in Fig. 9.8(b). The assignment of variables to the map defines which parts of the

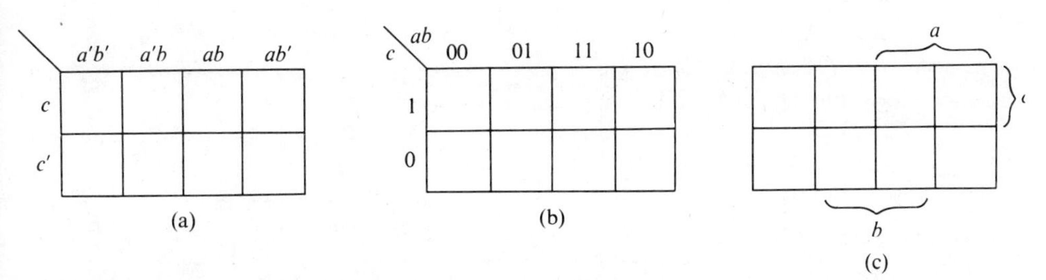

Fig. 9.8. The three-variable Karnaugh maps: (a) Variable assignment to squares; (b) Boolean assignment to squares; (c) The parts which belong to each variable

map "belong" to a given variable (represent the true state of that variable) x, as shown in Fig. 9.8(c); a is true in the right two columns, b is true in the center two columns, c is true in the top row. Each of the eight possible combinations is represented in only one square. This is therefore simply another way of drawing a truth table.

Unit distance adjacencies

As you examine the column and row headings in Fig. 9.8(a) or (b), it is immediately obvious that each row is a unit distance from its physically adjacent row. But what about the end columns? They are adjacent in unit distance to each other! For consideration of adjacency of columns, the map may be considered a vertical cylinder.

The four-variable map

The four-variable map is a square divided so that two variables are displayed along the top and two along the side. In the binary number system, two variables define four states; therefore, the square is subdivided four-by-four into 16 squares. In the map of Fig. 9.9(a), the variables a,b,c,d are shown assigned to the squares. As in Fig. 9.8, the shorthand notation is shown in Fig. 9.9(b) and the assignment of variables to the map defining which parts of the map "belong" to a given variable is shown in Fig. 9.9(c). As shown, a is true in the right two columns, b is true in the center two columns, c is true in the bottom two rows, and d is true in the middle two rows. Each of the 16 possible combinations is represented in only one square. In the four-variable map, as in the three-variable map, columns are treated as adjacent because the map may be considered a vertical cylinder. Similarly, the top row is considered adjacent to the bottom row, thus creating a horizontal cylinder. When one is dealing with both column and row adjacency, a bit of mental juggling is necessary to visualize vertical and horizontal cylinders simultaneously.

Numbering the map

Each small square within the Karnaugh map may be numbered with the decimal representation of the binary number formed by ordering the variables alphabeti-

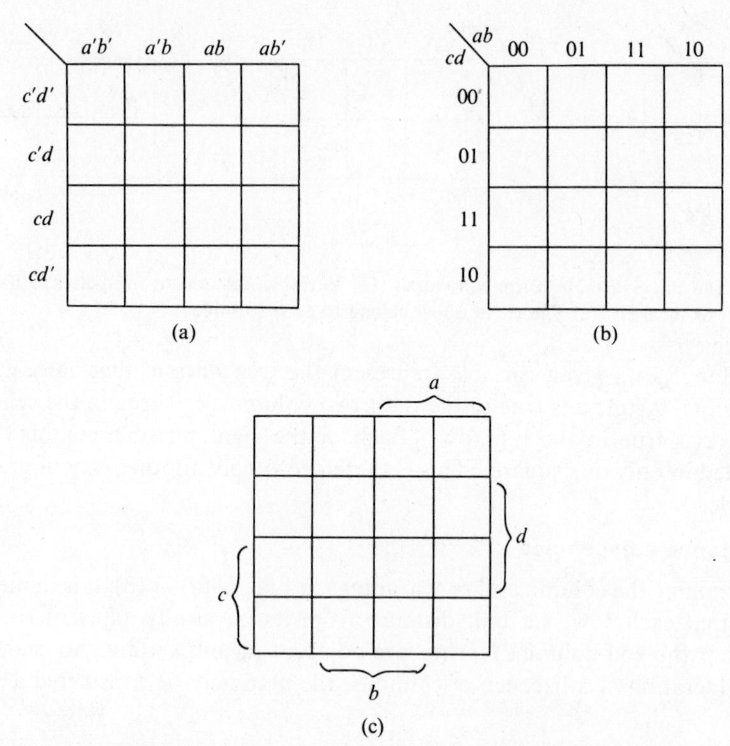

Fig. 9.9. The four-variable Karnaugh map: (a) Variable assignment to squares; (b) Boolean assignment to squares; (c) The parts of the map which belong to each variable

cd \ ab	00	01	11	10
00	0	4	12	8
01	1	5	13	9
11	3	7	15	11
10	2	6	14	10

Fig. 9.10 Minterm numbers represented in the Karnaugh map

cally: *abcd*. The binary numbers produced, from 0 to 15, represent the 16 minterms of four variables. The squares are numbered in the map of Fig. 9.10.

Entering terms on the map

The standard sum of products and the truth table are ideal forms from which to enter terms on the Karnaugh map. A 1 is entered in the appropriate square for

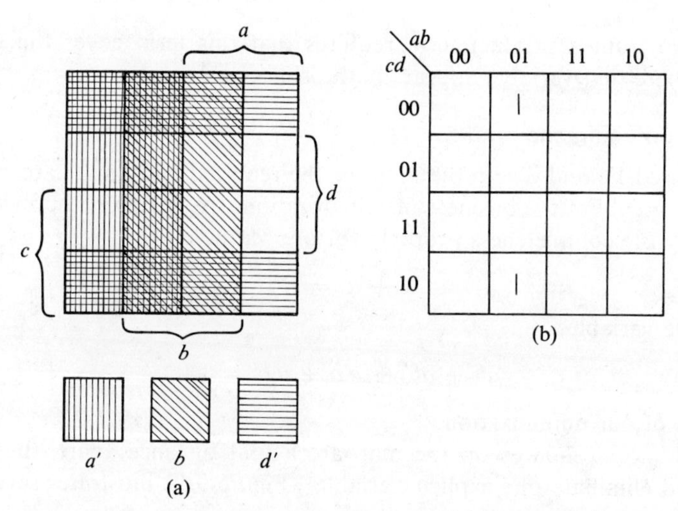

Fig. 9.11. Map entries as intersections: (a) Intersections; (b) Resultant map entry of $a'bd'$

each minterm which is part of the desired function. Similarly, a ϕ is entered for each don't care condition. The squares left empty are those which represent minterms not part of the function; those squares, therefore, along with the ϕ terms (if any), constitute the complement function.

If you are given the desired function in other than standard form, it is possible to enter it directly on the map without performing the expansion to canonical form. The interpretation of the map given in Fig. 9.9(c) is best suited for such entry. Simply consider each product term (in the sum of products form) to see which squares on the map belong to, or are covered by, the product term. Since a product term is a logical **AND**, the portion of the map covered by a product term is the intersection of the individual portions delimited by the variables constituting the product term. By way of example, consider the product term

$$a'bd'$$

which is entered in the map in Fig. 9.11. Part (a) shows the intersection of the variables, part (b) the resultant map.

Variable count/squares covered

There is a relationship between the number of variables appearing explicitly in a product term and the amount of the map covered by that term. Understanding this relationship is valuable both in entering terms on the map and in using the map for minimization.

The smaller the number of variables in a product term, the greater the number of squares covered by that term. This statement should be clear, since the smaller the number of variables, the more the product term will be expanded when converted to standard form; and each minterm covers exactly one square. The absence of a variable from a product term requires that this term cover the minterms (squares) where that variable appears both primed and unprimed.

Minimization using the map

Having entered 1's and ϕ's on the map for the required and don't care minterms, we can now go about the business of minimization. As mentioned previously, we shall employ the completeness property for one variable

$$a + a' = 1$$

and multiple variables

$$ab + ab' + a'b + a'b' = 1$$

as the basis of our minimization.

Since adjacent squares on the map are a unit distance apart, they may be combined to eliminate one explicit variable. Figure 9.12 illustrates several maps

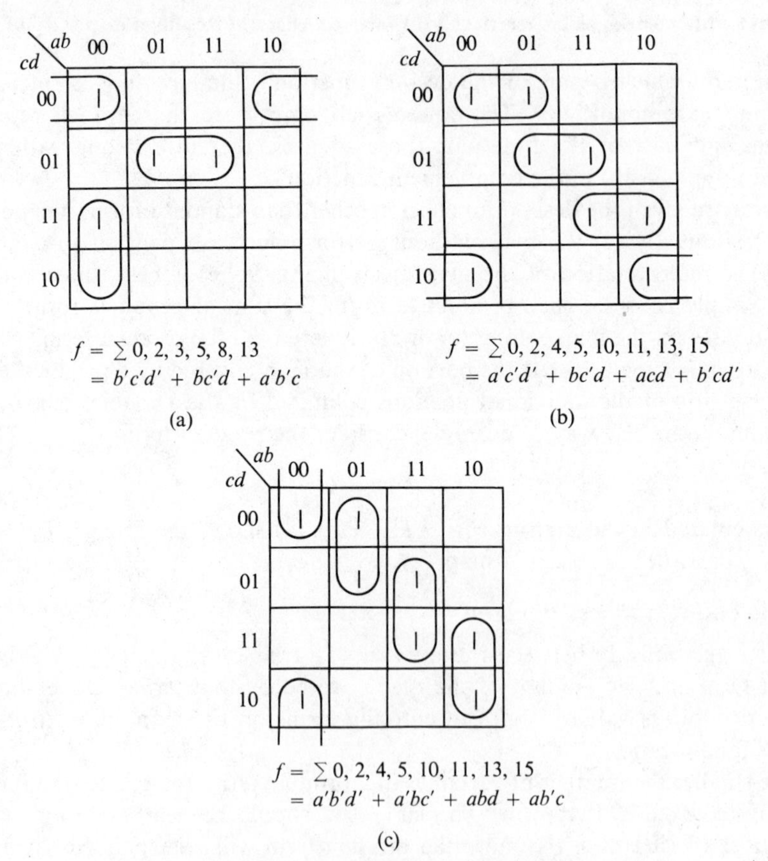

Fig. 9.12. Minimization by groups of two

where groups of two have been formed. Note that a grouping is indicated by encircling the minterms being combined. Adjacencies which occur over edges are indicated by open-ended encirclements.

The map of Fig. 9.12(a) represents the function entered on the map as

$$f = a'b'c'd' + a'b'cd' + a'b'cd + a'bc'd + abc'd + ab'c'd$$

It is simplified by the map to

$$f = b'c'd' + bc'd + a'b'c$$

Note that we could have included an additional grouping of two,

$$a'b'd'$$

but this grouping would not have added anything new. This product term would have provided duplicate coverage of terms already covered; therefore, it was not included.

The same original function could have been used to create the identical minterms to be covered in Fig. 9.12(b) and (c). The original function is expressed as

$$f = \sum 0,2,4,5,10,11,13,15$$

From Fig. 9.12(b), the simplification is

$$f = a'c'd' + bc'd + acd + b'cd'$$

but from Fig. 9.12(c) it is

$$f = a'b'd' + a'bc' + abd + ab'c$$

Clearly these two solutions are nonidentical, but both cover the necessary minterms. Both also require the same number of gates, and indeed, they require the same number of complemented values as inputs. The availability of these complemented values might be the deciding factor between the two solutions.

The next-largest-sized group that may be created during the simplification process is the group of four. There is no group of three minterms because 3, not being a power of 2, could not possibly be an expression of completeness. Groups of four take the form of rows, columns, or squares. Several examples are shown in Fig. 9.13. Note in Fig. 9.13(a) that two groups of four were chosen in preference to a group of four and a group of two, because the use of a group of two would have involved an additional variable. There is no disadvantage to covering some minterms twice. This is expressly permitted by the absorption property.

Figure 9.13(b) illustrates the optional use of don't care conditions: a don't care condition is employed only when it makes possible a larger grouping. As shown, the solution to this map is also nonunique.

Rules for map simplification

Rather than continue with examples of various sizes and shapes of grouping for minimization using map methods, we can collect our experiences in a few rules or procedures.

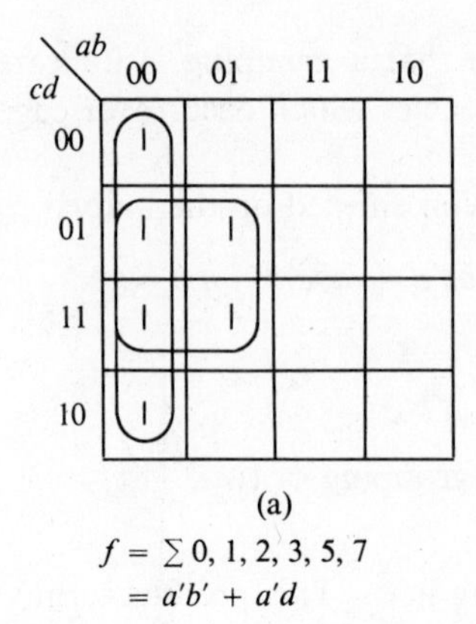

(a)

$f = \sum 0, 1, 2, 3, 5, 7$
$= a'b' + a'd$

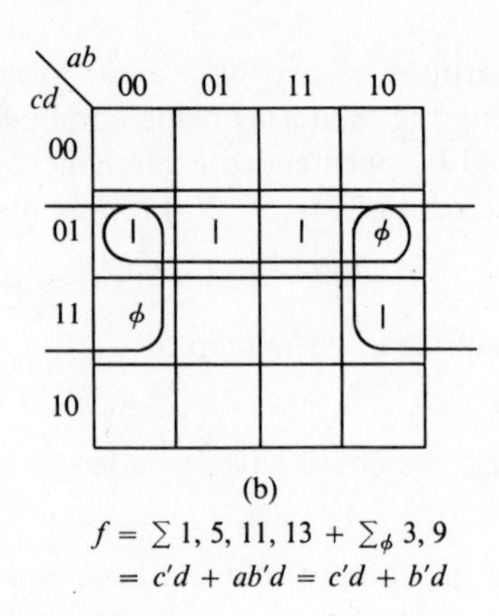

(b)

$f = \sum 1, 5, 11, 13 + \sum_{\phi} 3, 9$
$= c'd + ab'd = c'd + b'd$

Fig. 9.13. Groups of four minterms

The foremost consideration in forming a grouping is to get all the minterms covered, and to get them covered in the largest grouping possible. As previously mentioned, grouping sizes are powers of 2, i.e., 2, 4, 8, 16, ... The largest possible grouping covering a minterm is called a *prime implicant* and is formally defined to be a grouping (which covers the minterm) which is not itself covered by a larger grouping.

Our final selection of groupings on a map is seen to be composed exclusively of prime implicants. But not every prime implicant possible will be part of the final cover. Some, as we have shown, are included as a matter of choice; but others are essential. A prime implicant that covers a minterm not otherwise covered is called an *essential prime implicant*. To make sure that no essential prime implicant gets overlooked and to properly provide a minimum cover, it is wise to choose the essential prime implicants first. One way of detecting essential prime implicants is to look for those minterms that do not have a large number of neighbors. A minterm that is all alone, for example, must be covered by its own essential prime implicant. Once the essential prime implicants have been chosen, the remaining number of minterms to be covered by other prime implicants is reduced.

It may be helpful for the beginner to redraw the map after he has selected the essential prime implicants. All the terms covered by the essential prime implicants become don't care conditions for further simplification. Since these terms have once been covered, they may optionally be included in other groupings whenever necessary to increase the size of the grouping.

A classic example illustrating the necessity of choosing the essential prime implicants first is shown in Fig. 9.14. Note that the "obvious" grouping of four minterms is not an essential prime implicant.

Product of sums minimization

The discussion so far has all been directed toward using the map for minimization of Boolean expressions in the sum of products form. In many cases, however, it is

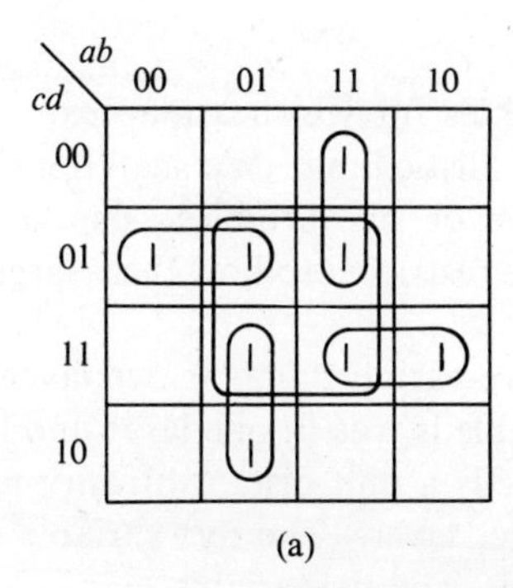

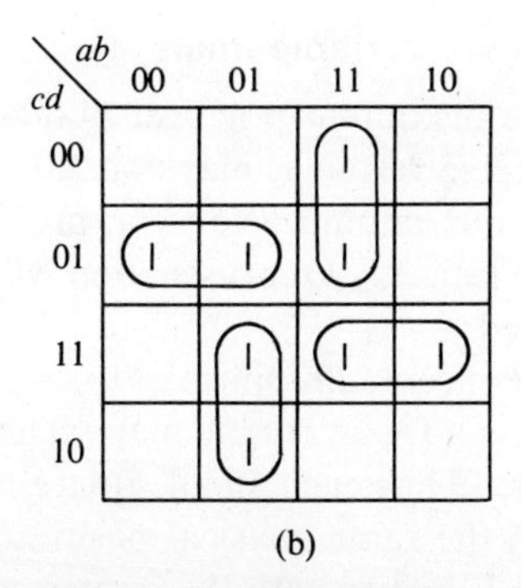

Fig. 9.14. "Starburst" map: (a) Poor solution; (b) Proper solution

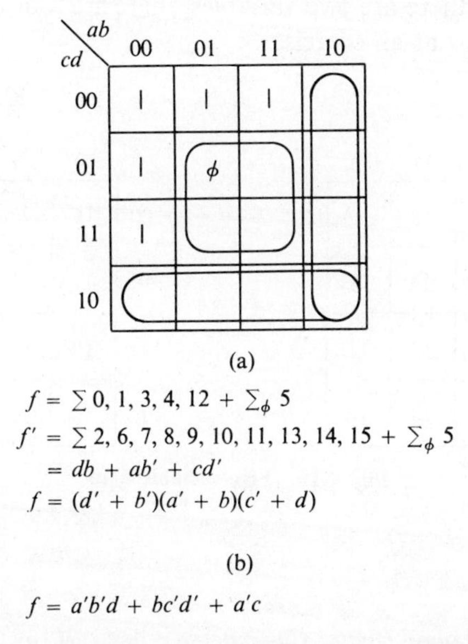

(a)

$$f = \sum 0, 1, 3, 4, 12 + \sum_{\phi} 5$$

$$f' = \sum 2, 6, 7, 8, 9, 10, 11, 13, 14, 15 + \sum_{\phi} 5$$

$$= db + ab' + cd'$$

$$f = (d' + b')(a' + b)(c' + d)$$

(b)

$$f = a'b'd + bc'd' + a'c$$

Fig. 9.15. Product of sums and sum of products groupings

more advantageous to minimize the product of sums. This procedure is best implemented on the map by application of DeMorgan's theorem. You minimize the complementary function on the map and then apply the theorem to obtain the result in true product of sums form. Minimization of the complementary function on the map consists of grouping the 0's instead of the 1's; the inclusion of ϕ's remains optional.

In the example of Fig. 9.15, the map is shown with three essential prime implicants of the complementary function. The reader should verify the correctness of the sum of products solution in (b). The choice between solutions (a) and (b) depends on the minimization criteria and the hardware implementation.

Five and six-variable maps

Six is the maximum practical number of variables for which a map can be constructed; five variables may constitute a more realistic limit. We shall not attempt to work out minimization for problems in five or six variables. Nothing new would be learned. Our discussion will instead be restricted to how these large maps are formed.

It is conceptually simplest to consider the five-variable map as two layers, each layer being a four-variable map. The fifth variable is true in one layer and false in the other. Thus each small square in one layer is a unit distance from the small square in the same physical position in the other layer. The five-variable map is shown in Fig. 9.16 with the minterms numbered. The six-variable map is likewise composed of layers, each of four variables. The other two variables designate the individual layer; since there are two variables, there are four layers. Constructing a six-variable map is left as an exercise.

$a = 0$

$de \backslash bc$	00	01	11	10
00	0	4	12	8
01	1	5	13	9
11	3	7	15	11
10	2	6	14	10

$a = 1$

$de \backslash bc$	00	01	11	10
00	16	20	28	24
01	17	21	29	25
11	19	23	31	27
10	18	22	30	26

Fig. 9.16. Five-variable map

REFERENCES

M. Phister, *Logical Design of Digital Computers*, Wiley, 1961.

See also references for Chapter 8.

QUESTIONS

1. Express the following verbal problem statement as a logical input/output relationship expressed in a truth table. Clearly indicate the proposition represented by each logical variable.

   A store has a burglar alarm system with sensors on all the windows, the doors, and the skylight, as well as a clock that indicates day of week and time of day. The alarm is to sound if the skylight is ever opened. The alarm is also to sound if the windows or doors are opened after 9 p.m., Monday through Saturday, or any time on Sunday.

2. Is there a completeness property, similar to Eq. (9.6), for three variables, for four variables? If so, what is the expression?

3. Using truth tables, prove the following.
   a) Eq. (9.5)
   b) Eq. (9.6)

4. Given the switching function

$$f = ab + ac' + b'c$$

   what would it cost to implement under **NAND/NOR** logic, where each two-input gate cost 7.5¢?

5. Given the cost information in problem 4, what is the cheapest way of implementing the following switching functions?
   a) $f = a'c'd' + ad + bcd'$
   b) $f = c'd + ab'd$
   c) $f = b'd + a'c'd$
   d) $f = b'cd + a'c'd + a'bc'$

6. Consider a simple intersection where two streets cross perpendicularly. The intersection is controlled by a traffic light, which changes in the sequence green-yellow-red-green. Automobiles may proceed through the intersection when the light facing them is green or yellow. Pedestrians may cross only when the light is green. Draw up a truth table expressing when an automobile and when a pedestrian may cross the intersection. Take note of ϕ conditions.

7. Draw up the truth table for a logical function of four input variables such that the output is true whenever two or three of the inputs are true, but the output is false if zero, one, or four of the inputs are true. Write out the algebraic switching function from the truth table.

8. Draw up the truth table and write the Boolean function for a binary subtractor. The inputs will be minuend bit, subtrahend bit, and borrow-in bit; the outputs are difference bit and borrow-out (to the next stage) bit.

9. Algebraically reduce the following switching functions to two-level minimum variable form.
   a) $f = a'b'c'd' + a'bc'd + abcd + ab'c'd + abcd' + a'bcd + ab'cd'$
   b) $f = abcd + ab'c'd' + a'b'cd' + a'bcd + abc'd + a'b'c'd' + ab'cd' + a'bc'd$
   c) $f = (a + b + c + d)(a' + b + c + d')(a' + b' + c + d)(a + b + c' + d)$
   $\cdot(a + b' + c + d')(a' + b' + c + d')(a' + b + c' + d')(a + b + c' + d')$
   d) $f = ab'c'd' + abcd + a'b'c'd' + ab'cd' + a'b'c'd + a'bc'd + ab'cd$
   e) $f = a'cd' + a'bd + bc'd + a'bc + ac'd + b'cd'$
   f) $f = abd' + a'cd' + bcd + a'bd + abc + bd + acd + bcd'$
   g) $f = bcd + abc + a'b'c'd + abd + abcd + a'bc'd' + abcd'$
   $+ [ab'cd + bc'd + a'b'cd]_{\phi}$
   h) $f = a'bc'd + abcd + a'bcd' + [a'b'cd' + abc'd + a'bcd + abcd']_{\phi}$

10. Using the Karnaugh map, reduce the following switching functions to two-level minimum variable form.

   a) $f = \sum 5, 6, 11, 15 + \sum_{\Phi} 4, 7, 13, 14$
   b) $f = \sum 1, 2, 3, 4, 5, 15 + \sum_{\Phi} 7, 9$
   c) $f = \sum 5, 9, 14, 15 + \sum_{\Phi} 1, 12, 13$
   d) $f = \sum 0, 2, 4, 5, 10, 11, 13, 15$
   e) $f = \sum 4, 5, 6, 7, 11, 12, 13, 14, 15$
   f) $f = \sum 2, 4, 5, 6, 9, 10 + \sum_{\Phi} 11, 13$
   g) $f = \sum 1, 5, 11, 14 + \sum_{\Phi} 3, 6, 9$
   h) $f = \sum 0, 4, 5, 6, 10, 12, 14, 15 + \sum_{\Phi} 2, 7, 8, 13$

11. Apply Karnaugh map techniques to the switching functions of problem 9.
12. Use the Karnaugh map to simplify the switching functions represented by the truth tables of problems 6, 7, and 8.
13. Repeat problems 6 through 10, designing for minimum cost. Two-input gates cost 75¢, three-input gates cost $1.10, and four-input gates cost $1.50.
14. As a logic designer, you are constrained to one prefabricated logic card composed of four **NAND/NOR** (positive logic/negative logic) and eight **AND/OR** (positive logic/negative logic) gates. Implement the switching functions of problems 9 through 13 using a minimum number of these cards.
15. Give the block diagram for a gate network, using **AND** and **OR** gates and *exactly one* **NOT** element, which provides the output

$$h(w, x, y, z) = w'x'yz + wy' + wz' + xy' + xz'$$

   Complements of the input variables are *not* available.

16. For the following switching function,

$$t(A, B, C, D) = \sum 3, 8, 10, 11, 13$$
$$\text{don't cares} = 0, 2, 4, 7, 15$$

   a) Find all prime implicants.
   b) Assume that both complemented and uncomplemented inputs are available. Find both two-level **NAND** gate circuits requiring the smallest number of inputs.

17. Design and specify components for hardware modules to perform the given arithmetic operations in each of the following representations of numbers: (a) sign-magnitude, (b) 1's complement, and (c) 2's complement. Assume a 36-bit word and floating-point storage of 1 bit for the sign, 27 bits for the fractional part, and 8 bits for the exponent.

   i) Divide integer
   ii) Floating-point multiply
   iii) Integer multiply and round

10
DIGITAL COMPUTER HARDWARE: I

At the heart of every digital computer is a group of electronic circuits designed to perform the various operations in the machine language instruction set. As we become more intimately involved with machine language and logic, we find it difficult to fully understand how to program so to take best advantage of the hardware without some background in electronics.

10.1 EVOLVING TECHNOLOGY

In the late 1950's, *transistors* and *junction diodes* overtook and replaced the *vacuum tube*, bringing great savings in space, power, heat dissipation, and reliability. Along with these solid-state devices came different design techniques (for example, the use of static voltage levels instead of pulses to represent binary numbers). In the late 1960's, *integrated circuits* (IC's) had begun to replace individual transistors and diodes with entire circuits in one small package. In the early 1970's, the trend was toward *large-scale integration* (LSI) of basic logical building blocks into single complex units. The CPU-on-a-chip was no longer an impossible fantasy. The entire field of logic design was restructured by the new technology.

The single-transistor circuits of the early 1960's, called *discrete-component logic*, were built of penny resistors, dime capacitors, and dollar transistors. Additional circuits and stages of amplification were significantly expensive and great efforts were made to reduce the complexity of designs. Today, several hundred transistors can be made and interconnected for a dollar. They are now cheaper than resistors or capacitors. The cost of sockets and printed wiring board to connect to an IC may be several times the cost of the circuit. Because today's logic hardware is so small and cheap, it has become no longer worthwhile to perform extensive simplifications. The major problem is one of *partitioning*—dividing a design into several individual building blocks in such a way as to minimize interconnections and maximize generality of each block.

Discrete-component circuits were also slow. Magnetic memories were available with cycle times on the order of two microseconds, and most logic circuits were operable no faster than one or two megahertz (10^6 cycles per second, 10^{-6} seconds per operation). For this reason, to make an efficient computer it was necessary to use much circuitry and complicated parallel techniques to keep the CPU from lagging behind the capabilities of the memory. Today's ICs are much faster, however. They normally operate at speeds between 2 and 40 megahertz. It is therefore possible to get many operations done by the CPU during each cycle of main memory. This has made possible many elegant simplifications in the design of computer hardware.

10.2 COMPUTER HARDWARE CIRCUITS—INTRODUCTION

Two overwhelming features are immediately clear when one inspects the design of a digital computer. The first is that all the circuits seem to fall into a small number of different types, and the second is that there are very many of each type. In

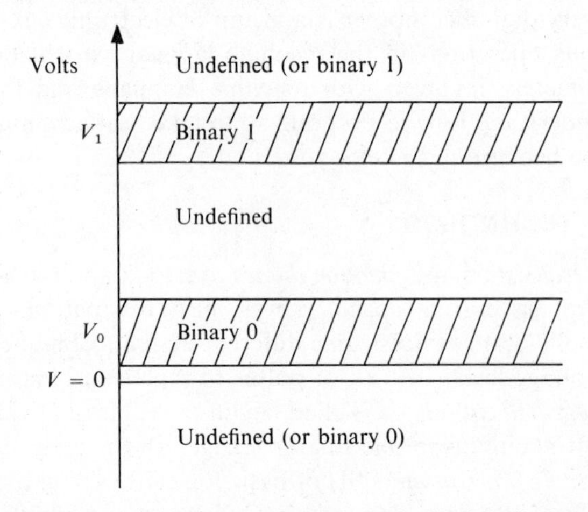

Fig. 10.1 Voltage ranges corresponding to logic levels (positive logic, positive voltages)

order to quickly transfer a machine word of 36 bits from one register to another, 36 (or sometimes 72) wires must connect them. Each of the registers must have 36 storage elements. A transfer between the same two registers in the opposite direction requires at least 36 more wires. A typical large computer might have 50 or 60 thousand gates.

Almost all of the functions desired in the detailed performance of digital computer instructions can be implemented by the construction of logic gates to perform the **NAND**, **NOR**, and other functions described in Chapter 8. If the number of different types of circuits is kept small and they are made in large quantities, the problems of size, complexity, and cost can be reduced to manageable proportions.

Designing digital circuits is easier than designing analog† circuits. The signals

† Active electronic circuits, such as are employed in radios, televisions, tape recorders, etc.

	Positive logic convention	Negative logic convention
Positive voltages (levels)	$V_1 > V_0$ $V_0 \geq 0$ volts	$V_1 < V_0$ $V_1 \geq 0$ volts
Negative voltages (levels)	$V_1 > V_0$ $V_1 \leq 0$ volts	$V_1 < V_0$ $V_0 \leq 0$ volts
Bi polar voltages (levels)	$V_1 > V_0$ $V_1 > 0$ volts $V_0 < 0$ volts	$V_1 < V_0$ $V_0 > 0$ volts $V_1 < 0$ volts

Fig. 10.2. Logic conventions

handled are allowed to occupy only two different states or voltage levels. Most digital systems have signal voltages, called *logic levels*, defined as follows:

$$\begin{array}{ll} \text{Binary 0} & V_0 \pm \Delta V_0 \text{ volts} \\ \text{Binary 1} & V_1 \pm \Delta V_1 \text{ volts} \end{array} \qquad (10.1)$$

with appropriate tolerances (ΔV) for deviations allowed at both levels (see Fig. 10.1). Any other signal voltage may be erronously interpreted as one of the above, or it may cause ambiguous results.

The convention $V_1 > V_0$ is called *positive logic* because binary 1 is a more positive voltage than binary 0. A similar definition applies when the logic levels are currents rather than voltages. (Of course, current and voltage are always both present; the difference is simply whether the logic level has been defined in terms of the voltage or the current.) Confusion results from a blurry distinction between positive logic and positive voltages. Figure 10.2 explains this distinction, and it also defines the negative logic convention. The levels $V_0 = 0$ volts, $V_1 = +4$ volts are almost universally used with digital ICs. They will be used exclusively here, but keep in mind that other values are common with some ICs, with discrete components, and with vacuum tube logic. The circuit diagram conventions for logic used here are explained in Appendix B.

Basic electronics

Any reader with a background in electrical circuits should skip the discussion in this section and go directly to the next section, on solid-state components.

The movement of electrons (small negatively charged particles) in any material takes place under the influence of a force, which is in turn due to an electrostatic potential difference called a voltage. The entire process is completely analogous to the flow of water through a pipe. The voltage represents the pressure difference between the inlet and the outlet of the pipe.

The quantity of water flowing through the pipe in one second has an analog in an electric current, a measure of the number of electrons per second flowing

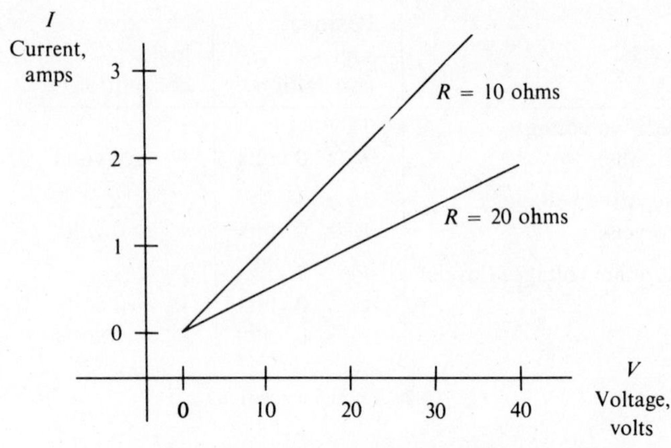

Fig. 10.3. ***I-V*** **graph for a resistor**

past a point in a wire. To express the relationship between voltage (pressure) and current (flow), we need another quantity. Just as pipe size can restrict water flow, electrical resistances can restrict the flow of current. The defining equation is called Ohm's law.

$$V = IR \tag{10.2}$$

where V is voltage in volts, I is current in amperes, and R is resistance in ohms. In hydraulics, the resistance of a pipe is usually difficult to calculate, and may not be a constant, due to turbulence, surface finish, etc. Electrical resistors, however, exhibit the very linear *ohmic* behavior with direct current expressed in Eq. (10.2). Figure 10.3 is a graph of this relationship, where voltage is expressed as the ordinate and current as the abscissa. The slope of the line is the resistance. Figure 10.3 is of very little interest in itself, but we will soon encounter *nonohmic* devices for which Eq. (10.2) is most emphatically not true. Then *I-V graphs* such as Fig. 10.3 will be useful.

As an illustrative example; consider Fig. 10.4, which shows separately the electrical symbols for a battery (source of voltage), a resistor, and current. Figure 10.4(b) shows the simplest electrical circuit, in which current from a battery is sent through a resistor and back to the battery. All circuits must be complete, as this one is, for any current to flow. By Ohm's law, the current is

$$I = V/R = 5/25 = 0.2 \text{ amperes or } 200 \text{ milliamperes (mA)} \tag{10.3}$$

Figure 10.5(a) shows two resistors connected so that all the current passing through one must pass through both. This is known as a *series* circuit. Since the current is the same in both, the voltage across the individual resistors must add up to the whole battery voltage, as is clearly true. Note that Ohm's law applies individually to *each* resistor for calculating the *voltage drop* across it, as well as to the whole

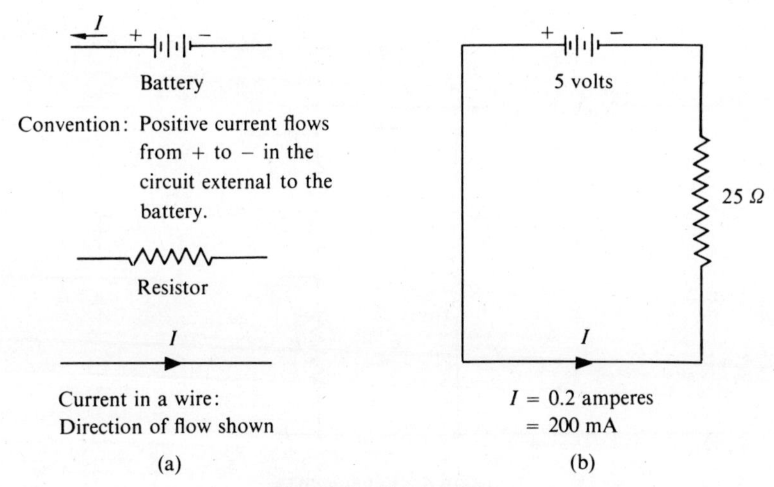

Fig. 10.4. Current flow

circuit for calculating the current. In general, for n resistors in series, the resistance, R, of the combination may be expressed as

$$R = R_1 + R_2 + R_3 + \cdots + R_n \tag{10.4}$$

The configuration of Fig. 10.5(b), in which current passes separately from the battery to each individual resistor and back to the battery, is known as a *parallel* circuit. The total current from the battery is the sum of the individual currents, and the presence of one resistor has no effect on the current through the other. The equivalent resistance of the combination as seen by the battery, for n resistors, is

$$R = \frac{1}{\dfrac{1}{R_1} + \dfrac{1}{R_2} + \dfrac{1}{R_n}} \tag{10.5}$$

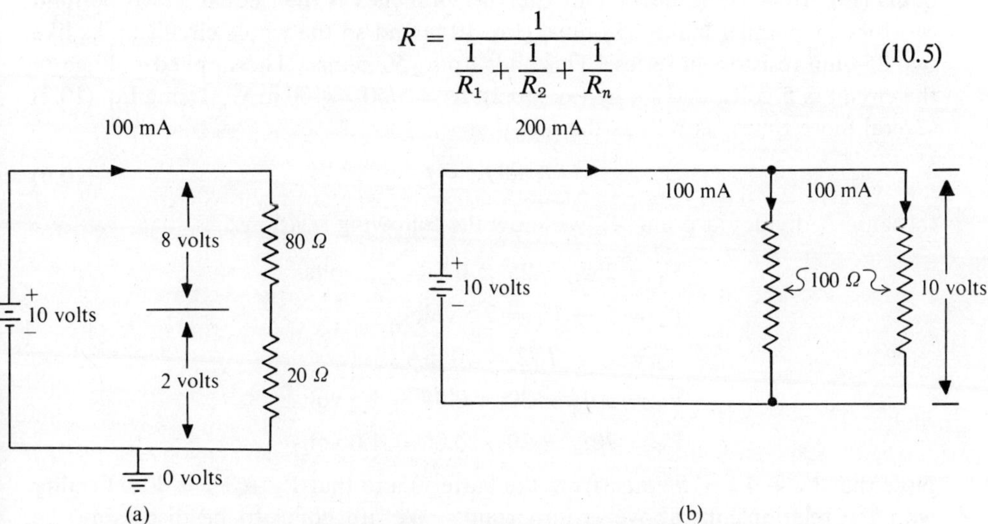

Fig. 10.5. Resistor combinations: (a) in series; (b) in parallel

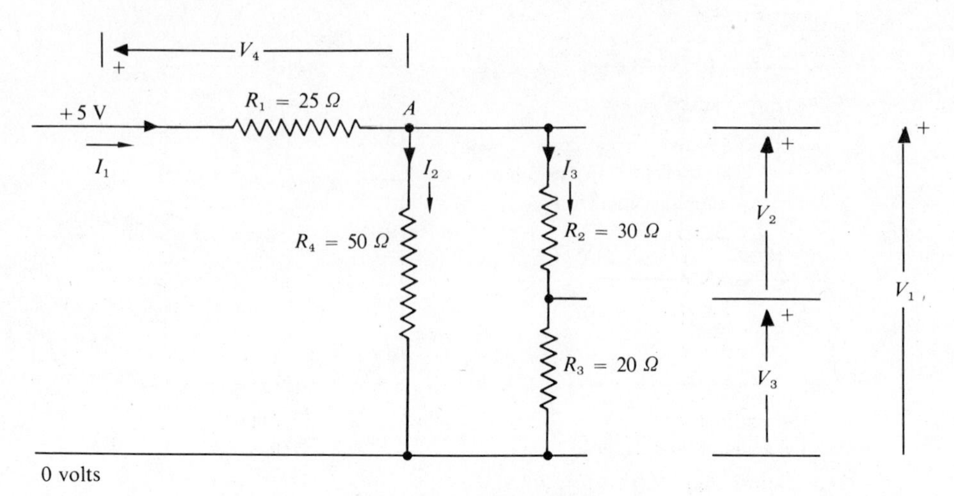

Fig. 10.6. Voltage divider

which, in the special case of two equal resistors, as in Fig. 10.5(b), reduces to simply half the resistance of either.

Figure 10.6 is a simple network circuit of the type which will be useful in designing digital circuits. It contains both series and parallel components. All the current flowing in the circuit I_1 must pass through the 25-ohm resistor, R_1. Then it divides, some (I_2) going through the 50-ohm resistor, R_4, the rest (I_3) through the other two, R_2 and R_3, which are in series. The 30-ohm and 20-ohm resistors, R_2 and R_3, may be thought of as being replaced by their equivalent resistance, 50 ohms (Eq. 10.4). The current in the two branches is then equal. Two 50-ohm resistors in parallel make 25 ohms, (Eq. 10.5) and so the whole circuit looks like two 25-ohm resistors in series. This adds up to 50 ohms. The applied voltage in the circuit is 5 volts, and so, by Eq. (10.2), $I_1 = 5/50 = 100$ mA. Using Eq. (10.2) several more times, as well as the fact that

$$I_1 = I_2 + I_3 \tag{10.6}$$

(because I_1 divides at point A), we show the following relations.

$$V_4 = 25I_1 = 25 \times 0.1 = 2.5 \text{ volts}$$
$$V_1 = 5 - V_4 = 2.5 \text{ volts}$$
$$I_2 = I_3 = I_1/2 = 50 \text{ mA}$$
$$V_2 = 30I_3 = 30 \times 0.05 = 1.5 \text{ volts}$$
$$V_3 = 20I_3 = 20 \times 0.05 = 1.0 \text{ volts}$$

Note that $V_1 + V_4 = 5$ volts (from the battery) and that $V_2 + V_3 = V_1$. Facility with the relationships above is important. We are going to be discussing the operation of digital logic circuits in terms of them.

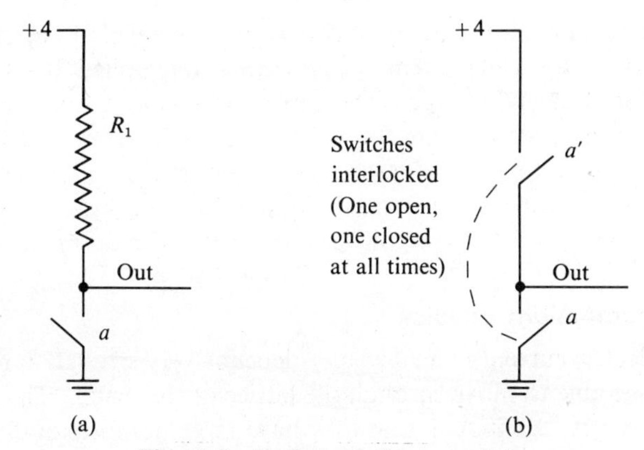

Fig. 10.7. Logic levels from a switch

Figure 10.7 shows two different ways in which a switch contact, such as the one in Chapter 8, can be translated into a logic level. If switch *a* is open, current from the +4 volt power supply is free to flow toward the wire marked OUT and from there to some other circuits which constitute a *load* on this source. Normally, the load is a large resistance compared to R_1, so that the voltage on OUT (V_{out}) is substantially the same as the power supply ($V_{out} = +4$). If switch *a* now closes, all the current through R_1 is shunted to ground, and $V_{out} = 0$. No current then passes through the load. The second circuit (Fig. 10.7b) accomplishes the same results but requires two switches. The OUT line is connected either to the power supply or to ground, never to both, and never to neither. This circuit is better than that in Fig. 10.7(a) only in that more load current can be carried by switch *a′* than by R_1. More loads may therefore be connected to Fig. 10.7(b) before its output voltage drops below the defined logic level.

Figure 10.8 shows the effect of connecting a capacitor to the output of a logic

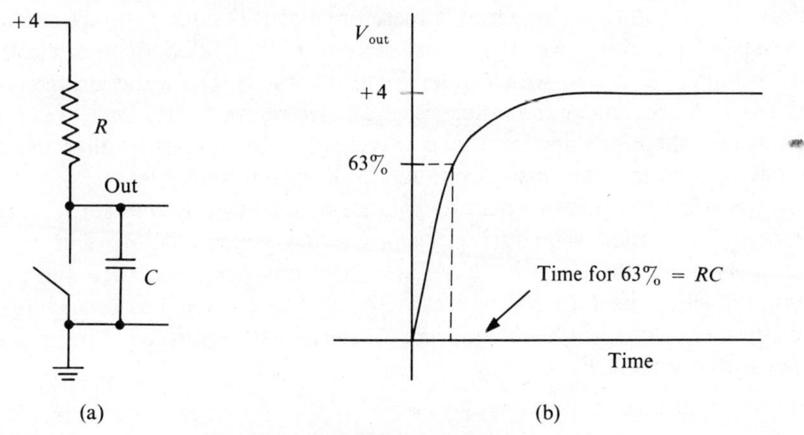

Fig. 10.8. Capacitive time constant

switch. A capacitor consists simply of two wires or metal plates insulated from each other but in proximity, as its schematic representation implies. It acts as a temporary storage unit for electrical charge. The capacitor acts as a transient or temporary load on the circuit and prevents the voltage on OUT from reaching the defined logic level until some time has elapsed. This effect might be important if two logic signals were required to be present simultaneously during a computation. Obviously, capacitance effects are significant in limiting the operating speed of digital circuits.

Solid state component physics

The flow of electric current in a *conductor* depends very strongly on the availability of free electrons able to move through the lattice of the metal. The internal energy structure of most metals is such that they have large numbers of electrons that are very loosely bound, and so the electrostatic forces due to low voltages are more than enough to accelerate these electrons and cause them to become, collectively, an electric current. Copper is an excellent example of such a material. Other materials, called *insulators*, have few, if any, free electrons. To make them conduct current requires large forces which actually remove bound electrons from the atomic structure of the materials. Any insulator will have some voltage at which these bound electrons will break loose, but there are many practical materials for which this voltage is so high that they are nonconductors in everyday use. Glass is a typical insulator. Our major concern here is with materials in a gray area between conductors and insulators, called *semiconductors*. A highly refined semiconductor material, such as silicon or germanium, is actually a very poor conductor. At extremely low temperatures, around 0 K (or −273 C), there are few free electrons. energy states (available places for electrons) are full. This means that there is no room for any more electrons before the material will conduct. At room temperature (300 K or 27 C), the additional energy due to temperature is enough to cause some electrons to leave their bound state and contribute to conduction. A piece of semiconductor material is therefore a sensitive temperature sensor.

When an electron leaves the bound state to become a conduction electron, it leaves behind a *hole* into which other electrons may go. As the electrons move away from an area under the influence of an electric field, they leave a surplus of holes, which therefore appear to be traveling in the opposite direction. The material has two mechanisms of conduction: electrons and holes.

It is possible to upset this delicate balance by introducing very small quantities of impurities into the crystal lattice structure of the semiconductor. This process is called *doping*. If the impurity has fewer electrons per atom than the pure or *intrinsic* material, its presence will cause extra holes. If it has a surplus, extra conduction electrons will be introduced. Material with a surplus of holes is called *p-type*; with a surplus of electrons, *n-type*.

Junctions and bias

Transistors, diodes, and integrated circuits are made by building *junctions* between regions of *p*-type and *n*-type material. Since the *n*-type has more electrons in a high

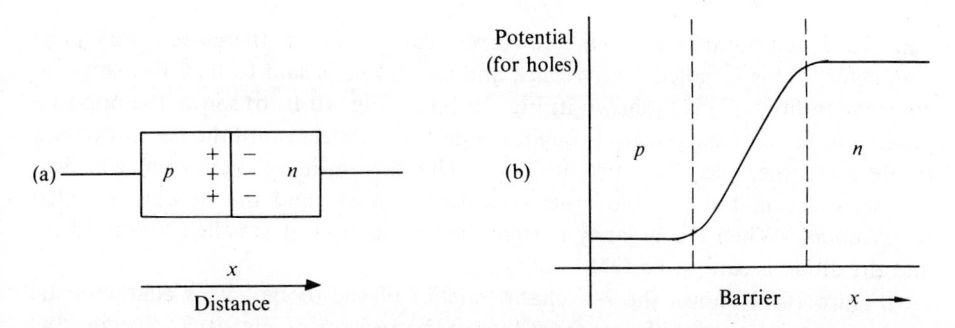

Fig. 10.9. ***p-n*** **junctions**

energy state than the *p*-type, such a junction will have a natural potential difference across it. When the junction is first created, this potential causes a momentary current to flow, moving charges through the junction into a configuration whose electrostatic potential just negates the junction potential. This results in an electronic no-man's land in the region of the junction, where all free charges, both electrons and holes, have been swept away. This region is then called a *barrier*. The *n* region, by itself, is uncharged but has an excess of free electrons balanced by bound holes. In order to maintain the barrier, electrons cross the junction, leaving the *n* region with a surplus of holes and a net positive charge. The reverse is true for *p*. The junction and barrier are illustrated in Fig. 10.9. The potential distribution is shown in Fig. 10.9(b). The actual barrier potential difference is dependent on the amount and type of doping, and is on the order of three-tenths of a volt for germanium and seven-tenths of a volt for silicon (as the doped material).

If we apply an external potential to this situation, say from a battery, we can modify the barrier. Adding a potential in the same direction as the natural one tends to pull the holes and electrons further away from the barrier, and current does not flow in the circuit. The material in the barrier region, much like intrinsic

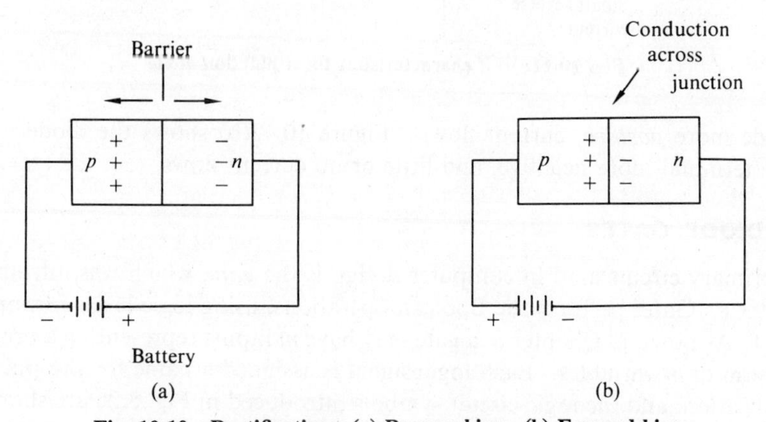

Fig. 10.10. Rectification: (a) Reverse bias; (b) Forward bias

material, is temporarily without free current carriers and is therefore a very poor conductor. This is called *reverse bias*, and the device is said to be OFF, since no current can flow. This is shown in Fig. 10.10(a); Fig. 10.10(b) shows the opposite polarity applied. Charges are brought closer to the barrier until the natural potential is overcome, and then current flows. This device, in which current will flow in one direction but not the other, is called a *diode*, and the process is called *rectification*. When the polarity is right for conduction, it is called *forward bias*, and the diode is said to be ON.

Figure 10.11 shows the *I-V* characteristics of the diode. This characteristic curve is certainly very different from that of the resistor of Fig. 10.3. Specifically, it is neither linear nor symmetric. Note that the conduction curve does not cross the *V* axis at zero; it crosses at the natural offset voltage of the junction. In Fig. 10.12(a), the symbol for a diode is introduced, and the circuit shown has the diode biased ON, so that current flows. Note that when the anode terminal of the diode

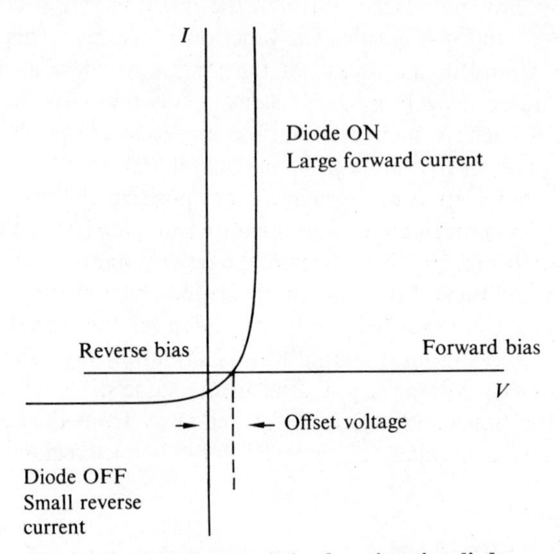

Fig. 10.11. ***I-V*** **characteristics for a junction diode**

is made more *positive*, current flows. Figure 10.12(b) shows the diode with the anode terminal more negative, and little or no current flows.

10.3 DIODE GATES

The primary circuit used in computer design is the *gate*, which was introduced in Chapter 8. Gates perform the Boolean operations, using logic levels for input and output. As noted in Chapter 8, a gate may have *m* inputs representing a product or sum term in *m* variables. Each logic signal is assigned a name for the purpose of identification, and the logic circuit symbols introduced in Fig. 8.26 are shown with

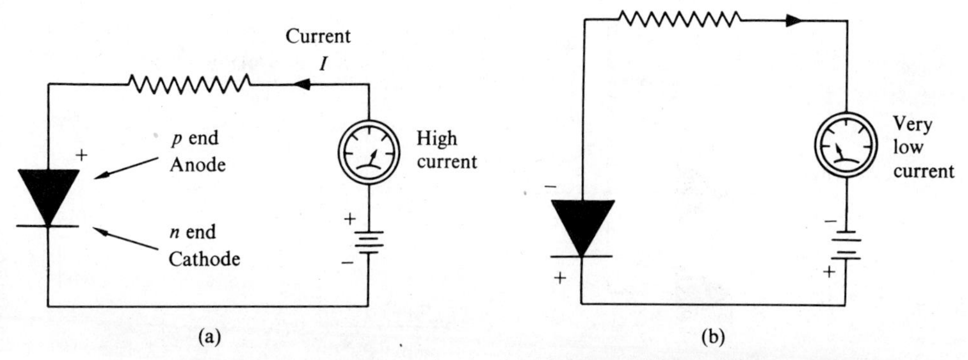

Figure 10.12. Diode symbol shown in circuit: (a) forward biased; (b) reverse biased

signal names on input and output. A typical use of an **AND** function would be transferring data into a register. Figure 10.13 shows four **AND** gates with their outputs **OR**ed together to form the input to a register stage. The top **AND** gate, for example, is always looking at register *B* bit 6. When the binary value of the signal "load *A* from *B*" is 0, the contents of bit 6 of register *B* are not loaded into register *A*. The **OR** gate performs the function of allowing *any* of the appropriate signals from register *B*, *C*, *D*, or *E* to be transferred into *A* simply by raising the proper "load" signal from 0 to 1. This structure is repeated for every bit of register *A*. To load *A* from *B*, the line "load *A* from *B*" is raised so that it is binary 1. This line goes to a similar **AND** gate on every bit of register *A*, so that when it

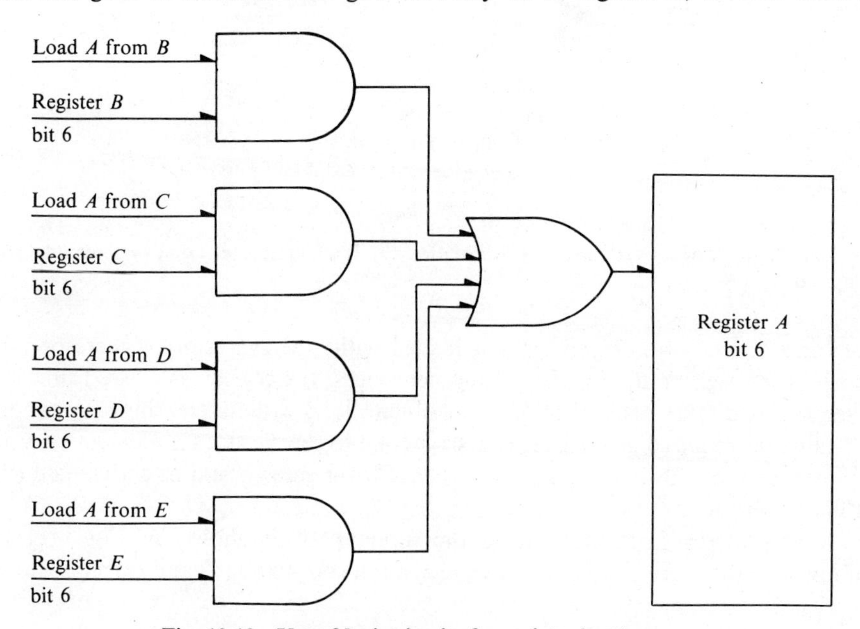

Fig. 10.13. Use of logic circuits for register inputs

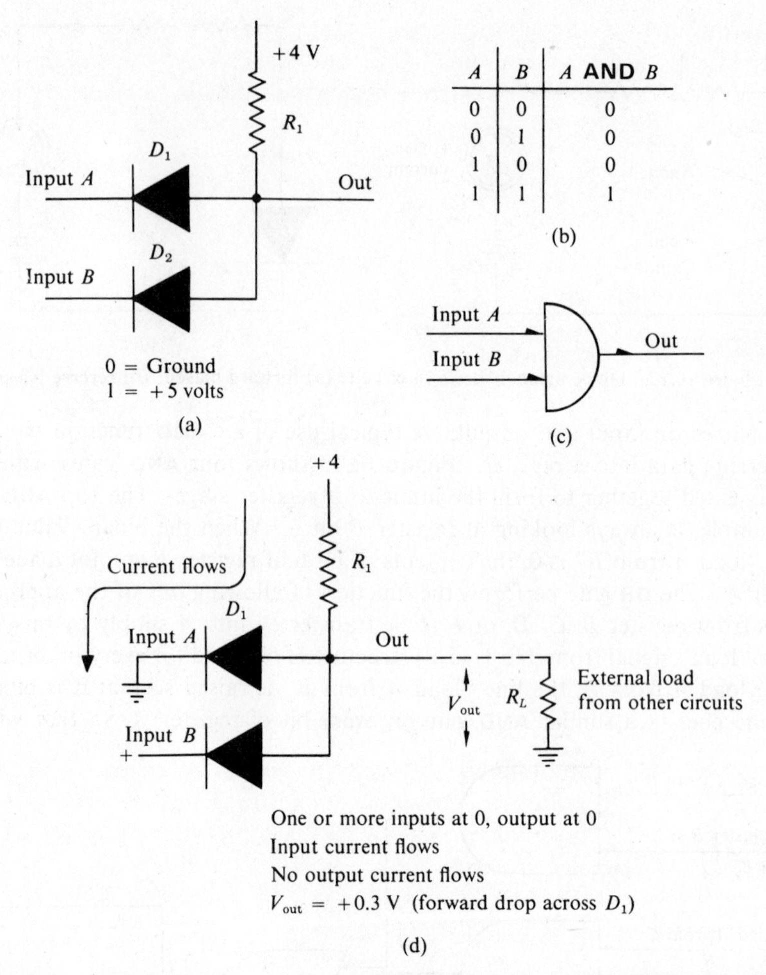

Fig. 10.14. Diode AND gate: (a) Schematic; (b) Truth table; (c) Logic symbol; (d) Any input 0

becomes 1, every bit of register A is loaded with whatever is in the corresponding position of register B. For actual register stages, the register is cleared first, and then the load takes place. It is also possible to load without clearing, but two sets of gates are required at each register stage, one to set the storage element to 1 and the other to reset it to 0. One or the other of these gates would be activated during a load operation.

The electrical construction of the diode AND is shown in Fig. 10.14(a), along with the truth table (b) and logic symbol (c). Parts (d) and (e) show how it works.

If either input is grounded (connected to 0 volts), as in Fig. 10.14(d), current will flow through R_1, through the diode for that input, to ground. The voltage on the output (V_{out}) is held near ground *by current going toward the gate input.* Little or no current flows through the external load provided by the circuit to which this gate's output is connected.

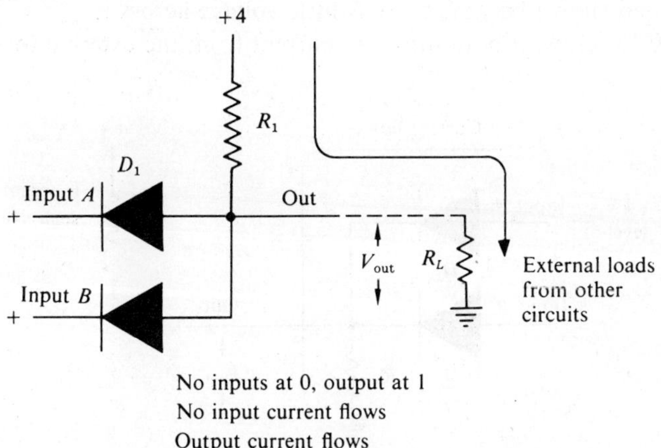

No inputs at 0, output at 1
No input current flows
Output current flows

$$V_{out} = 4\,\text{V}\left(\frac{R_L}{R_1 + R_L}\right)$$ R_1, R_L are a voltage divider for the +4 supply

(e)

Fig. 10.14.(e) All inputs 1

In Fig. 10.14(e), if both inputs are held at 1, current is free to flow from the power supply through R_1 to the external load, and the diodes are reverse-biased. The voltage on the output is determined by the divider created by R_1 and the load R_L. If R_1 is small and R_L is large, the resulting V_{out} will be within the defined values for binary 1. The student should verify for himself that the electrical properties described above fulfill the requirements for the **AND** function.

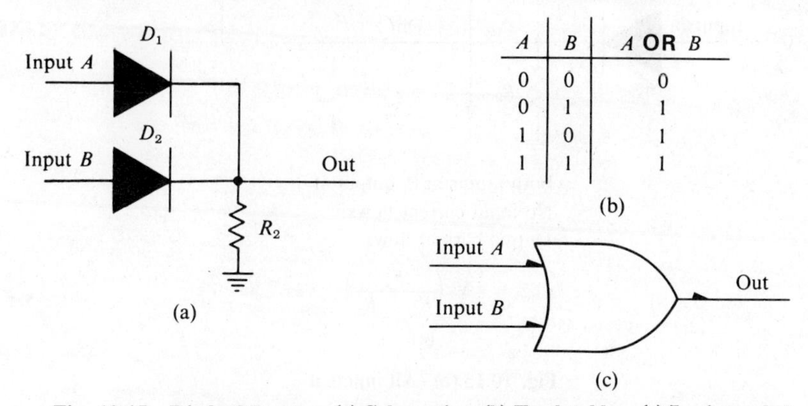

A	B	A **OR** B
0	0	0
0	1	1
1	0	1
1	1	1

Fig. 10.15. Diode OR gate: (a) Schematic; (b) Truth table; (c) Logic symbol

The electrical construction of the diode **OR** is shown in Fig. 10.15(a), along with the truth table (*b*) and logic symbol (c). Parts (d) and (e) show how it works.

In Fig. 10.15(d), with either input at 1, current *from the input* flows through the diode and causes the output to be at binary 1. Little or no current flows through the external load shown because there is little voltage across it.

In Fig. 10.15 (e), with both inputs 0, current from the external load, R_L flows

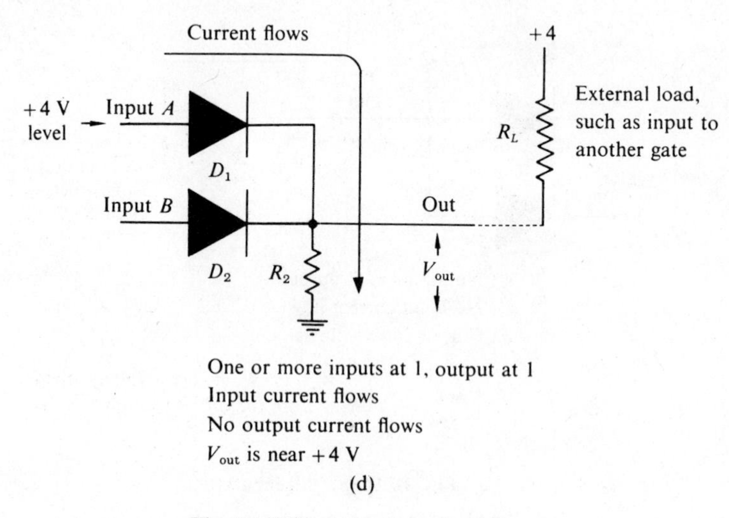

Fig. 10.15.(d) At least one input 1

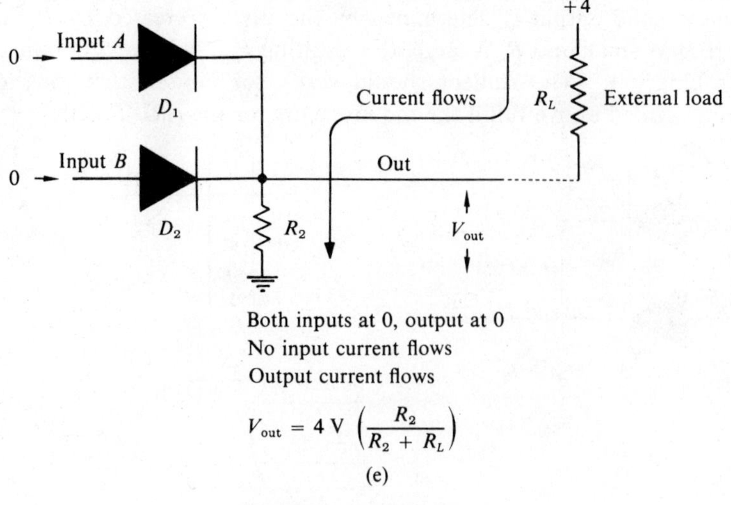

Fig. 10.15.(e) All inputs 0

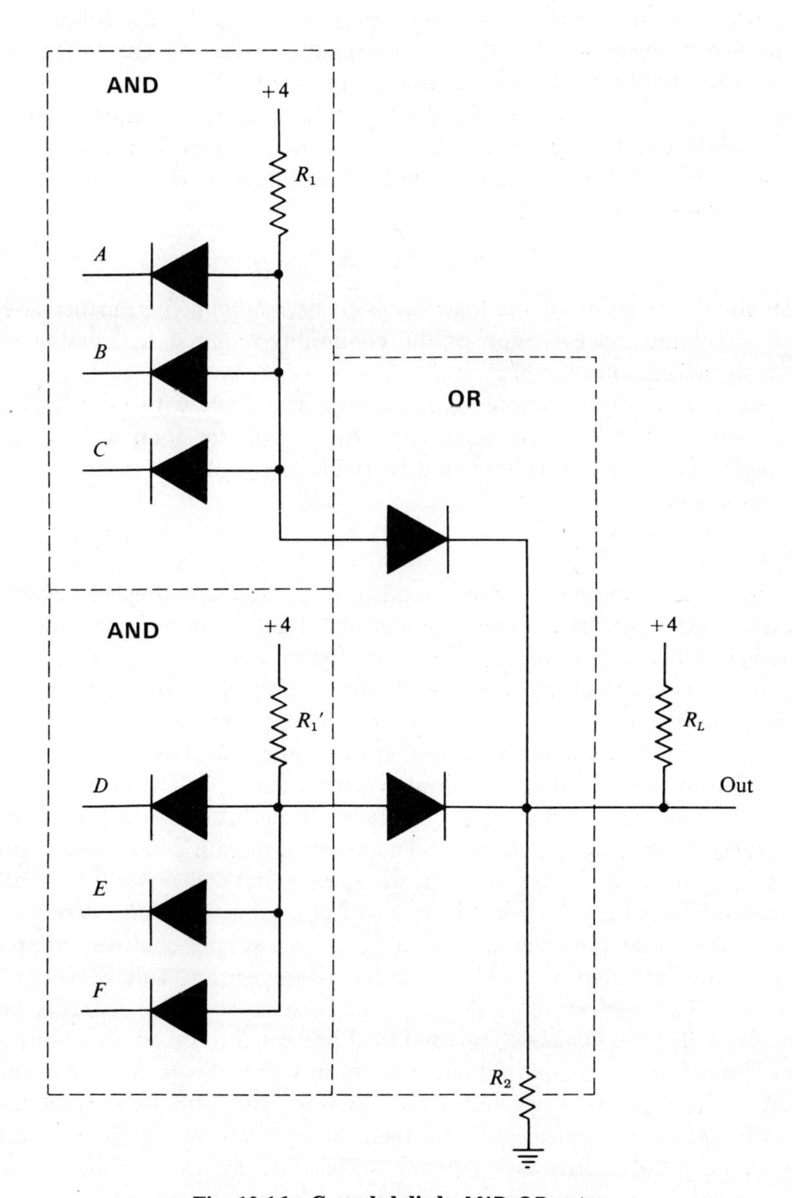

Fig. 10.16. Cascaded diode AND-OR gates

through R_2 to ground, and V_{out} is again given as the result of a voltage divider from the 4-volt power supply. If R_2 is much smaller than R_L, the voltage on the output of the circuit will be within the defined range for binary 0.

Figure 10.16 shows a cascaded **AND/OR** which performs part of the function shown symbolically in Fig. 10.13. The R_2 of the **OR** gate is the *load* resistor (called R_L in Fig. 10.14) for the **AND** gate. The values of R_L, R_2, and R_1 must satisfy the relation

$$R_L \gg R_2 \gg R_1 \dagger \qquad (10.7)$$

in order for the integrity of the logic levels to be maintained. Further levels of diode logic require an extension of the conditions required by relation (10.7), which soon becomes intolerable.

In the early days of computers, most logic was done with multilevel diode gates like those we have been discussing. As we will see soon, it is no longer economically necessary to do this, and therefore the problems mentioned above in (10.7) are now rare.

Loading

When one is designing a computer or other large electronic logical device, it is simplest and least expensive to package a group of gates into a single unit, usually a printed circuit card; and to make up entire logical units of the equipment, such as a register bank, out of many identical cards. It would be difficult or even impossible to make an analysis such as the elementary one above for each time an **AND/OR** or other multilevel situation arose. For this reason, circuits were designed with certain maximum allowable output current. The current for each gate input was also specified, so that the number of inputs that could be connected to each output was easy to calculate. The lowest common denominator, usually the input current for a single gate term, was chosen and called a *unit load*, and all other currents were then described in terms of unit loads. Circuits were designed so that in the worst possible case (loading, power supply changes, component variations) the integrity of the logic levels within certain voltage ranges was maintained. The number of loads that a given gate can drive from its output terminals is called the *fan-out*. The total load presented to other circuits by all of a gate's inputs is called *fan-in* (although this term sometimes refers to the number of discrete gate inputs). These terms are still employed, although it is no longer usual to be extensively concerned with them in logic design. Circuits with large fan-out and capacitive drive capability, as in Fig. 10.8, are now routinely available so that the propagation time for a signal through a gate (called *delay*) is usually a more important design parameter.

† The analysis supporting this statement is a straightforward but complex application of Ohm's law. It will not be presented here.

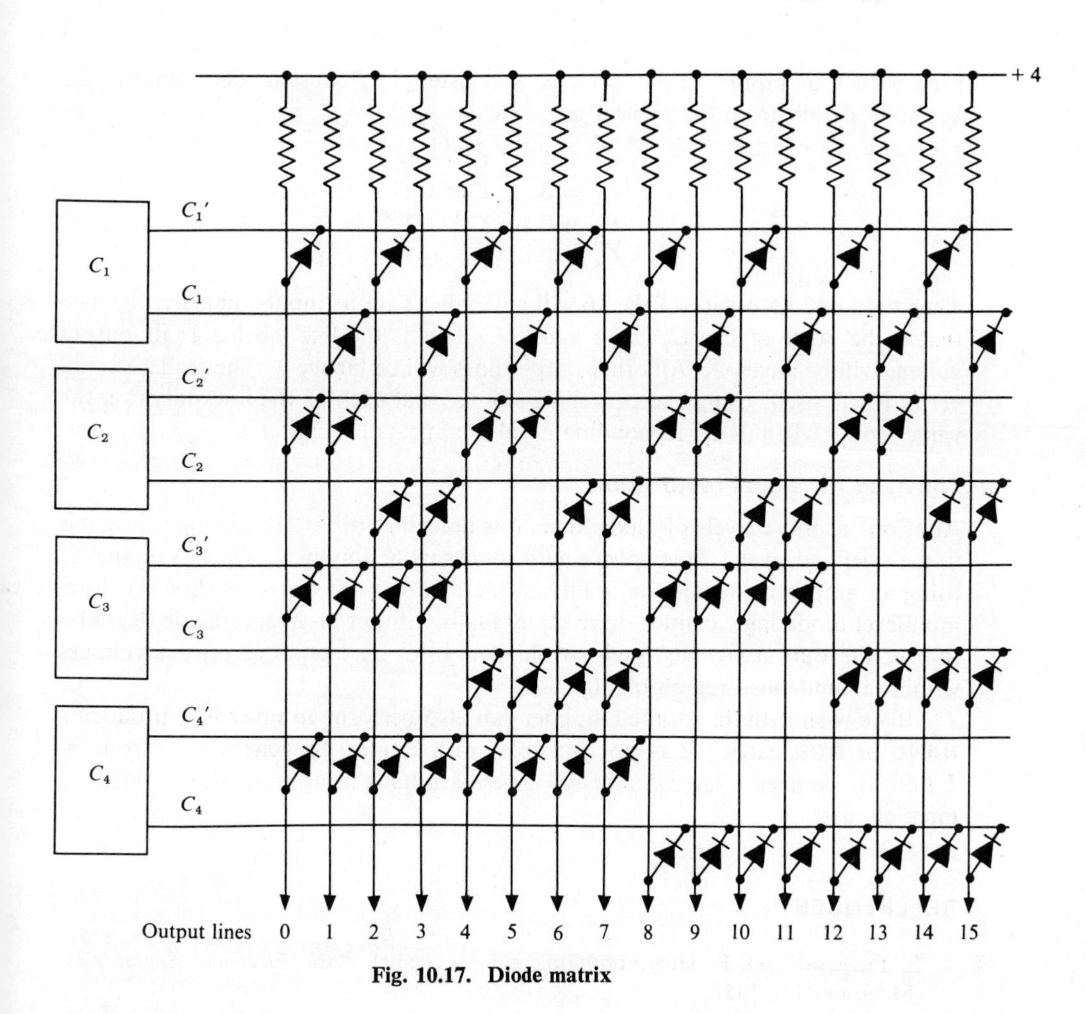

Fig. 10.17. Diode matrix

The diode matrix

Figure 10.17 shows a circuit with wide application inside the computer being used as an op-code decoder. During the execute phase of an instruction, the portion of the instruction register set aside for storing the op code must be interpreted so that the rest of the control section knows which operation to perform. The circuit takes four register stages, C_1 through C_4, with their complements C_1' through C_4' as inputs; and it produces 16 lines of output. One of these lines will be binary 1, according to the binary number in bits 1 through 4 in register C. All other output lines will be binary 0. This single line that is now active may be used to enable the desired control sequences for the instruction.

This circuit is nothing but 16 **AND** gates of the type shown in Fig. 10.14(a), each with four inputs. For example, if register C_{1-4} contains 0001 binary, the available signals from the register are:

$$\begin{array}{ll} C_1 = 1 & C_1' = 0 \\ C_2 = 0 & C_2' = 1 \\ C_3 = 0 & C_3' = 1 \\ C_4 = 0 & C_4' = 1 \end{array}$$

Then only one **AND** gate of the 16 will have all 1's on its inputs, namely, the gate that is the **AND** of C_1, C_2', C_3', and C_4'. This is the line labeled 1; its output voltage will be binary 1. All other output lines will be binary 0. The reader should verify for himself, using the procedure above, that each of the lines takes on the value 1 only when its corresponding number appears in register C.

The need for signal restoration

After one or more levels of diode logic, it is necessary to restore the logical signals to their defined levels. This is done with some sort of amplifier. The advantages of using an amplifier include the ability to increase fan-out since, as shown before, multilevel diode logic cannot drive many loads without an unacceptable degradation of the output signal. Such overloaded gates may produce output voltages within the undefined region of Fig. 10.1.

If we wish it to do so, the amplifier can also perform an inversion, producing **NAND** or **NOR** gates. It is not possible, with diodes, to generate A' from A. To do so, we need a *logical inverter*. These are built using transistors, which we must discuss.

REFERENCES

A. E. Fitzgerald, D. E. Higginbotham, and A. Grabel, *Basic Electrical Engineering*, McGraw-Hill, 1957.

D. L. Schilling and C. Belove, *Electronic Circuits: Discrete and Integrated*, McGraw-Hill, 1968.

P. A. Ligomenides, *Information-Processing Machines*, Holt, Rinehart and Winston, 1969.

W. H. Hayt, Jr., and G. W. Hughs, *Introduction to Electrical Engineering*, McGraw-Hill, 1968.

R. J. Smith, *Circuits, Devices, and Systems*, Wiley, 1966.

QUESTIONS

1. In the following table of logic levels, identify those that are positive logic and those that are negative logic.

Binary 0	Binary 1
0 volts	−4
−3	−7
−4	0
+12	+30
0	+5

2. What would be the effect on the output of an **AND** gate if one of the inputs was at an undefined voltage?
3. Draw the *I-V* graph for three ohmic devices with the resistances of 10, 50, and 720 ohms in the voltage range of 0 to +100 volts.
4. Several simple series circuits consisting of one battery and one resistor are to be set up. In each of the following cases, two of the three values, current, voltage, and resistance, are given. Find the third.

Voltage	Current	Resistance
?	10 mA	1,000 ohms
11 volts	?	22 ohms
21,322 volts	1 A	?
5 volts	?	(10^6 ohms)

5. Two resistors, one 120 ohms and one 240 ohms, are placed in series in a circuit with a 10-volt battery. Calculate the current in the circuit and the voltage drop across each resistor.
6. Two resistors, one 120 ohms and one 240 ohms, are placed in parallel in a circuit with a 10-volt battery. Calculate the current through each resistor.
7. What is the equivalent resistance of three resistors in parallel, one 100 ohms, one 200 ohms, and one 400 ohms?
8. What is the equivalent resistance of three resistors in series, one 1000 ohms, one 2000 ohms, and one 4000 ohms?
9. For the diagram of Fig. 10.6, the resistance values are: R_1 = 75 ohms; R_4 = 100 ohms; R_2 = 100 ohms; R_3 = 400 ohms. Calculate the values for I_1, I_2, I_3, V_1, V_2, and V_3. Repeat for resistance values: R_1 = 1000 ohms, R_4 = 500 ohms, R_2 = 20 ohms, R_3 = 50 ohms.
10. In Fig. 10.7(a), if R_1 is 50 ohms and an external load of 100 ohms between the output line and ground is connected, what will the output voltage be with the switch open? With the switch closed?
11. Repeat Question 10 for Fig. 10.7(b).
12. Repeat Question 10 for Fig. 10.8(a).
13. What features differentiate insulators, conductors, and semiconductors?
14. If an electronic field is applied across a piece of semiconductor material that has both holes and electrons as carriers, will the holes and electrons move in the same

or opposite direction? Will the current caused by hole motion add to or subtract from the current caused by electron motion.

15. A capacitor consists of two conducting metal plates separated by an insulator. When the capacitor is charged, equal negative and positive charges appear on opposing plates. Contrast this charge distribution with the positive and negative charges appearing inside semiconductor material across a barrier.
16. Contrast the *I-V* characteristic of a diode with the *I-V* characteristic of a resistor. Draw an *I-V* graph showing both.
17. In Fig. 10.13, what would be the result if two load lines, load *A* from *B* and load *A* from *C*, were raised at the same time? Assume that bit 6 in register *B* and bit 6 in register *C* are both binary 1.
18. Figure 10.14(e) shows a diode **AND**. If all inputs were set at 1, there would be a 1 at the output. If R_1 is 100 ohms and R_2 is 1000 ohms, what is the output voltage?
19. Figure 10.15(e) shows a diode **OR** with all zero inputs. If the external load resistor, R_L, is 100 ohms and the output resistor, R_2, is 500 ohms, what is the output voltage of the circuit?
20. In Fig. 10.16, if R_1 is 100 ohms, R_2 is 1000 ohms, and R_L is 10,000 ohms, what is the output voltage for each of the following conditions: (a) all inputs *A* through *F* are binary zero; (b) all inputs *A* through *F* are binary one; (c) inputs *A*, *B*, and *C* are binary zero; inputs *D*, *E*, and *F* are binary one.
21. The diode matrix shown in Fig. 10.17 decodes four binary inputs into 16 single-line outputs. Draw a diode matrix that converts 8421 BCD into the 0 through 9 decimal outputs. Draw a diode matrix that converts 10 lines of decimal information into 8421 BCD.

11
DIGITAL COMPUTER HARDWARE: II

In the earliest days of the computer, the logical inversion and signal restoration functions were as essential as they are now. These functions were performed by vacuum tubes or gas-filled tubes. Because each stage of restoration was far more unreliable and expensive than the diode logic preceding it, it was routine to maximize the number of functions performed by diodes. With transistors available (by about 1963) in inexpensive plastic cases, this optimization was no longer essential. By 1970, integrated circuits had become so extensively applied that the routine use of individually packaged components was no longer economically justifiable. Transistors and diodes are produced in such profusion on an integrated circuit that the price distinction between them is obliterated. It is unthinkable to have even a single level of diode logic without one or more transistors following. The result is greater reliability, more noise immunity, vastly greater speed and fan-out capability, and availability of an almost unlimited repertoire of logic functions.

Despite this progress, it is still very important for the computer designer to understand the properties of transistors and their uses. ICs are, after all, simply compact interconnected networks of transistors and diodes located in a single sealed package. Designing products with them entails many of the same pitfalls that a discrete-component designer will encounter, and a thorough understanding of the circuits is necessary.

11.1 TRANSISTORS

Figure 11.1(a) shows the semiconductor construction of a transistor, which in this example consists of a very thin layer of *p*-type material sandwiched between two pieces of *n*-type material. This is called an *npn transistor*. The *pnp* type is also possible (Fig. 11.1b) and almost as common except in digital circuits, but we will limit discussion to the operation of the *npn* type. Figure 11.2 shows how the *npn* transistor is connected in a circuit. The transistor consists simply of two diode junctions. The first, called the collector-base junction, is shown reverse-biased (*n* more positive than *p*) and it seems that little or no current would flow across it. The emitter-base junction is shown forward-biased (*p* more positive than *n*), and therefore current flows here. Suppose, though, that of the current crossing from

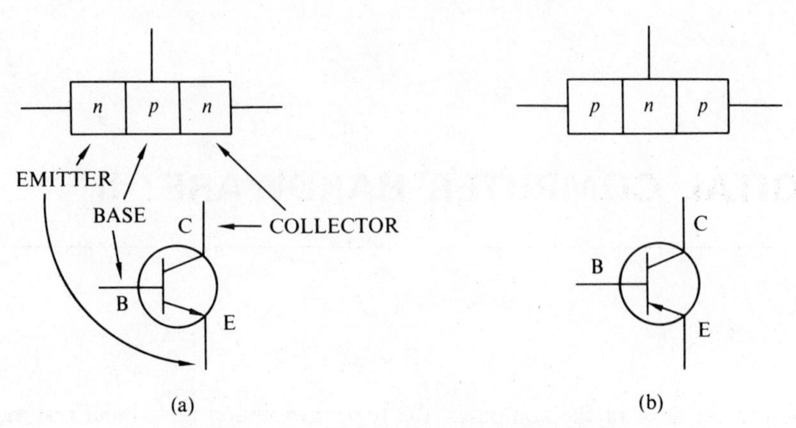

Fig. 11.1. Transistor constructions and symbols

emitter to base, some electrons were able to fully cross the base region and enter the barrier between collector and base. This is common, since the base region is very thin. Upon entering the base-collector barrier region, the electrons would encounter a large accelerating force due to the voltage on the collector, and a large collector current would flow. Normally no collector current flows because there are no electrons in the collector-base junction region. If electrons are injected by the emitter-base current, though, collector current will flow. The collector current that does flow is proportional to the emitter-base current that causes it and is usually much larger. Because of this proportionality, the transistor can act as an amplifier.

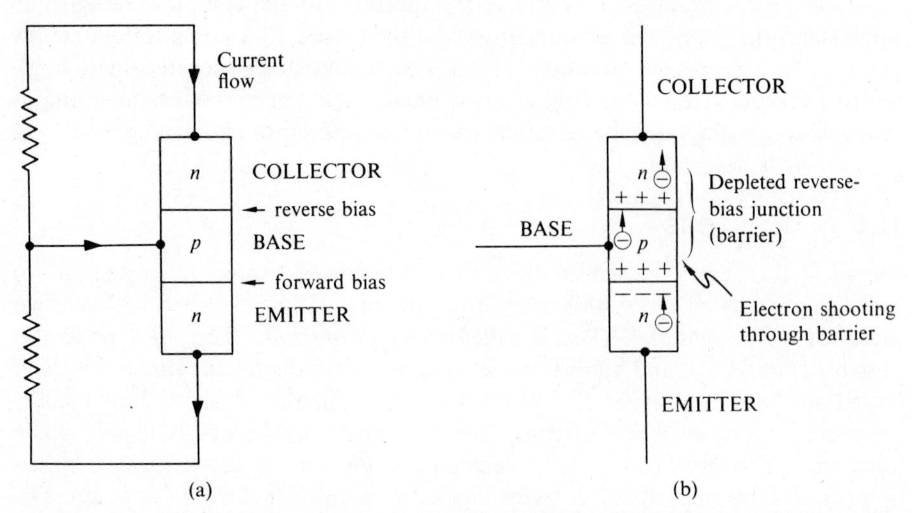

Fig. 11.2. An *npn* transistor: (a) biased for normal operations; (b) carrier distribution when transistor is biased as in (a)

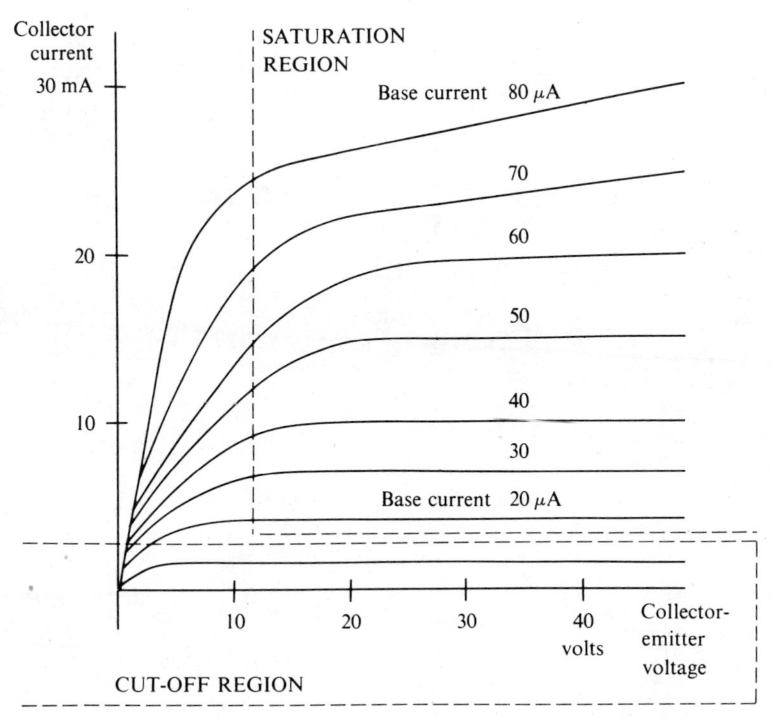

Fig. 11.3. Transistor *I-V* characteristics with emitter grounded

Figure 11.3 shows the *I-V* characteristics of an *npn* transistor for different values of base current. Note that for small collector-to-emitter voltages, the transistor looks like a resistor, and the collector current is roughly proportional to the collector-emitter voltage. Above a certain voltage, however, more collector voltage does not result in more collector current, because the number of electrons entering the collector region is being limited by the base current. The transistor is then said to be *saturated.* This state is of special interest to digital designers because large variations in circuit parameters and power supply voltages will not produce proportional changes in output current. This is an ideal way of generating a logic level. Note also that if the base current is very small, or if the emitter-base junction is reverse-biased so that no base current flows, no collector current flows. This condition, called *cut-off,* is also very insensitive to changes in circuit values and input voltages, and it is therefore commonly used to generate the other logic level. A saturated transistor will conduct a lot of current between collector and emitter and is therefore said to be ON. A cut-off transistor is OFF and does not conduct.

If the collector current of a saturated transistor is limited by a series resistor, R_C, as shown in Fig. 11.4(a), the collector voltage will be near zero, since the transistor is conducting heavily and looks like a low resistance. The voltage

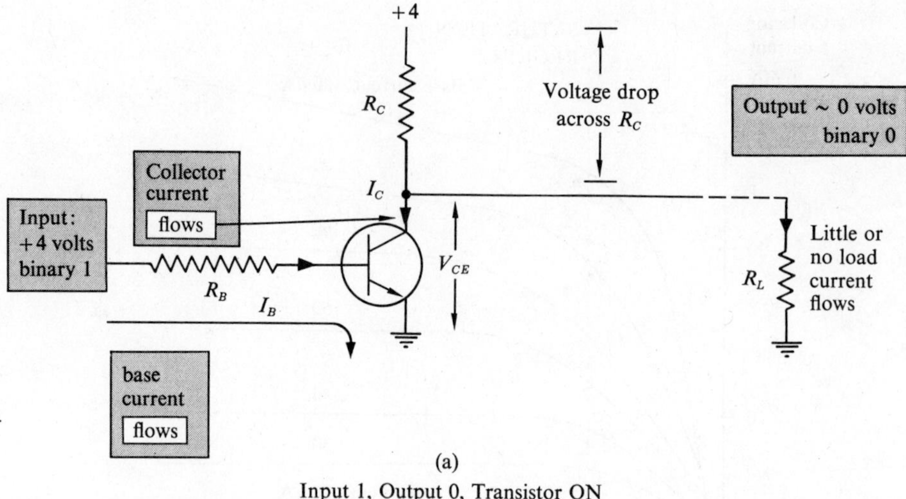

(a)
Input 1, Output 0, Transistor ON

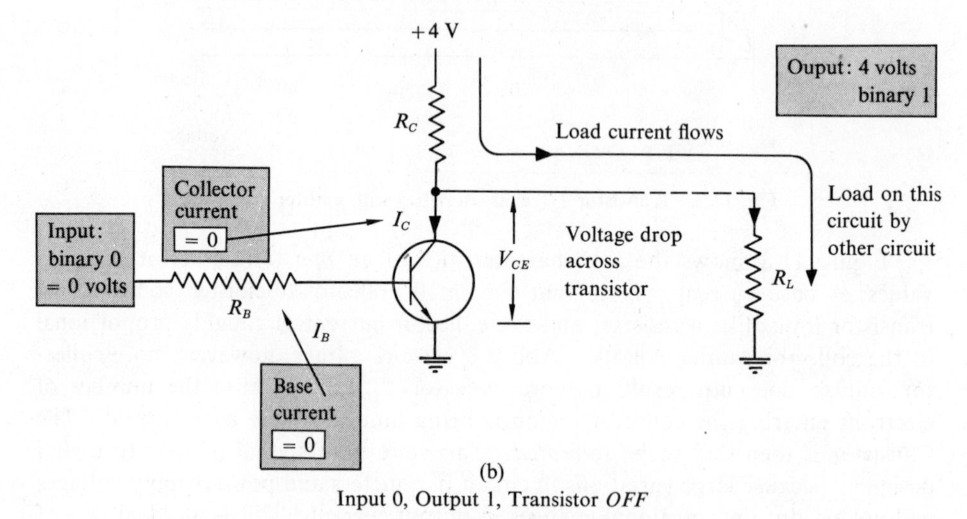

(b)
Input 0, Output 1, Transistor *OFF*

Fig. 11.4. Transistor inverter

across a saturated transistor may be as low as 0.1 volt if the base current is more than enough for saturation. This voltage, known as $V_{CE(sat)}$, is a common parameter for digital (switching) transistors. Conversely, if the transistor is cut off, as in Fig. 11.4(b), there is little current, there is no voltage drop across the resistor, and the collector voltage is equal to the supply voltage. The transistor therefore acts exactly like the logic level switch of Fig. 10.7(a), and the discussion about capacitive loading and Fig. 10.8 apply. In addition, a two-transistor switch circuit can be built which looks like the double switch of Figure 10.7(b).

11.2 TRANSISTOR LOGIC CIRCUITS

The inverter

The circuit of Fig. 11.4 is, in fact, a simple logical inverter. If the base (input) is held at logical 0 (ground), the transistor is cut off and the collector voltage (output) is logical 1, near the power supply. If the base is held at logical 1, the transistor is saturated, and the output is logical 0, within $V_{CE(sat)}$ of ground. The normal tolerances of the logic levels are wide enough to handle variations in $V_{CE(sat)}$, etc. The truth table is

In	Out
0	1
1	0

so the circuit performs a logical **NOT**, or inversion.

NAND circuits

Figure 11.5 is a complete **NAND** circuit, comprising a diode **AND** and a transistor **NOT**. The output of the diode network is no longer directly connected to the input of the next stage, and so problems of fan-in and fan-out have been greatly reduced. The **AND** function can be found at node *A*. If any of the inputs are at ground, as in Fig. 11.5(a), current will flow from the power supply through R_B, through any of the diodes $D_{1,2,3}$ that are ON, and to ground. Node *A* will be close to ground but will be a bit positive because of the few tenths of a volt offset

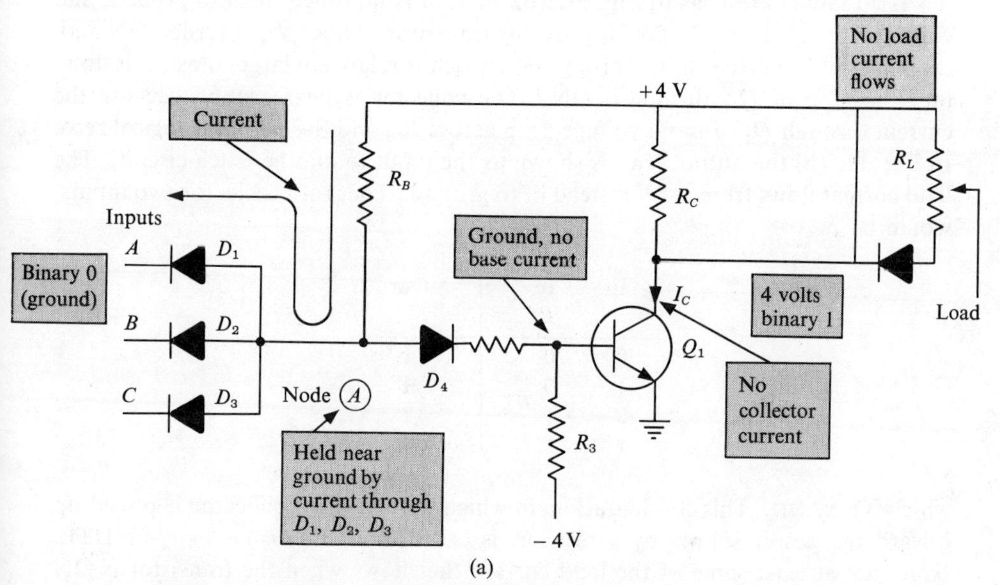

Fig. 11.5.(a) Diode-transistor logic (DTL) NAND: at least one input at 0

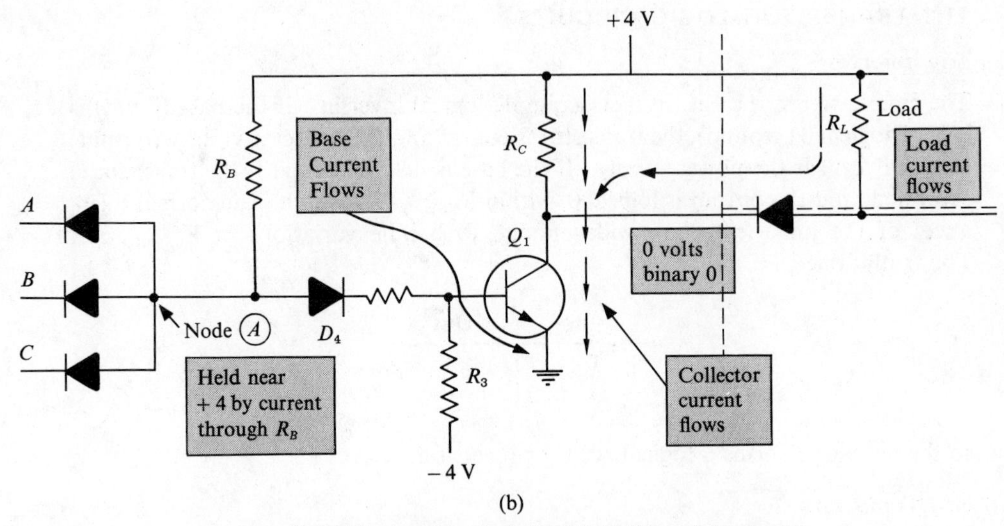

(b)

Fig. 11.5. (b) Diode-transistor logic (DTL) NAND: all inputs at 1

that is characteristic of the diode. If node A is a few tenths of a volt above ground, the voltage offset of D_4, which is in the opposite direction, will effectively be subtracted. Current from the negative power supply, through R_3, is enough to turn both D_4 and the transistor, Q_1 OFF. Since no collector current flows, there is no drop across R_C, and the output is logical 1.

If all inputs are 1, as in Fig. 11.5(b), node A is no longer held to ground, and current from R_B is free to flow toward the transistor. Diode D_4 is turned ON and, though a bit of current flows through R_3, which is relatively large, most of it flows into the base of Q_1, turning it ON. The collector is near ground because the current through Q_1 causes a voltage drop across R_C, and the output is logical zero. In Fig. 11.5(b) the output load is shown as the input to another such circuit. The load current flows from +4V instead of to ground. The truth table, for two inputs, would be

In A	In B	Out
0	0	1
1	0	1
0	1	1
1	1	0

which is a **NAND**. This configuration, in which the transistor collector is pulled up toward the power supply by a resistor, is called *diode-transistor logic*, or DTL. Note that at least some of the load current that flows when the transistor is ON comes through the resistor which is used as R_B for the next gate stage, and that this

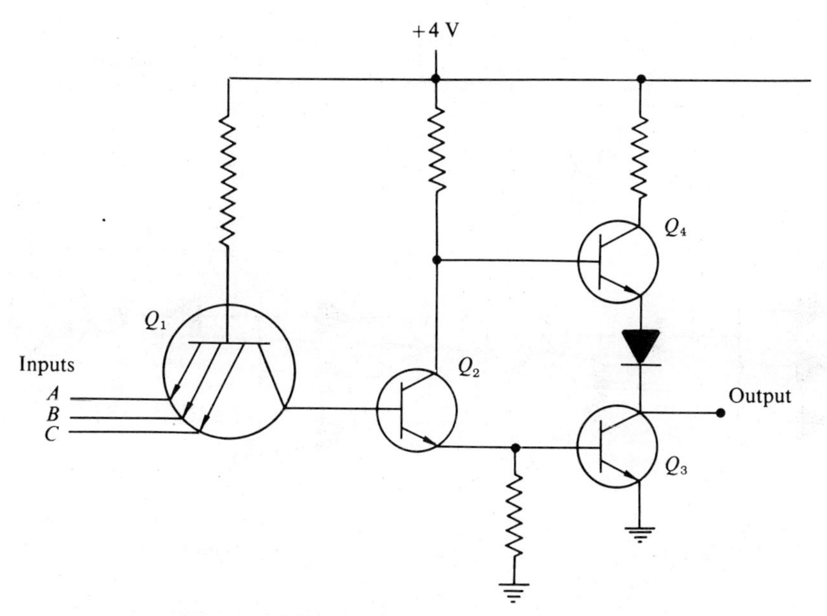

Fig. 11.6. TTL NAND with multiple-emitter transistor input

current flow is the same as the one shown in Fig. 11.5(a) through R_B and out toward the input connection. Input current therefore flows to ground through a saturated transistor located in the stage preceding the one shown. If we replaced $D_{1,2,3}$ with resistors, which would be cheaper but would not isolate the signals connected to inputs A, B, and C from each other, it would be *resistor-transistor logic* (RTL).

Another commonly used circuit is that of Fig. 11.6. Just as in Fig. 10.7(b), a switch (in this case a transistor) is used to connect the output either to the power supply or to ground. This is called *transistor-transistor logic* (TTL). The input logical elements may be diodes, with the dual-transistor output shown; or a transistor with multiple emitters like that in Fig. 11.6 can be used. The *multiple-emitter transistor*, Q_1 in the figure, allows any one of the multiple isolated inputs to turn on the transistor whenever any of the emitter-base junctions is conducting.

The NOR

The **NAND** circuit (Fig. 11.5) functions as a **NOR** if the input levels are defined as for negative logic. It is therefore possible to construct circuits which form the sum of products (The **OR** of **AND**s) using only **NAND** circuits. The products are formed and inverted by the first **NAND**s. The inverted products are equivalent to the products expressed in negative logic. The next **NAND** circuits, electrically the same as the first, perform a **NOR** function: a logical **OR** plus a return to positive logic.

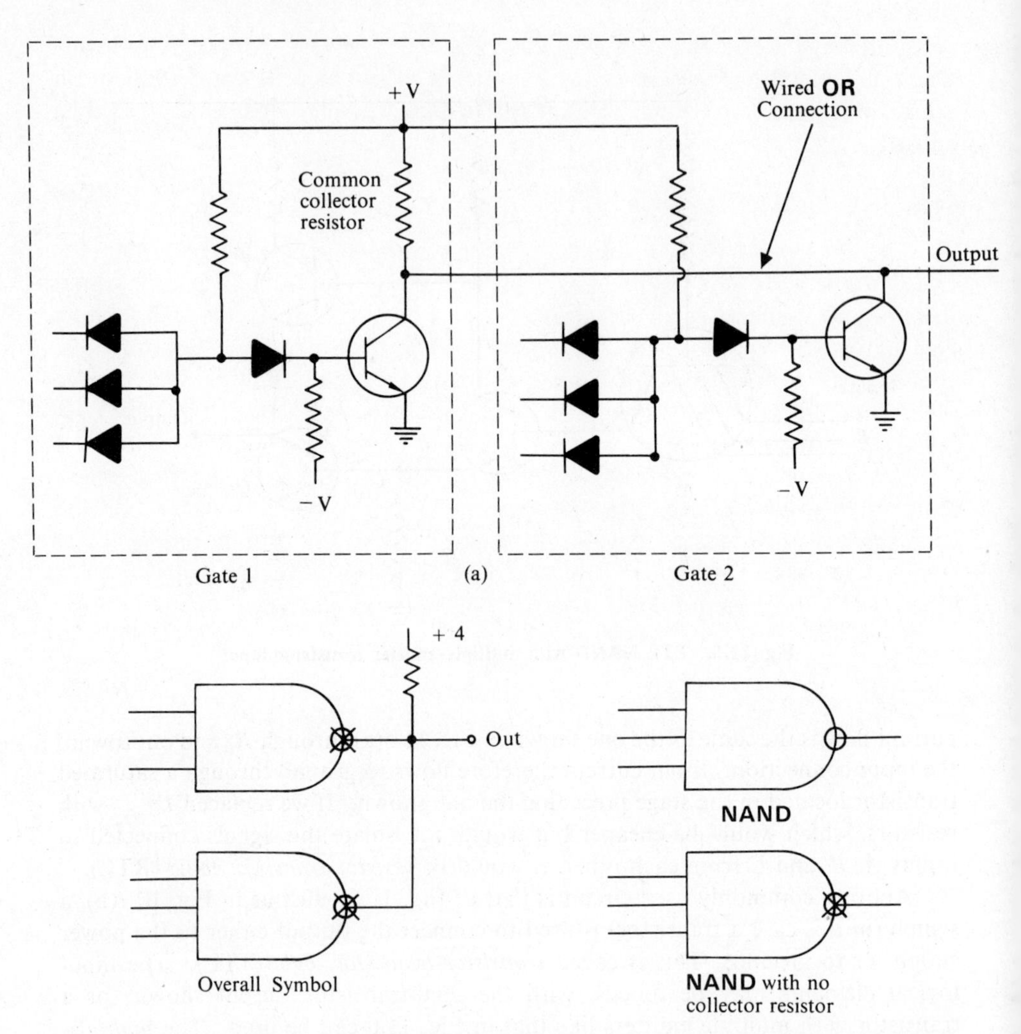

Fig. 11.7. (a) Wired OR of two NANDs with no collector resistor; (b) logic symbol representation

The Wired OR

In Fig. 11.7, two DTL **NAND** circuits are shown with their outputs wired together, sharing a common collector resistor. If either gate has all of its inputs 1, then its transistor will be saturated and it will ground the output, causing a 0, and current will flow through the common resistor. Since either transistor or both may cause current flow, the function performed is an **OR**, and the circuit is called a *wired* **OR**. The wired **OR** performs the **OR** function at no expense; no components are involved. It is necessary only that the gates be manufactured without a collector load resistor, and that that resistor be supplied externally. This function cannot be performed by the circuit of Fig. 11.6, since the logic 1 level is produced by a saturated transistor, not a resistor. The performance of the wired **OR** is dependent on the positive logic convention. With negative logic the wired connection would be an **AND**.

11.3 THE BINARY ADDER

Most of the operations a computer performs involve some kind of arithmetic. Even an operation not explicitly calling for addition may require indexing, and therefore the contents of an index register must be added to the operand address field of the instruction register to obtain an effective address. The arithmetic circuits perform this procedure. Some machines have separate address adder logic; others use the same circuits that do user-specified arithmetic. The principles involved are the same.

Any binary addition of two variables must involve, in each bit position (column), consideration of three input operands: the addend, the augend, and the carry from the previous column. It must also produce two output operands: the sum and the carry to the next column. Figure 11.8(a) shows the truth table for binary addition. Figure 11.8(b) is a typical example of such addition.

Inputs			Outputs	
Carry in C	Addend B	Augend A	Sum S	Carry out T
0	0	0	0	0
0	0	1	1	0
0	1	0	1	0
0	1	1	0	1
1	0	0	1	0
1	0	1	0	1
1	1	0	0	1
1	1	1	1	1

(*a*)

```
1111010000 Carry
 110101100 Addend
 101101010 Augend
1100010110 Sum
```

(*b*)

Fig. 11.8. Binary addition: (a) truth table; (b) example

The logical equations for binary addition can be reduced to the form

$$S = (A \oplus B) \cdot C' + (A \oplus B)' \cdot C$$
$$T = A \cdot B \cdot C' + (A + B) \cdot C$$

These equations may be routinely implemented in logic circuits, one for each bit position in each operand. Such a circuit, keyed to the equations above, is shown in Fig. 11.9.

An ADD instruction would proceed by presenting both operands to the inputs shown in Fig. 11.9. The operands would most likely be stored in registers, and the

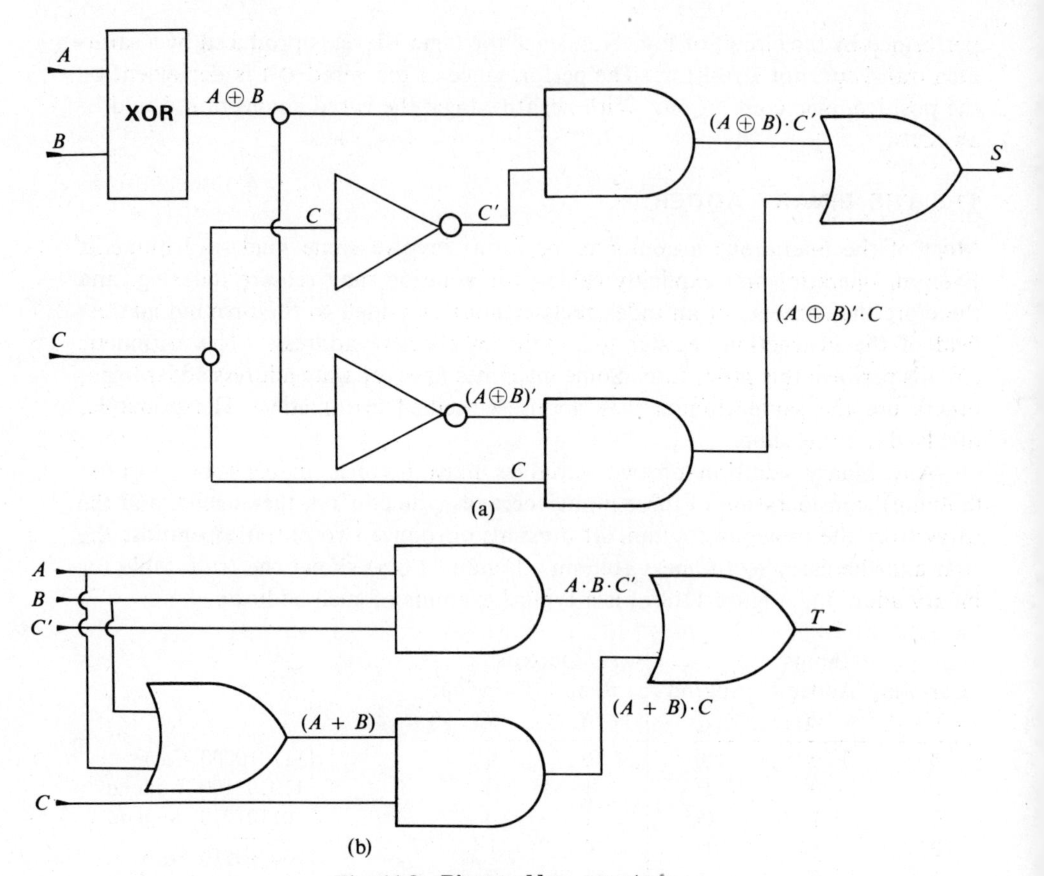

Fig. 11.9. Binary adder—one-stage

register outputs would be gated to the adder circuits by a signal received from the op-code decoder. After waiting some period of time for the output of the adder to settle to its final value, the op-code decoder would send a signal to load the sum into some destination register. The carry from the leftmost place is usually loaded into some overflow or carry-sensing flip-flop, which may then be tested under program control.

Carry anticipation

Although it may not be clear from the preceding elementary presentation, a serious difficulty is inherent in this simple adder. Consider the addition example in Fig. 11.10.

When the addend and augend of Fig. 11.10 are first presented to the adder circuits, each of the bit positions except the rightmost will see no carry and will set up a sum and carry out accordingly. The first circuit, however, will produce a carry, which will change the carry input of the second position (from the right).

$$\begin{array}{r} 1111111111 \\ + \qquad\quad 1 \\ \hline 10000000000 \end{array}$$

Fig. 11.10. Addition example with lots of carries

This second position will then be required to set up a sum and carry different from the original one. This change will then affect the third position, and the changes will propagate (ripple) in this way toward the most significant bit. The final result will be correct, but it will take a relatively long time for the final answer to be formed. To alleviate this problem, circuits have been devised which anticipate the need for a carry.

Let us look at a circuit which will anticipate the need for a carry from the first two places into the third place. Figure 11.11 is the truth table for such a circuit. The subscripts on each position indicate bits from the corresponding bit number position in each variable or output.

As Fig. 11.11 shows, the carry input for bit position 3 is calculated immediately, without ripple, based on the contents of both positions 1 and 2. With much more in the way of hardware, the same construction can be carried out for each subsequent position. Usually a compromise is reached between speed and cost in which the adder is broken into groups of bits. Each group has carry anticipation within itself, and carries propagate between groups.

11.4 BINARY STORAGE ELEMENTS—THE FLIP-FLOP

The internal circuitry of computer systems consists mostly of two types of logical elements. The first type, the gate, has already been discussed. The other type is a

$$\begin{array}{cccc} & C_3 & & \\ & A_3 & A_2 & A_1 \\ + & B_3 & B_2 & B_1 \\ \hline S_4 & S_3 & S_2 & S_1 \end{array}$$

(*a*)

A_2	B_2	A_1	B_1	C_3
0	0	0	0	0
0	0	0	1	0
0	0	1	0	0
0	0	1	1	0
0	1	0	0	0
0	1	0	1	0
0	1	1	0	0
0	1	1	1	1
1	0	0	0	0
1	0	0	1	0
1	0	1	0	0
1	0	1	1	1
1	1	0	0	1
1	1	0	1	1
1	1	1	0	1
1	1	1	1	1

(*b*)

Fig. 11.11. (a) addition example—bit positions labeled; (b) carry anticipation truth table for (a)

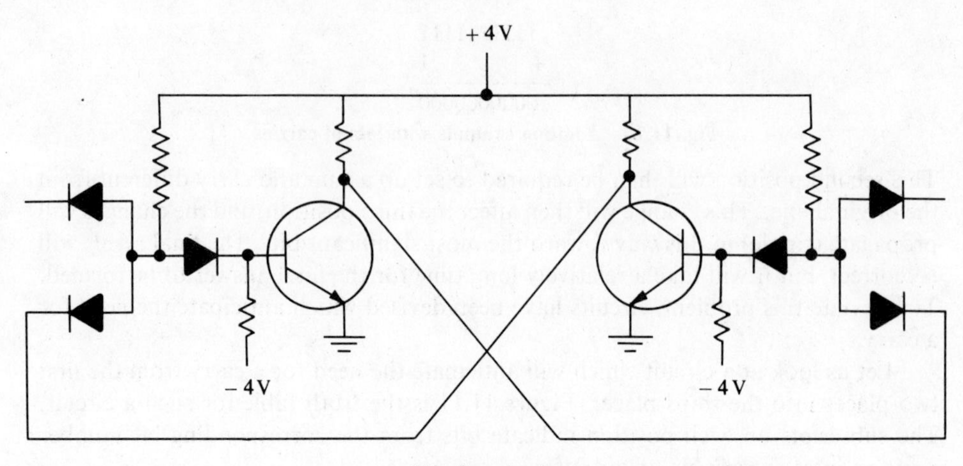

Flip-flop circuit

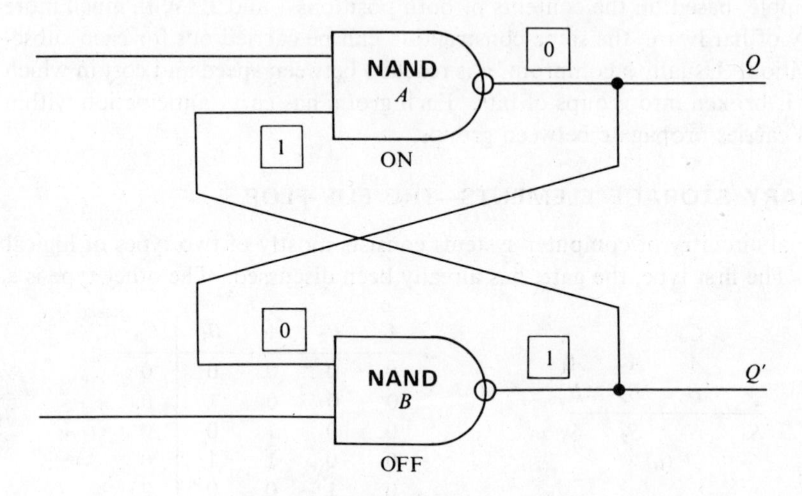

Flip-flop, logically, from gates

Fig. 11.12. Storing a bit in a flip-flop (numbers in boxes represent present state)

storage element. Registers are constructed of these storage elements, and transfer of data or instructions between registers is controlled by gates. The most common binary storage element is called a *bistable multivibrator*, or *flip-flop*. Figure 11.12 shows a flip-flop circuit and its equivalent, a circuit with two **NAND** gates.

R-S flip-flops

Two **NAND** gates connected in the circuit of Fig. 11.12 perform the binary storage operation as follows: Suppose that gate A is ON, so that its output (labeled Q) is binary 0. This output is in turn used as an input to gate B. Since B inverts it is

OFF, and the output of B is 1, which is applied to the input of A, turning it ON. Since we assumed that A was ON anyway, this situation is seen to be stable. Another stable state can be inferred by symmetry: when A is OFF and B is ON. This circuit therefore has two stable states and can be used to store one bit of binary information. One of the pair of states is arbitrarily called 0 and one called 1, and the bit stored in the flip-flop at any time corresponds to the internal state assigned.

Two further functions are required of the flip-flop. We must be able to determine the internal state and report it to the rest of the system as a logic level, and we must be able to externally cause the state of the flip-flop to change by application of a logic level. These functions allow us to construct registers and transfer data among them.

Since the components of a flip-flop are gates, the gate outputs may be used to provide logic levels which indicate the internal state of the circuit. At any time, one of the two lines is at binary 1, and the other is at binary 0. The signal indicating the contents of the flip-flop, Q, and its complement, Q', are therefore both available to drive other circuits. Their fan-out is reduced by an amount equal to one gate input, since each output is connected to one input of the other **NAND** within the flip-flop. The formal definition of Q is that it represents the binary number stored: If the flip-flop is *reset* (contains a 0), the logic level on Q will be binary 0. If the flip-flop is set (contains a 1), the level on Q will be binary 1. The definition of Q' is exactly the reverse; its availability is especially useful for inputs to other gates where a complemented input is required. The direct availability of a complement obviates the necessity of an extra inverter. Figure 11.13(a) shows a way of changing the state of the flip-flop. By choosing the **NAND** with inputs of (1, 1) and output of 0, and by putting a logical 0 on its free input, the output is forced by the **NAND** operation to become 1. This changes the state of the other **NAND** gate, which then holds the first one in its new state. *It is now possible to remove the signal which caused the flip-flop to flip* (Fig. 11.13b). The wires going to the externally available gate inputs cause the flip-flop to change state, but the change occurs when the applied signal changes state from binary 1 to binary 0. The sense of this change is logically inconsistent with the principle that binary 1 is an "active" or "causative" state. For this reason, we place an inverter in each input wire. Figure 11.14(a) shows the standard logic symbol for this circuit; the device is called an *R-S flip-flop*. Figure 11.14(b) is the truth table, or *state table*, for the *R-S* flip-flop. This truth table is different from the others we have used because it includes time as a dimension, showing the storage property of the flip-flop. Note that the solid half-arrow heads on the outputs of the flip-flop indicate that the output level remains even after the input has been removed. If both the R and S inputs are made binary 1 simultaneously, both **NAND** gates will have outputs of 1. Both gates are ON, and Q is not the complement of Q! When the input signals are removed, random noise or small transistor differences will cause the flip-flop to seek one state, 1 or 0, but this state is not determined by the history of its inputs and is therefore shown in Fig. 11.14(b) as indeterminate.

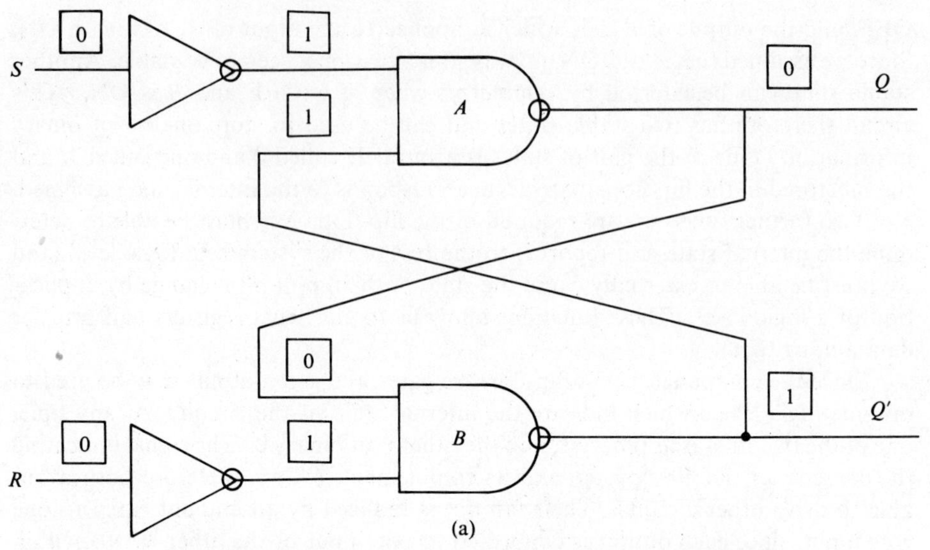

(a)

Flip-flop storing a binary 0 (numbers in boxes indicate present state)

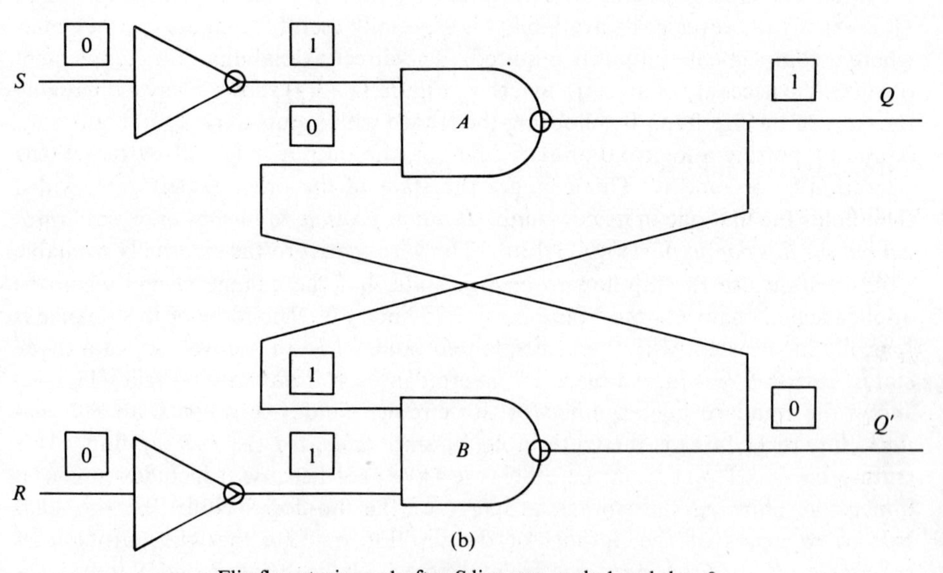

(b)

Flip-flop storing a 1 after *S* line was made 1 and then 0

Fig. 11.13. Changing the state of a flip-flop

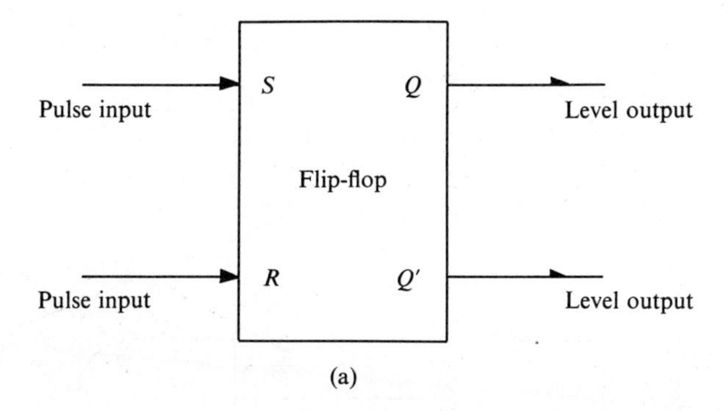

(a)

Present state $Q(t)$	Next state $Q(t+1)$ present input $R(t)S(t)$			
	00	01	11	10
0	0	1	?	0
1	1	1	?	0

Note: ? means undefined state

(b)

Fig. 11.14. *R-S* flip-flop: (a) logical representation; (b) state table

Information transfer to and from flip-flops

Figure 11.15(a) shows an *R-S* flip-flop in which the input inverters have been replaced by **NAND**s. This is easy enough to do, requiring only a few extra diodes. Now, both *R* inputs to the reset input **NAND** must be 1 to reset, and both *S* inputs to the set input **NAND** must be 1 to set. Connecting one *R* input and one *S* input together, we have a wire which, when made 1, causes the flip-flop to be either set or reset, depending on the state of the other inputs. If both extra *S* and *R* inputs are 0, nothing happens. If they are both 1, the ambiguous state results. If the extra *S* input is 1 and the extra *R* input is 0, the flip-flop will set, but this will happen only when the common wire is 1. The flip-flop will reset with 1 on the *R* line and 0 on the *S* line when the common wire is 1. This wire, which must be 1 for set or reset, is called the *common load* or *strobe line*. A normal means of loading data into a flip-flop is to place the data on the *R* and *S* lines, and then to raise the strobe line to 1 for a short time, causing the transfer to happen. This process is called a *jam transfer* since the data are "jammed" into the flip-flop by the strobe line. Figure 11.15(b) shows the contents of two flip-flops that are part of a register being jam-transferred into another pair of flip-flops by the action of one strobe line. In this fashion, register-to-register transfers take place in the machine. Such

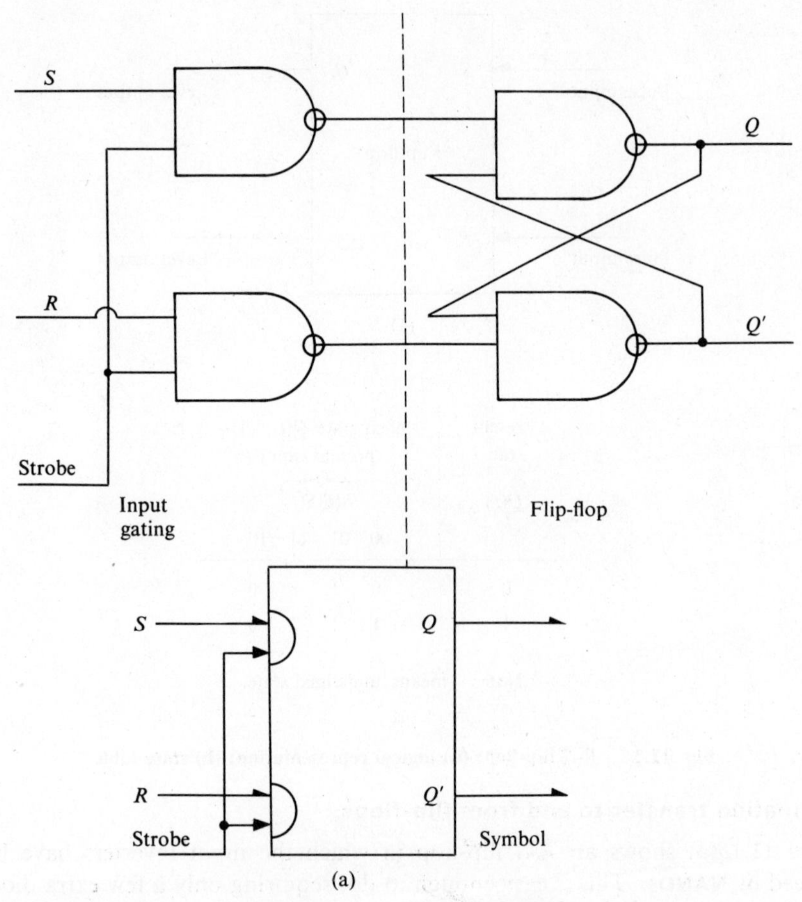

Fig. 11.15. (a) Strobed flip-flop

a transfer is called *double-rail logic*, since both Q and Q' are required. Figure 11.15(c) shows how one more inverter can recreate Q' at the register being loaded. Such a transfer is called *single-rail logic*; only one wire is needed to carry the data from the source flip-flop to the destination flip-flop. This is sometimes referred to as a D flip-flop or a *D latch*.

Figure 11.16 is a register stage which can be loaded from any of three data sources, single-rail. One strobe pulse is used for each data source, and the **NAND**s used for input strobe gating are connected together with a wired **OR**. Any **NAND** output will pull the flip-flop gate input to ground, causing the circuit to change state.

Shifting

A common machine instruction is the *shift*, which causes data to move within a register one or more positions to the left or right. The left shifting of a binary

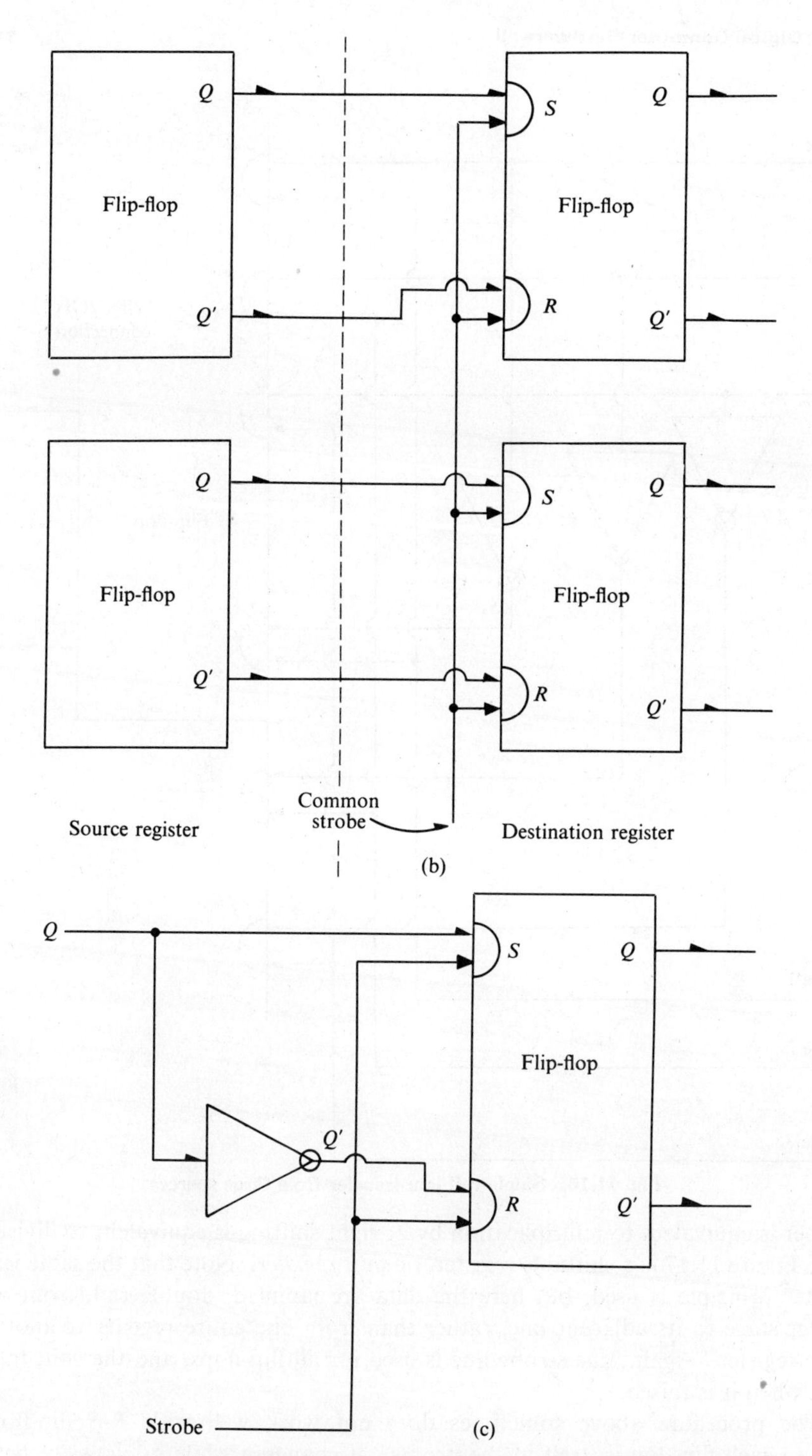

Fig. 11.15. (b) Jam transfer between registers; (c) single-rail input to an *R-S*

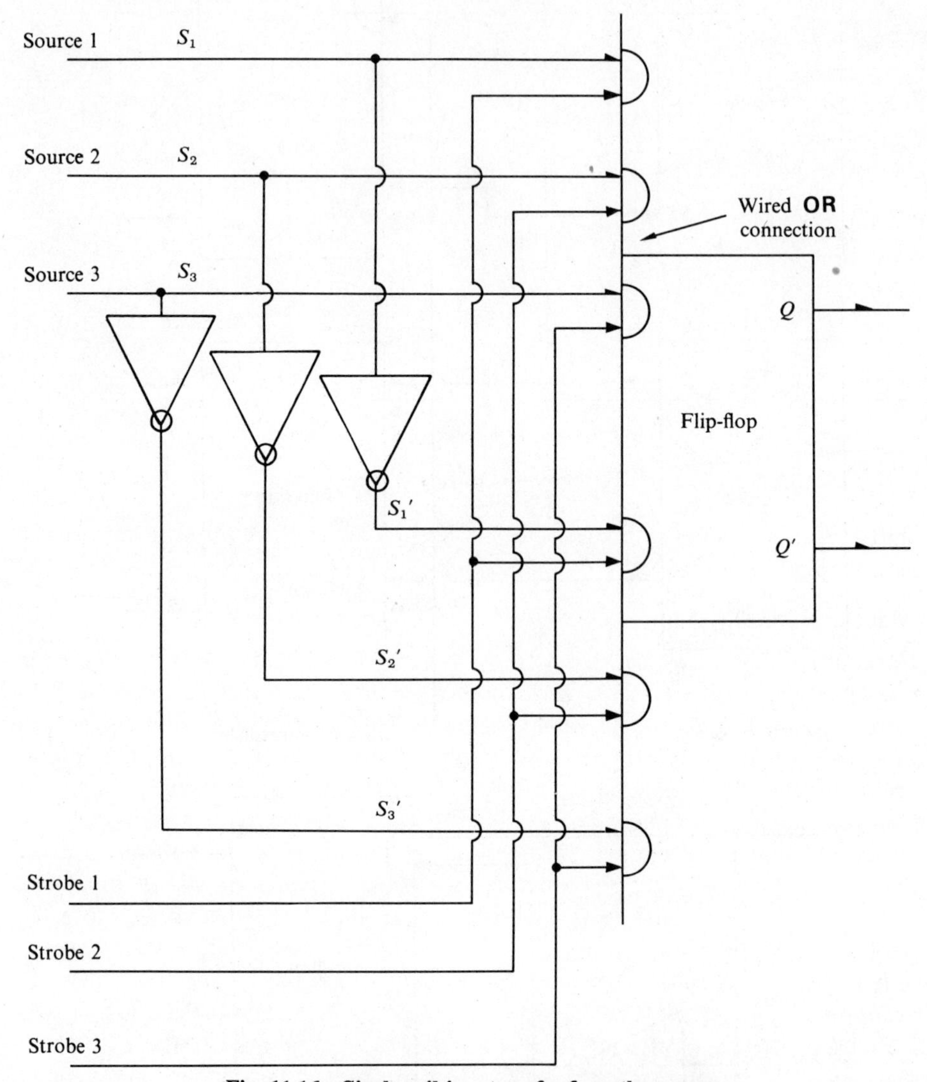

Fig. 11.16. Single-rail jam transfer from three sources

number is equivalent to multiplication by 2; right shifting is equivalent to division by 2. Figure 11.17 is a shiftable register, or *shift register*. Note that the same jam-transfer principle is used, but here the data are jammed, double rail, from one register stage to its adjacent one, rather than from one entire register to another entire register. Again, one strobe line is used for all flip-flops, and the shift takes place when it is raised.

The procedure above sometimes does not work well with *R-S* flip-flops, because each flip-flop is itself in the process of changing while its data are being

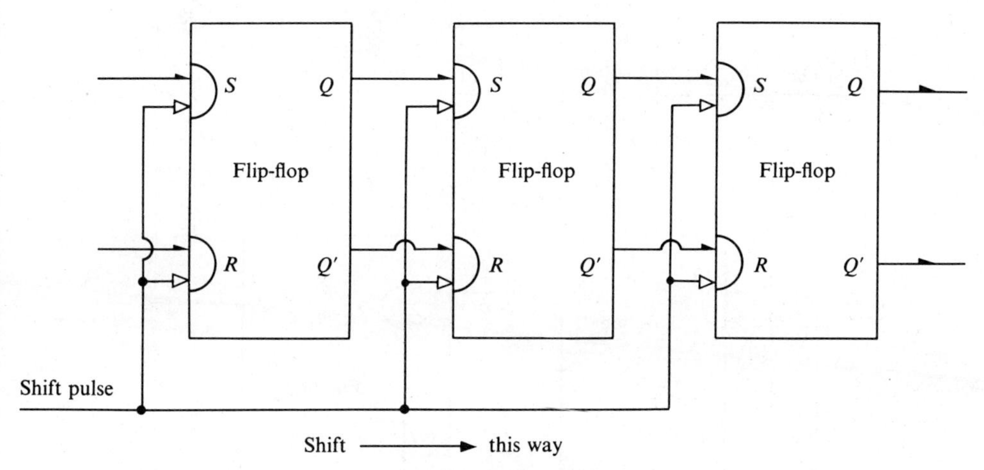

Fig. 11.17. Shift register

shifted into the next one down the line. A circuit that obviates the problem appears as Fig. 11.18.

Master-slave flip-flop

Conceptually, it is necessary to provide some sort of temporary storage between flip-flop stages so that data on the R and S inputs may change while the permanent storage element, the flip-flop itself, is being loaded. This storage would be needed in a shift register since all stages are strobed at once. In Fig. 11.18, this goal is accomplished by using another flip-flop for the intermediate storage, and the configuration is called a *master-slave* flip-flop. Here the pulse property of the strobe signal is used to activate first one flip-flop, the master; and then the other, the slave. The flip-flop strobe input is designed to respond to a change in voltage from ground to $+V$ and to ignore steady levels and other changes. Since the strobe signal is in the form of a pulse, both the first transition, from ground to $+V$ (called the *leading edge*), and the second, from $+V$ to ground (called the *trailing edge*), are used. When the strobe goes positive, data on the R and S inputs are loaded into the master flip-flop where they are temporarily stored and where they cannot be sensed from the outside (there are no external connections to the Q and Q' outputs of the master). When the strobe goes back to ground, the slave is loaded because the **NOT** in the strobe line inverts the polarity of the pulse and the slave responds only to positive-going changes. Because the output of the entire circuit changes after the clock returns to ground, the process is called *trailing-edge triggering*.

AC and DC triggering

If the strobe inputs to the flip-flops responded to steady voltage levels, the master could continue to accept changes in data from the R and S inputs during the time

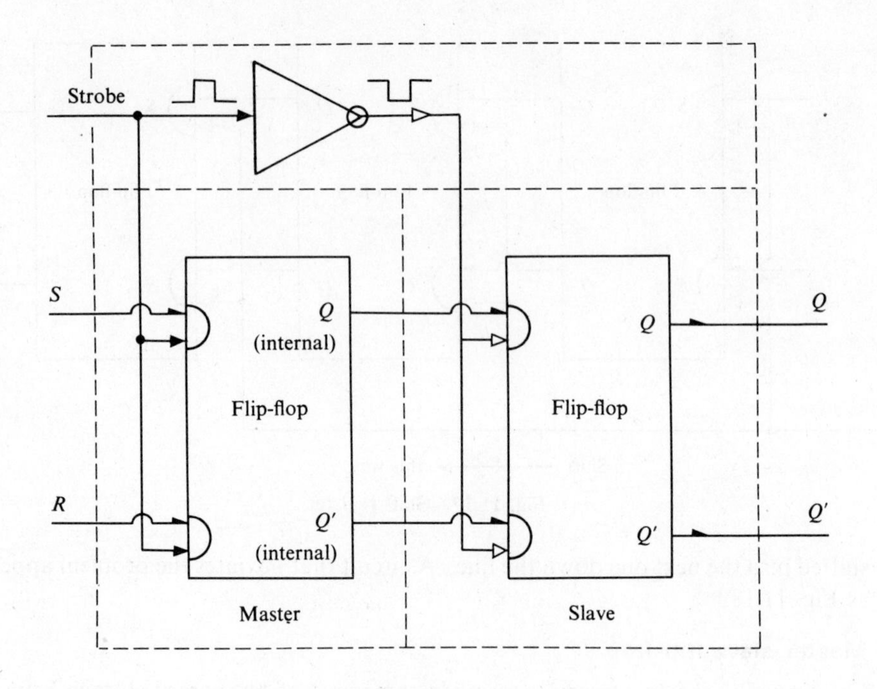

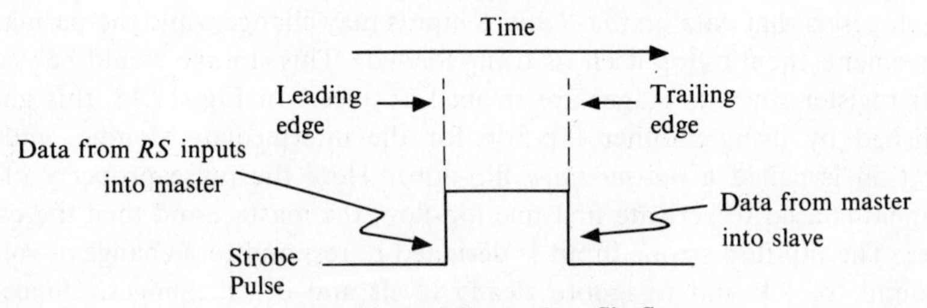

Fig. 11.18. Master-slave flip-flop

that the strobe was positive. It is usually safer to have the actual transition between ground and $+V$ cause the transfer of data into the master, and to have the transition between $+V$ and ground transfer the data into the slave. This modification is called *pulse triggering*, or *AC triggering*, and the process which accepts levels rather than transitions is called *level sensing*, or *DC triggering*. Some flip-flops have both AC and DC inputs, with the DC inputs usually used for functions like a common reset wire for all of the elements in a register.

Integrated-circuit flip-flops are usually of the master-slave type whenever intermediate storage is required. In discrete-component circuits, capacitors are commonly used as the temporary storage elements. In both methods, the principle of utilizing the leading and trailing edges of the strobe is the same.

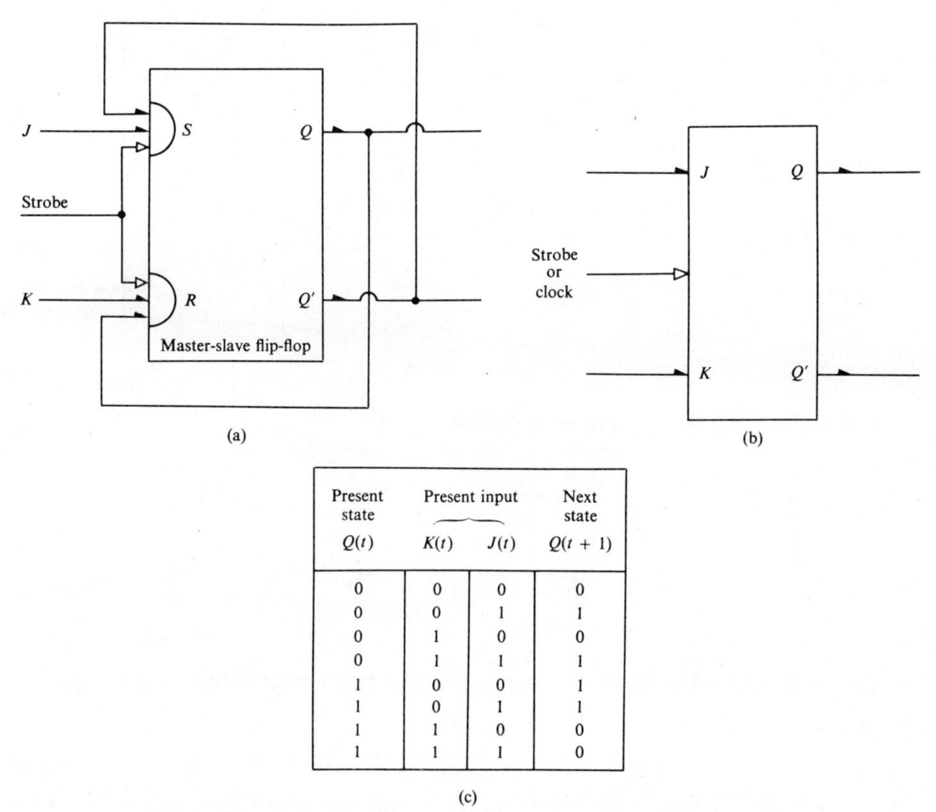

Present state $Q(t)$	Present input $K(t)$	Present input $J(t)$	Next state $Q(t+1)$
0	0	0	0
0	0	1	1
0	1	0	0
0	1	1	1
1	0	0	1
1	0	1	1
1	1	0	0
1	1	1	0

(c)

Fig. 11.19. ***J-K* flip-flop: (a) logic; (b) representation; (c) state table**

J-K flip-flops

The major disadvantage of the *R-S* flip-flop is the presence of undefined states in the state table (see Fig. 11.14b). This problem is solved by the development of the *J-K* flip-flop, shown in Fig. 11.19. If an extra input is added to the *R-S* **NAND** gates, the Q and Q' outputs may be logically combined with the inputs. This is done to eliminate the ambiguous state possible with an *R-S* flip-flop.

Let us examine the operation of the *J-K* flip-flop. Suppose that the slave flip-flop is in the 0 or reset state. This means that Q is 0, and the R gate, to which Q is wired, will not accept data or a strobe from the R input wire. Since Q' is 1, the S gate will accept data and strobe. If both the R and the S input wires are 1 when the strobe pulse arrives, only the S pulse will reach the flip-flop. The R pulse is blocked by a closed input gate. Since a similar argument would hold if the slave flip-flop started at 1, the ambiguous case has been eliminated. The operation of the *J-K* flip-flop has been formalized in the state table of 11.19(c). The logical representation of the *J-K* flip-flop is in 11.19(b).

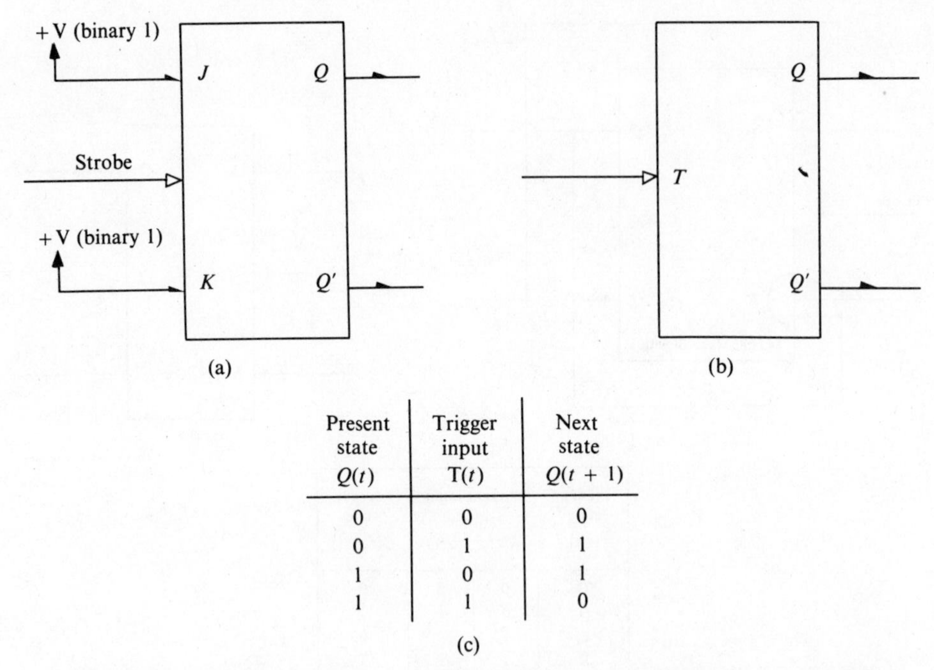

Present state $Q(t)$	Trigger input $T(t)$	Next state $Q(t+1)$
0	0	0
0	1	1
1	0	1
1	1	0

(c)

Fig. 11.20. ***T* flip-flop: (a) from a *J-K*; (b) in logical representation; (c) as a state table**

T flip-flop

Further inspection of the state table of Fig. 11.19(c) reveals that if both the J and the K inputs are binary 1, the flip-flop will change state each time a strobe pulse is received. This special case is referred to as a *toggle* or *T flip-flop*. Figure 11.20 shows the T flip-flop as derived from a *J-K*, in its logical representation, and as a state table.

The T flip-flop has the logical property of counting modulo two. If two input pulses are placed sequentially on the T input, the first will cause a change of state from 0 to 1, the second from 1 to 0. This means that the Q output of the flip-flop will have a pulse from ground to $+V$ and back whose width is the spacing of the input pulses. This is formalized in Fig. 11.21(a), which is called a *timing diagram*. The horizontal axis represents time, increasing to the right. Two or more logical signals are plotted in separate graphs on the same time axis. Each graph shows the state, whether 0 or 1, of a particular signal at any given time. Broken lines connect simultaneous events on different graphs. Thus the temporal correlation of several signals can easily be plotted.

Binary counting

Since a T flip-flop counts by 2, and since the Q output is a single, wide pulse, we may make a simple binary counter of any desired length by cascading stages, as in Fig. 11.21(b). The Q output of each stage is wired to the T input of the next.

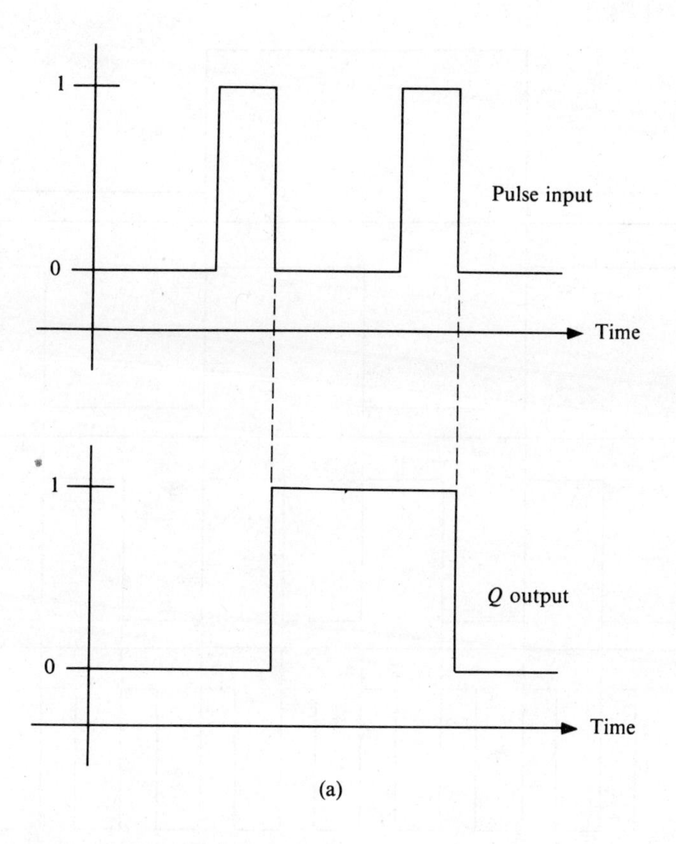

(a)

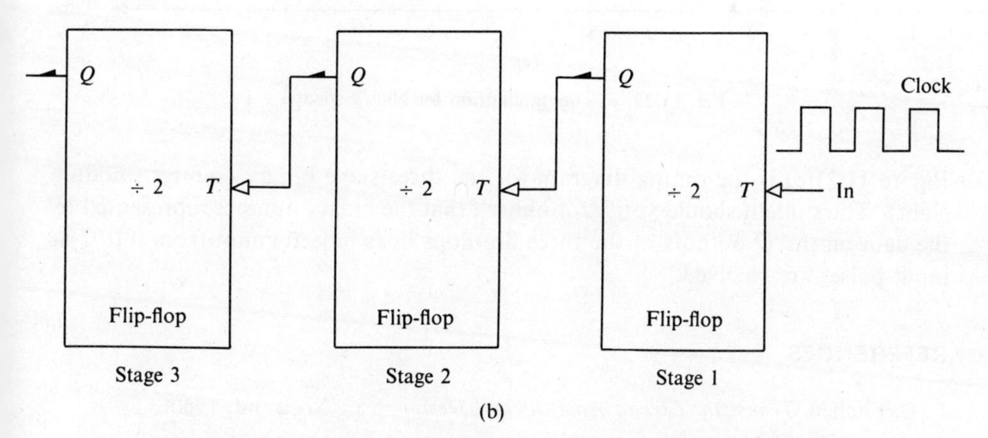

(b)

Fig. 11.21. ***T*** **flip-flop: (a) dividing by 2; (b) three-stage binary counter**

(continued)

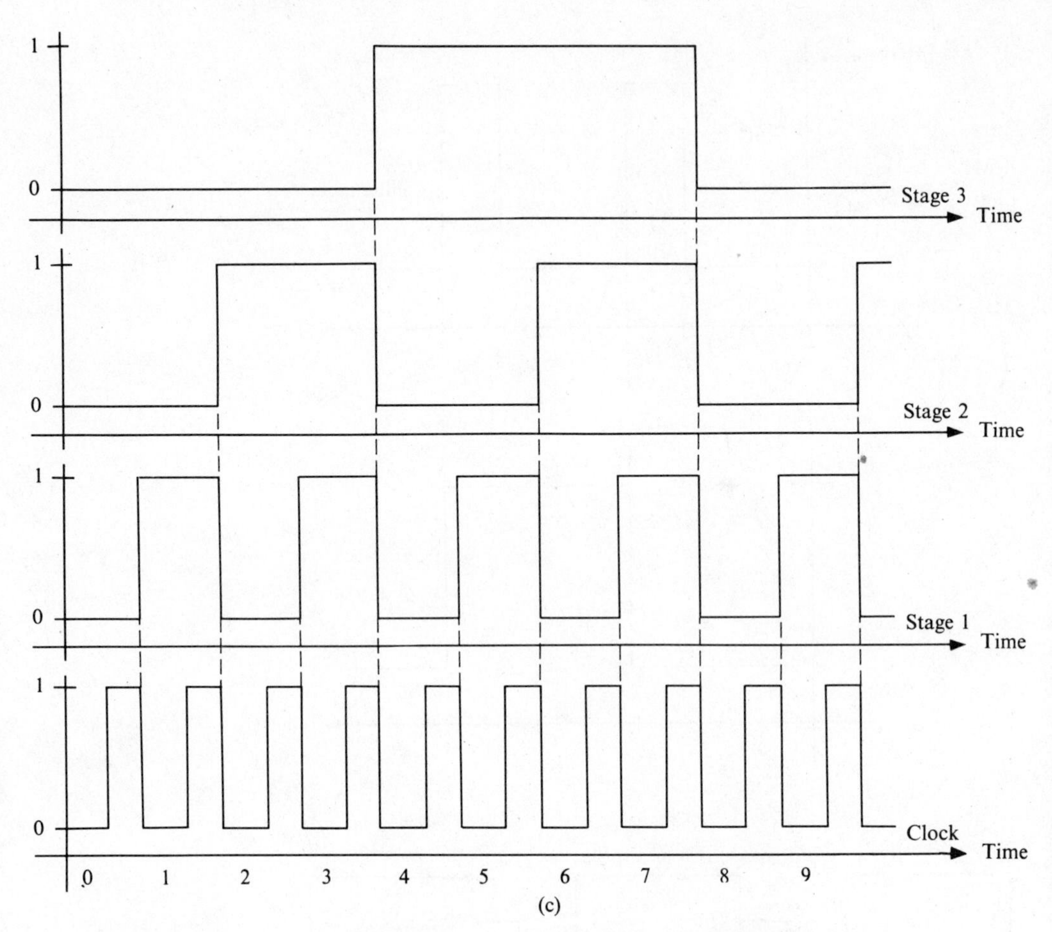

Fig. 11.21.(c) Timing diagram for binary counter

Figure 11.21(c) is the timing diagram for the three-stage binary counter (modulo eight). The student should verify for himself that the binary number represented by the data on the Q outputs of the three flip-flops does indeed count from 0 to 7 as input pulses are received.

REFERENCES

F. C. Fitchen, *Transistor Circuit Analysis and Design*, Van Nostrand, 1960.

J. Millman and C. C. Halkias, *Electronic Devices and Circuits*, McGraw-Hill, 1967.

J. M. Pettit and M. M. McWhorter, *Electronic Switching, Timing, and Pulse Circuits*, McGraw-Hill, 1970.

R. Littauer, *Pulse Electronics*, McGraw-Hill, 1965.

R. H. Mattson, *Electronics*, Wiley, 1966.

Y. Chu, *Digital Computer Design Fundamentals*, McGraw-Hill, 1962.

Burroughs Corporation, *Digital Computer Principles*, McGraw-Hill, 1969.

J. Millman and H. Taub, *Pulse, Digital, and Switching Waveforms*, McGraw-Hill, 1965.

L. V. Azaroff and J. J. Brophy, *Electronic Processes in Materials*, McGraw-Hill, 1963.

QUESTIONS

1. Is it possible to make a transistor out of two discrete diodes, simply by connecting their two anodes, or *P* regions, together? Explain.
2. Draw a circuit involving a transistor with applied voltage polarity for the base-emitter junction to be (a) forward-biased, (b) reverse-biased. Show the carrier distribution and the current flow.
3. Explain why the transistor conditions of saturation and cut-off are extremely important in the design of digital circuits. Explain why it is important to avoid them for linear circuits such as audio-amplifiers.
4. In what ways is a transistor like a switch? In what ways is it unlike a switch? What mode of transistor operation do you assume?
5. In the inverter circuit of Fig. 11.4, what would cause the output voltage to be (slightly) less than 4 volts when the transistor is OFF?
6. Using the complete circuits of Fig. 11.5(a) and (b), draw in its totality the circuit diagram for a cascaded **NAND/NOR** analogous to the cascaded **AND/OR** of Fig. 10.16. Be careful with definitions of positive and negative logic.
7. Show how you might use a wired **OR** connection in loading a bank of register flip-flops. (*Hint:* See Fig. 10.13.)
8. Derive the logical equations for binary sum and carry out.
9. Derive the logical equations for sum and carry out from (a) a group of two, (b) groups of three, and (c) groups of four. Design a circuit implementation using this carry anticipation to reduce carry propagation time.
10. If a 1 bit is placed on both the set and reset lines of an *R-S* flip-flop at the same time the output condition is ambiguous. When the reset and set signals go away, what in fact determines the state of the flip-flop?
11. Using logic symbols, show how to build a *J-K* flip-flop out of some gates.
12. Explain the difference between single-rail and double-rail jam transfers.
13. Show how the register in Fig. 11.17 can be made a cyclic shift register, such as was described in Chapter 7.
14. Show how the register in Fig. 11.17 can be made to do an arithmetic shift, as described in Chapter 7.
15. Why would the introduction of integrated circuits make it much easier to build master-slave flip-flops than before?

16. If a flip-flop employs DC triggering and the strobe line stays high or at binary 1, what will happen to the internal state of the flip-flop as the data change? What kind of circuit can be used to obviate this problem?
17. What kind of flip-flop is produced if the *J* and *K* inputs to a *J-K* flip-flop are wired together?
18. Design a giant multipurpose register with four flip-flop stages. The register should be able to count in binary, shift left, and be jam-transferred from another source. Each function is to be controlled by pulses on external strobe lines, one for each function.

12
MEMORY TECHNOLOGY

Memory is one of a digital computer's most expensive resources; it is also scarce because buying enough is prohibitively costly. Much talent is therefore devoted to decreasing the cost per bit of storage and/or increasing memory speed. Since memory technology is constantly changing, we will concentrate our discussion on fundamental concepts that are basic to the understanding of more advanced memory technology.

Just as the semiconductor is the basis for the active circuits and flip-flop memory devices discussed in Chapter 11, the ferromagnetic device is fundamental to *passive* memory technology. Passive memory devices possess two stable states which require no external energy to maintain them. The two states have a *hysteretic* characteristic; i.e., the quantity called *magnetic induction* has nonzero values even in the presence of no energization. The sign of this residual quantity is dependent on the sign of the energizing force (magnetizing force) most recently applied, provided that the magnitude and duration of that force were sufficient to change the state of magnetic induction.

Intensive development effort has been expended on *active* memory systems. These employ elements which require external energization to maintain the stored state, as opposed to passive memory devices which require energy only to change states. Under the rapidly changing conditions that exist in the memory marketplace, only the expert is interested in the details of each technology and their comparative advantages. In this chapter we first present the fundamentals of the established core-memory technology and follow with a similar treatment of the chief competitor, semiconductor memory technology. We close the chapter with a brief introduction to other devices used for secondary memory.

12.1 FERROMAGNETIC PHENOMENA

The mechanism most commonly used in computer memories is ferromagnetism. The ferromagnetic effect is used in core, drum, disk, and tape memory systems. Constant striving for technological improvement and commercial advantage may uncover other phenomena as useful as ferromagnetism. They will also have to be economically competitive, however, in order to replace those currently in use.

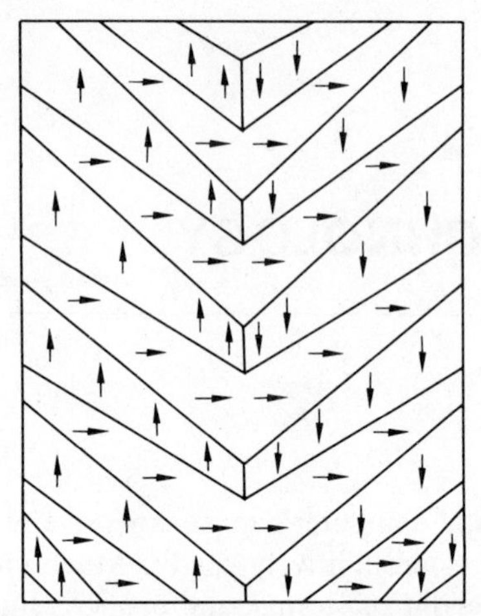

Fig. 12.1. Schematic representation of ferromagnetic domain structure

From the viewpoint of the early 1970's, it appears that ferromagnetic techniques will continue to be used for a long time. A discussion of the physical principles of ferromagnetism will assist us in understanding these computer memory devices.

The atoms of ferromagnetic material possess a magnetic dipole moment, which is the nonzero sum of the magnetic dipole moments of its spin and orbital electron motion. Only certain materials are ferromagnetic and have this property because of their atomic electron configuration. When one examines bulk ferromagnetic material in detail, one discovers small regions where most of the ferromagnetic dipole moments are aligned in the same direction. These regions are known as ferromagnetic domains. An unmagnetized sample is actually composed of a very large number of randomly oriented ferromagnetic domains. Figure 12.1 shows such an orientation. If these domains were aligned so that their magnetic moments all pointed in the same direction, then the bulk material would exhibit a ferromagnetic moment. The determination of the orientation of ferromagnetic domains is a study in energy minimization.

When a ferromagnetic material exhibits a net magnetic moment, we know that the domains are organized to point in one particular direction. On the other hand, when there is no net magnetic moment, we know there is complete disorganization. Figure 12.2 illustrates the relationship between organization and energy. In this qualitative representation you will note that, although there is a lower energy involved in an organized state than in a disorganized state, there is also an energy barrier or potential which must be overcome in going from a disorganized to an organized state. You must put energy into the system to go from a nonmagnetic state to a magnetic state, but when the magnetic state is reached, no further energy is required to maintain that situation. A more common graphical representation of the state of magnetization of ferromagnetic material is given by

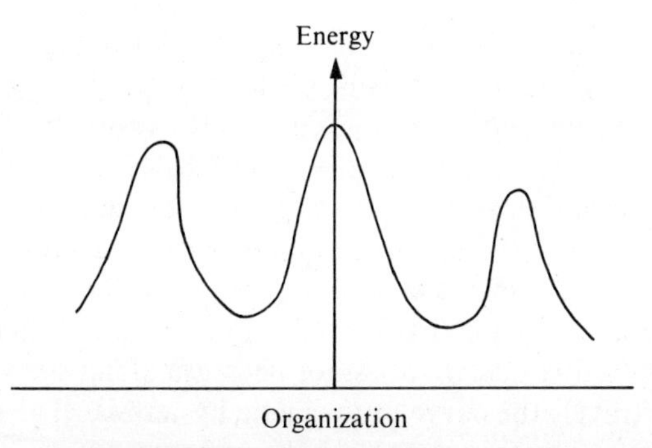

Fig. 12.2. Qualitative relationship between energy and organization

the curve shown in Fig. 12.3, which is known as a *B-H* loop, or hysteresis loop. The ordinate of this curve is labeled *B*, which is magnetic flux density. Its units are webers/meter^2 (magnetic flux per unit area) *B* is a normalized representation of the degree of alignment or organization of the magnetic domain. The abscissa is labeled in units of *H*, the magnetizing force, which has the units amperes/meter. *H* is a measure of the external force being applied to the magnetic material tending to align the magnetic dipoles.

In Fig. 12.3, the section of the curve labeled *virgin magnetization curve* describes the relationship between *B* and *H* for a material which has previously been demagnetized and is being newly exposed to a magnetizing force. Note that the stable point on this curve is the intersection of the *H* and *B* axes; in other words, when there is no *H* field applied, there is no magnetic moment from the material. Note also the arrow heads pointing in both directions along this curve,

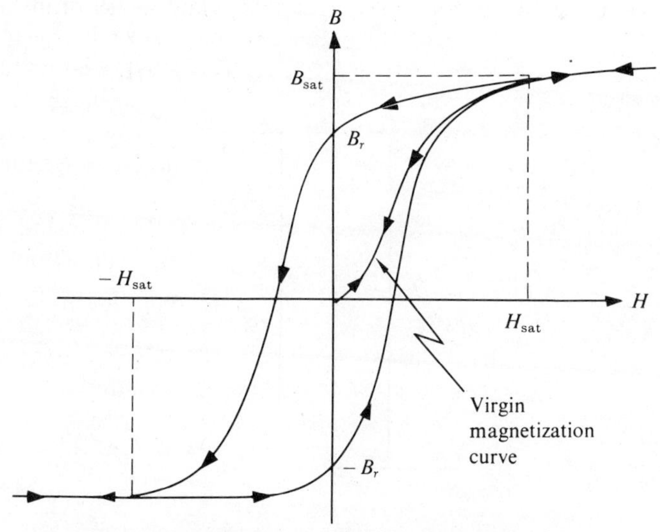

Fig. 12.3. Hysteresis loop and virgin magnetization curve

indicating that the process is reversible. If the magnetizing field is removed, then the magnetization of the material will go to zero, according to the relationship described by the virgin magnetization curve. Reversibility disappears, however, when we go over the energy hump at the point labeled H_{sat}, meaning H saturation. Saturation implies that all of the ferromagnetic domains have parallel alignment. It is possible to increase the magnetic flux density by increasing the magnetizing force above H_{sat}, but the increase becomes very much smaller.

If the magnetizing field is first brought up to H_{sat} and then reduced from this value, the magnetic flux density does not decrease along the virgin magnetization curve but rather travels the curve of magnetic hysteresis. The arrows on the curve for hysteresis show that this is not a bidirectional process. As the magnetizing force H is decreased to zero, having been brought up to H_{sat}, the magnetic flux density B does *not* go to zero but rather remains at the point we have labeled B_r, the residual flux density. This is a stable point; for no applied magnetic field ($H = 0$) we *do* obtain a magnetic flux density ($B = B_r$). If we change the direction of the applied magnetizing force through zero down to a value of $-H_{sat}$ and then return the applied magnetizing field to zero, we will achieve the state of $-B_r$, which is equal in magnitude but opposite in direction to the residual flux density we first encountered. As we continue to cycle between positive and negative H_{sat} we will likewise go through positive and negative B_r.

It is the existence of the two stable states, B_r and $-B_r$, which makes ferromagnetic material usable as computer memory elements. There is a relationship between direction of magnetization intensity and direction of magnetic flux density. For computer use, however, the hysteresis curve of Fig. 12.3 is the wrong shape. The hysteresis curve of Fig. 12.4 is much preferred for computer applications. This curve is, of course, an ideal; but with a knowledge of metallurgy and

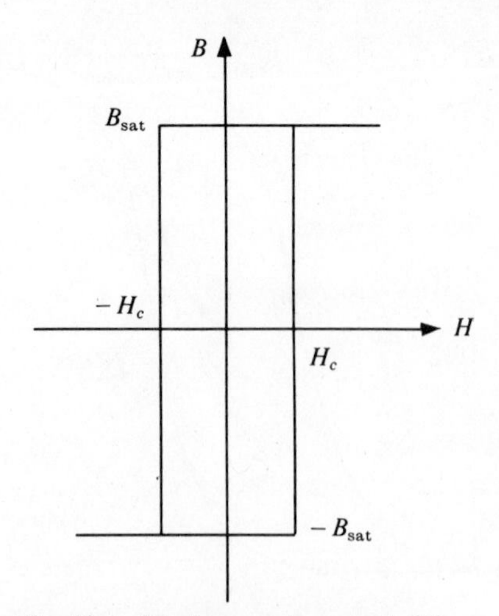

Fig. 12.4. Ideal square-loop material

material sciences, it is possible to design a ferro-magnetic substance that will approach a curve of this type, called *square-loop* material.

The curve for our ideal ferromagnetic material exhibits a large change in flux density for a small change in magnetization intensity. Furthermore, the switching from positive to negative residual magnetism occurs for an infinitesimal change in magnetization intensity (the sides of the curve rise very steeply). The magnetization intensity required for switching is labeled H_c, meaning H critical. The top and bottom of the curve are flat, so that there is negligible difference between the saturation flux density and the residual flux density. The sharp delineation of the two stable states makes it possible to determine without any ambiguity the state of the material. The determination of the state in which the material is residing is found from an application of Faraday's law, Eq. (12.1):

$$E = -\frac{d\Phi}{dt}, \tag{12.1}$$

where E is the induced voltage and $d\Phi/dt$ is the change in flux linkages per unit time. The Φ is related to the flux density B by Eq. (12.2), which is simply an integration of the flux density over the area of

$$\Phi = \int_s B \cdot da, \tag{12.2}$$

the conducting loop, which is used to sense the change in flux. A large $d\Phi$ is produced by the ideal material of Fig. 12.4.

There are two ways in which $d\Phi/dt$ can be produced. When the magnetic material is stationary, $d\Phi/dt$ is produced by changing the state of the material from positive to negative saturation or the reverse. If the magnetic material is in motion, however, it is possible to sense the $d\Phi/dt$ caused by the appearance and disappearance of a magnetic field and thus make it unnecessary to change the state of magnetization of the material. The introduction of mechanical motion, however, presents us with a myriad of problems, which include inertia and friction. All these problems produce an upper limit on the access speed available to the information. Therefore most contemporary computing equipment uses a stationary magnetic memory medium for primary storage.

12.2 MAGNETIC CORES

The magnetic material used in primary memory is formed into the doughnut-like shape called a *toroid*. Each of these toroids is referred to as a magnetic *core*, and the collection of toroids in a primary memory is referred to as a *core memory*. We may magnetize a core either clockwise or counterclockwise ($+B_r$ or $-B_r$) by causing a current to pass along a wire threaded through the hole in the core (Fig. 12.5). It is arbitrary whether clockwise or counterclockwise is defined as positive magnetization; furthermore, it is irrelevant. The definition of importance, though also in fact arbitrary, concerns whether clockwise or counterclockwise

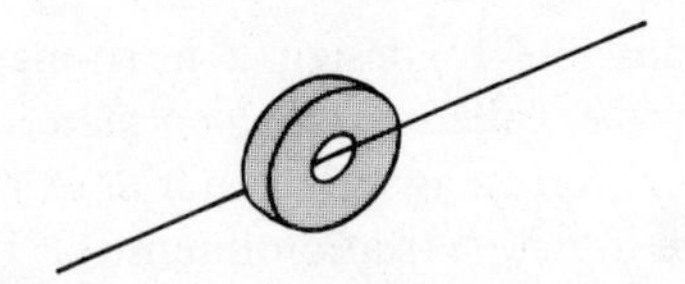

Fig. 12.5. Magnetic core with a single threaded winding

magnetization represents a binary 1. The other direction then becomes defined as binary 0. Each core can store one bit. A group of n cores will represent an n-bit word.

We may pass current through the wire shown in Fig. 12.5 and therefore write information into a magnetic core by saturating it, but with one wire there is no way to read it out (i.e., to determine the state of a core). To determine the bit stored in a core, we must provide at least one additional wire through the core. The winding along which magnetizing current is passed is the *drive winding*, and the second winding, used to determine the state of the core, is the *sense winding*. In principle, the sense winding could be connected to a voltmeter, as illustrated in Fig. 12.6. The determination of the state of a magnetic core is destructive; the state must be changed in order for us to read a voltage on the voltmeter. If we wish to retain the state of the core, we must restore it to its original state. The state of a core is read by passing a current along the drive winding. Let us assume that we pass a positive current. If the core was in the negative state, a positive current will cause it to switch to the positive state, and we will read an output voltage. If, on the other hand, the core was already in the positive state, then the presence of a positive drive current will produce no change in the flux, and therefore no voltage will be produced for our voltmeter to sense.

The sensing of a voltage means that core was in the 0 state but is no longer in that state; therefore, we must provide a negative drive current if and only if an output voltage is produced by the positive drive current. The positive drive current is the *read* half of the cycle, and the negative drive current (if present) is the *write* half of the cycle. This pair is referred to as the *read-write* cycle, or *read-restore* cycle. In the organization of a computer memory, there will always be a read-write cycle, even if the core had originally been in the positive state and it was therefore

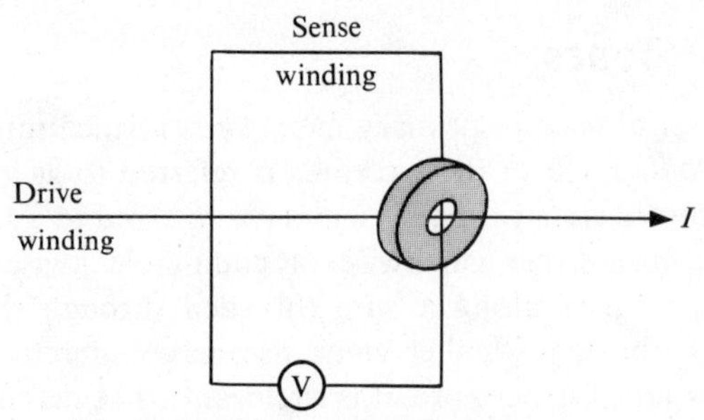

Fig. 12.6. Drive and sense windings

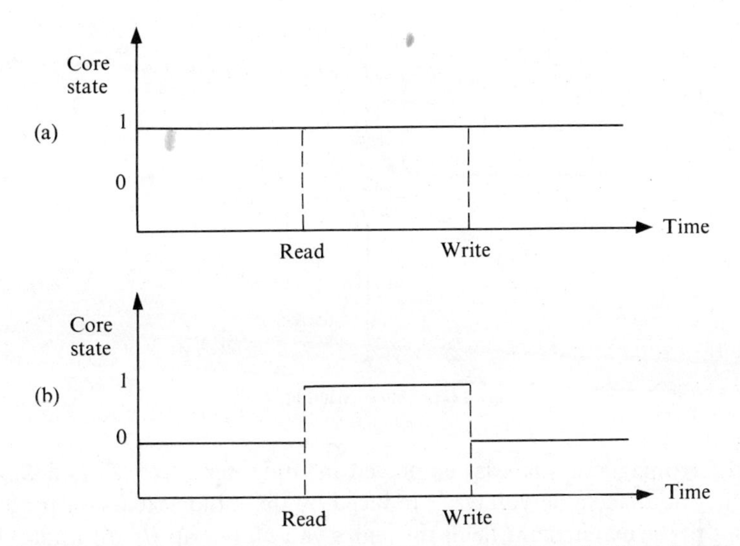

Fig. 12.7. Read-write cycle: (a) previously in 1 state; (b) previously in 0 state

not necessary to reset it to that state after its state had been read. A timing chart for the read-write cycle is presented in Fig. 12.7. Part (a) shows what would happen if the core had originally been in the 1 state; part (b) shows what the result would be if it had previously been in the 0 state.

12.3 COINCIDENT CURRENT SELECTION

If our computer memory were composed of only a few bits, there would be no problem in using a single drive winding and a single sense winding threaded through each bit. But a memory small enough to use only a single drive winding would be too small for any practical purpose. When a single drive winding is used there must be a drive wire for every single bit of every single word in the memory. Obviously, as the memory size grows, the number of drive windings becomes prohibitively large. In memories of any practical size, more than one wire is threaded through the hole in the core to carry a current that will be used to switch the state of the core from positive to negative saturation.

The hysteresis loop relationship describing the magnetic material is completely independent of how many wires are threaded through the core. It is concerned only with the net magnetizing field. If the net magnetizing field is of sufficient magnitude and direction to cause the core to switch from one steady state to another, switching will occur. If the net magnetizing field is of the correct direction but its magnitude is less than H_c, switching will not occur. Examine the hysteresis loop of Fig. 12.4, and note that if the magnetizing force for a core in negative saturation is increased from zero to some value less than H_c and then decreased back to zero, the steady state will not have changed; it will remain at $-B_{sat}$. For

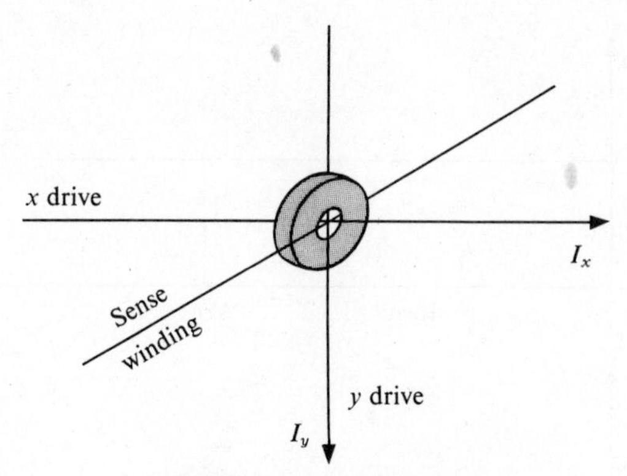

Fig. 12.8. Three-winding core

the ideal ferromagnetic material employed in computer cores, B_r and B_{sat} are so similar that no detectable voltage is induced by the minor excursion produced by the cycling of the magnetizing fields through a value less than H_c. Multiple windings threaded through the cores take advantage of this necessity for a critical magnetizing force. If current is sent along each of several windings, such that the vector sum of the magnetizing field is greater than H_c, then switching will occur. However, if the vector sum is less, switching will not occur. The importance of the vector sum is that current can be sent along wires in such a way that only in one core or group of cores does the vector sum exceed the critical magnetizing force.

In commonly used memory organizations, three wires are threaded through the core. Two of them are always drive wires, and the third can be used as either a sense or an inhibit winding. When that wire is functioning as a sense winding, it is exactly the same as the sense winding of Fig. 12.6. When current passes through only one of the two drive wires that thread the core, no switching occurs and no output is detected on the sense winding. If and only if current passes on both of the drive windings, will the core switch states. This is known as *coincident current* switching. Figure 12.8 shows the single core threaded by three windings, and Fig. 12.9 shows the output voltage that is sensed when only one and then when both drive windings are energized. The functioning of the third winding as a sense winding occurs in the read half of the cycle. In the write half of the cycle it is used for an *inhibit function*. The inhibit function operates by sending a current along the inhibit winding of opposite direction (that is, opposite polarity) to the current in the drive windings. Thus, if all three windings are energized, the current in the inhibit winding may be used to prevent the net magnetizing force from coming up to the critical H_c. Current is passed through the inhibit winding to prevent the state of a core from being switched.

Having thus selected a core, how do you cause the data in it to be read?

Let us suppose that the sense-inhibit winding is connected to a register. When we select a given word through a coincident current scheme, the contents

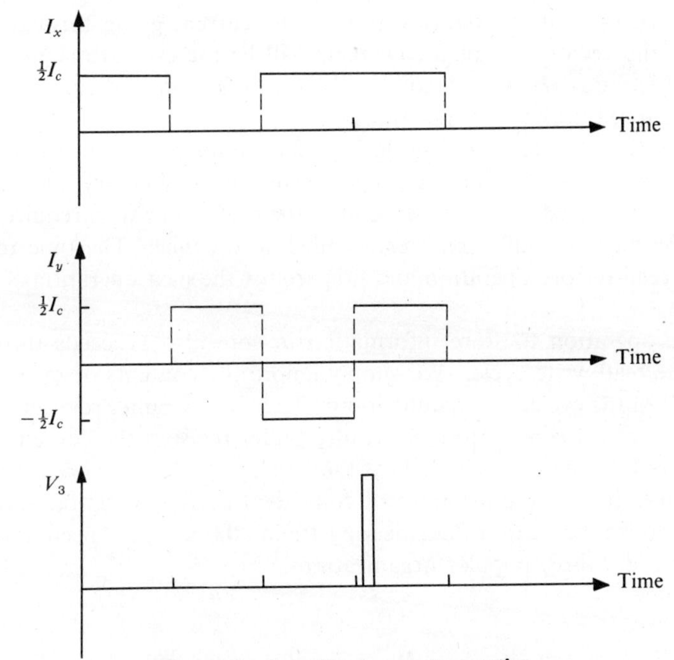

Fig. 12.9. Coincident current operation

of the word will be exhibited as the absence or the presence of a voltage on each of the sense windings. Let us say that we drove the current along the drive windings so as to force the word into the negative saturation, the 0 state. If a particular bit was already in the 0 state, no output voltage would occur; therefore the bit position in the memory buffer register to which the sense winding was connected would remain 0. On the other hand, if that bit was in the 1 state, driving it to the 0 state would produce a voltage on the sense winding which would set that particular bit in the memory buffer register to 1.

The physical geometry of such a memory has several *planes* of cores; one for each bit in the word to be stored. Drive wires are threaded similarly for each plane, and determine which core (one for each plane) is selected. One sense winding threads all cores in a plane, and reads out only the selected one. If the planes are visualized as being horizontal and stacked one on top of another, a given computer word can be found in a given vertical column of cores, one in each plane.

This readout is *destructive*. During the write half of the cycle, the contents of the word just read are restored. To accomplish this, current is passed along the drive windings, tending to switch all the cores in that word to the 1 state. If no current were passed along the inhibit winding we would obtain a word of all 1's. We wish to prevent switching of those bit positions which originally contained 0; therefore, in the write half of the cycle we will send a current along the sense-inhibit winding for all those bits which were originally 0. The direction of this current will be opposite to the direction of the current going through the drive winding, so the vector sum of the currents will be the critical magnetizing force insufficient to produce H_c. Those cores will not switch from 0 to 1, and the state of 0 will remain stored in those bit positions.

The result of the read-write cycle is to load into the memory buffer register the contents of the word whose address is given in the memory address register. The word is unchanged in core at the end of the cycle. The time required to bring data into the memory buffer register is called *access time*. The time required to complete a read-restore operation and prepare for the next operation is called the *cycle time*.

A write operation to store information in core also proceeds through both halves of the read-write cycle. We simply ignore the contents of memory during the read half of the cycle, not loading it into the memory buffer register. The word we wish to store is loaded into the memory buffer register, the contents of which are then stored during the write half of the cycle.

The application of the principle of coincident current selection and practical memory organization requires engineering trade-offs between speed and expense. We shall consider three popular organizations.

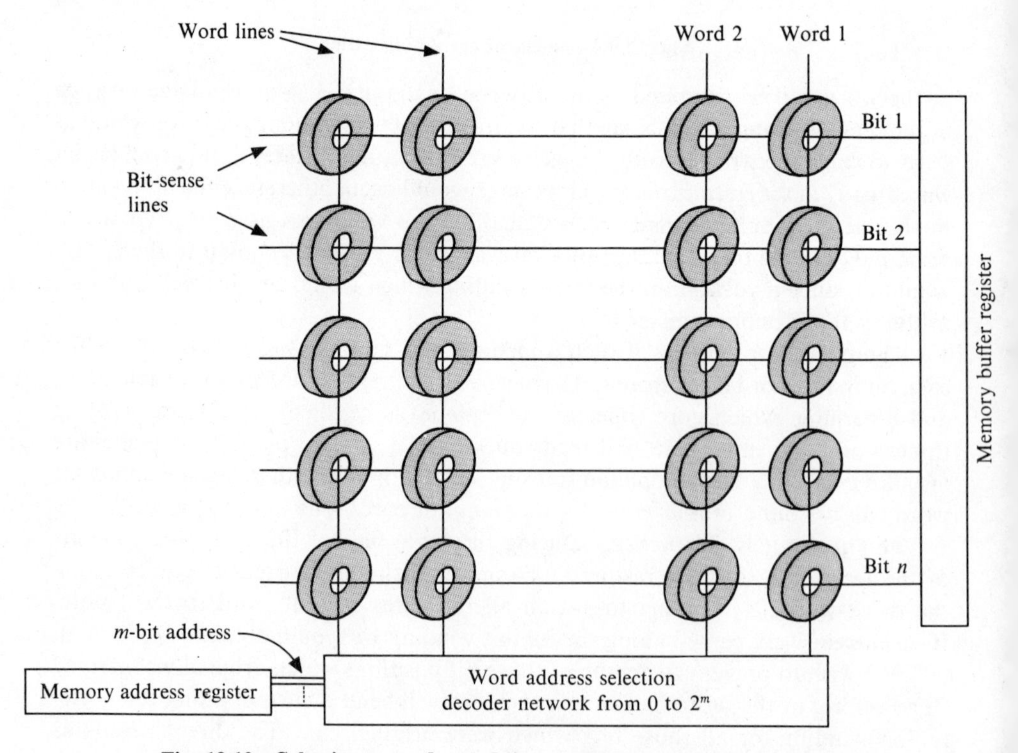

Fig. 12.10. Selection network, word lines, and bit lines in 2D organization

12.4 ADDRESSING ORGANIZATION

In current memory designs, there are three different organizational structures. For convenience, they are referred to as 3D, $2\frac{1}{2}$D, and 2D. Concise definitions of the three basic schemes follow.

3D—A memory system in which reading is accomplished by the coincidence of two current pulses on two orthogonal axes. Writing is accomplished by the coincidence of three current pulses on three orthogonal axes. This is the coincident current memory used in most conventional core memories today.

2D—A memory system in which reading is accomplished noncoincidentally by a single current pulse uniquely generated in the (word) group of cores selected to be read. Writing is accomplished by a coincidence of two current pulses, one on the unique "word" wire and the other on an orthogonal plane.

$2\frac{1}{2}$D—A memory system in which reading is accomplished by the coincidence of two current pulses on two orthogonal axes. Writing is accomplished also by the coincidence of two current pulses on the same two orthogonal axes. This system is a hybrid, having features of both the 3D system and the 2D system—thus the name $2\frac{1}{2}$D.

All these organizations produce *random-access memories* (RAM); the access time to any word is the same as to any other word.

2D memory organization

Conceptually, the simplest organization of a random-access memory consists of completely separating the address selection function from the storing function. Such memories are variously known as *word-organized*, *switch-driven*, or having *linear* or *end-on* selection. The memory elements are in a rectangular array, with each row line corresponding to a word and each column line to a bit, as illustrated in Fig. 12.10. If there are 2^m words, each composed of n bits, 2^m word lines are required, and n bit lines are required. The 2^m words are specified by an m-bit address; the 2^m word lines are driven by a decoder switch, which selects one of the lines uniquely from this m-bit coded address. To write, the bit lines are energized according to the n bits of the word to be stored. To read any word, the appropriate word line is selected, and the signals corresponding to the bits of the stored word appear on the bit lines.

Every bit in a word is pulsed during the read cycle. The contents of the cores are read from the sense lines. In writing, coincident current is employed, with partial current being applied to all cores of the selected word, and partial current being applied to the lines threading those bits which are being set to 1. The line which threads all bits of a single word is the *word line*; the one that threads all the bits in a plane is the *bit-sense line*.

3D memory organization

Great economy in decoder switching is obtained if the cores are made to participate in the address selection function. For example, consider that the core can respond

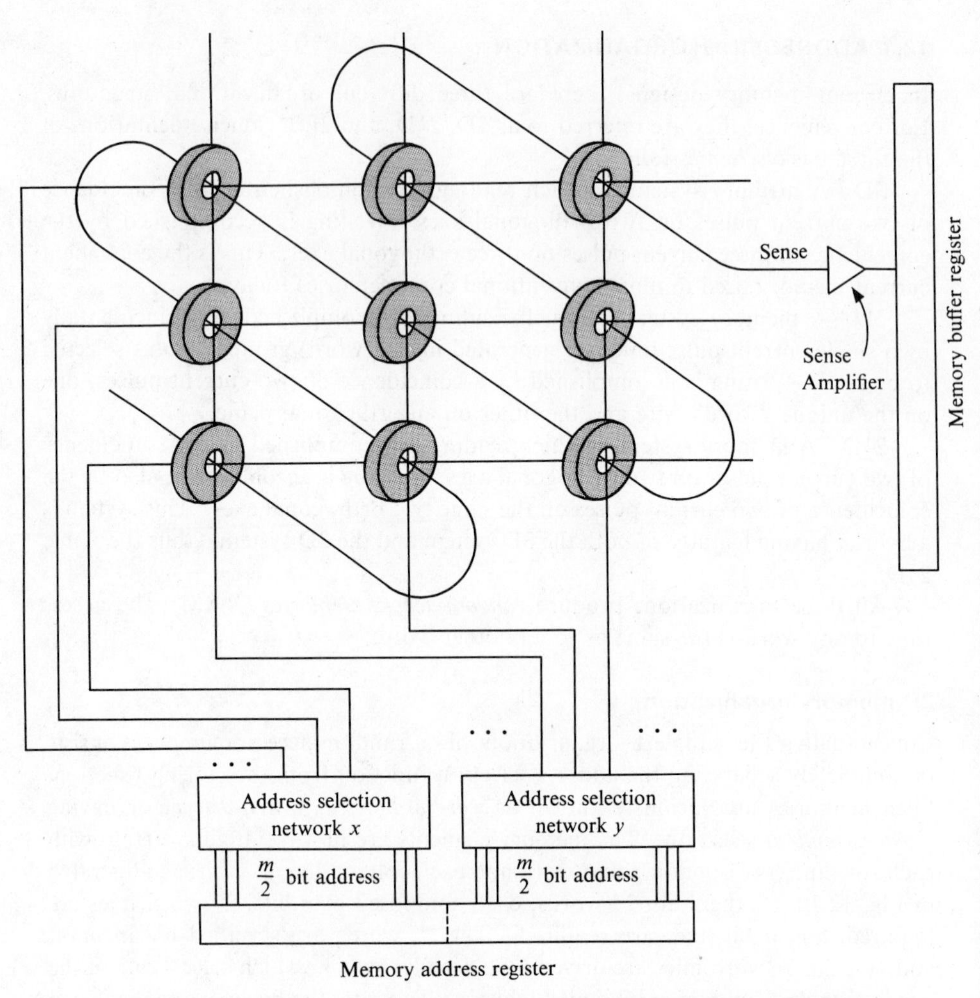

Fig. 12.11. Selection networks, drive lines, and sense lines in 3D organization

to two simultaneous address selecting signals instead of one. It is then possible to energize the 2^m words by two sets of selecting lines of size $2^{m/2}$ as shown in Fig. 12.11. The economy of the required decoder is evident, because the single m to 2^m decoder (for example, $m = 12$ to $2^m = 4096$) has now been replaced by two much smaller decoders of $m/2$ to $2^{m/2}$ (for example, $m/2 = 6$ to $2^{m/2} = 64$) capacity. Memories organized with such partial core participation in the selection are variously called *bit-organized*, *current-coincident*, or *coincident*.

Let us suppose that we have available the m-bit address of the computer memory word we wish to access. We shall first learn how to gain direct access to all bits of that word and no other word, and then we shall consider how to separate the data and switching signals on each of the individual bits.

The cores in a memory are arranged in planes, part of one of which is shown in Fig. 12.11. Through each core in a plane there are three windings threaded. One set of windings runs only horizontally and is known as the set of x-windings; another set of windings runs only vertically and is known as the set of y-windings. These x- and y-windings are the drive windings discussed above. The third wire, the sense-inhibit winding, threads all the cores in one plane. Figure 12.11 shows this selection mechanism. The m-bit address is stored in the memory address register. It is broken into two fields, each of which is connected to a drive line selecting network. This network, in turn, sends current through the drive windings that are threaded through the cores.

As a numerical example, consider a machine with an 8-bit address. There are $2^8 = 256_{10}$ words that can be addressed by the machine. If the 8-bit address is divided into two groups of four bits each, we have now the inputs to our drive line selector networks. Since 2^4 is 16, there will be 16 x-drives and 16 y-drives. One individual core will be selected by sending currents along only one of the x-drives and only one of the y-drives, which intersect at only one of the 256 locations.

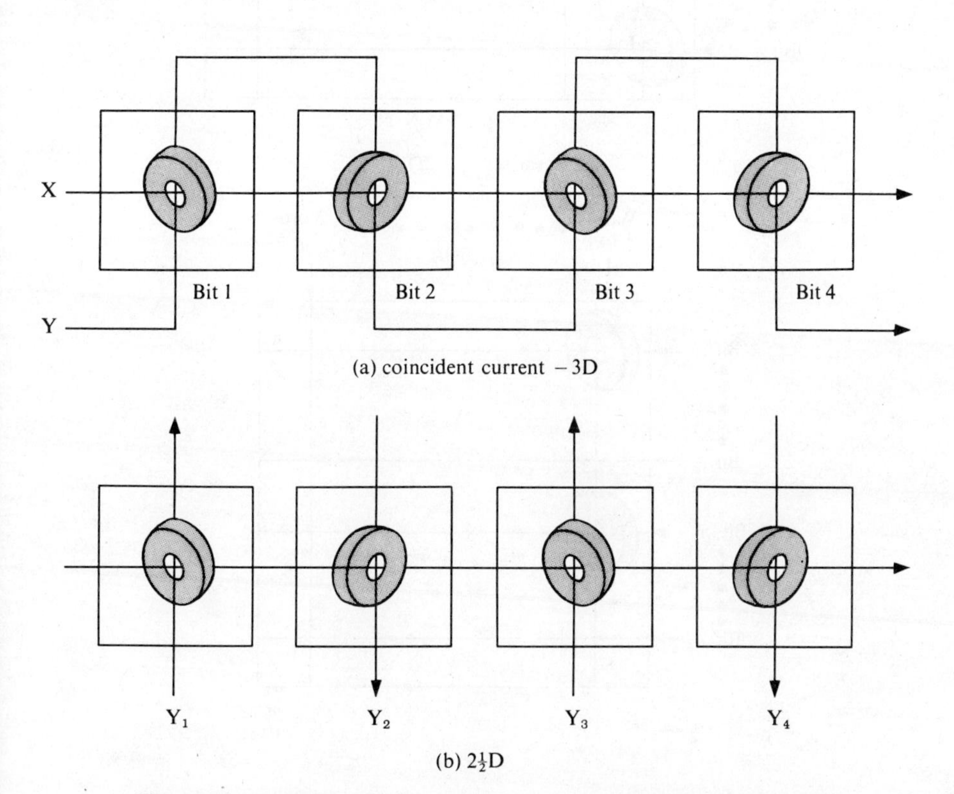

Fig. 12.12. Derivation of $2\frac{1}{2}$D system from coincident current (3D) system

$2\frac{1}{2}$D memory organization

In the 2D system, each word to be operated on is chosen individually by a word select line. The choice of word select line is external to the core stack. In 3D, the word in question is located at the intersection of x and y lines carrying partial currents. In the $2\frac{1}{2}$D scheme, a decoder creates word group select lines by decoding only, say, the x lines. Selection of a particular y line is therefore the equivalent of choosing a group of words which can be read or written into; the choice of which member of the group is active is made by the decoded x lines.

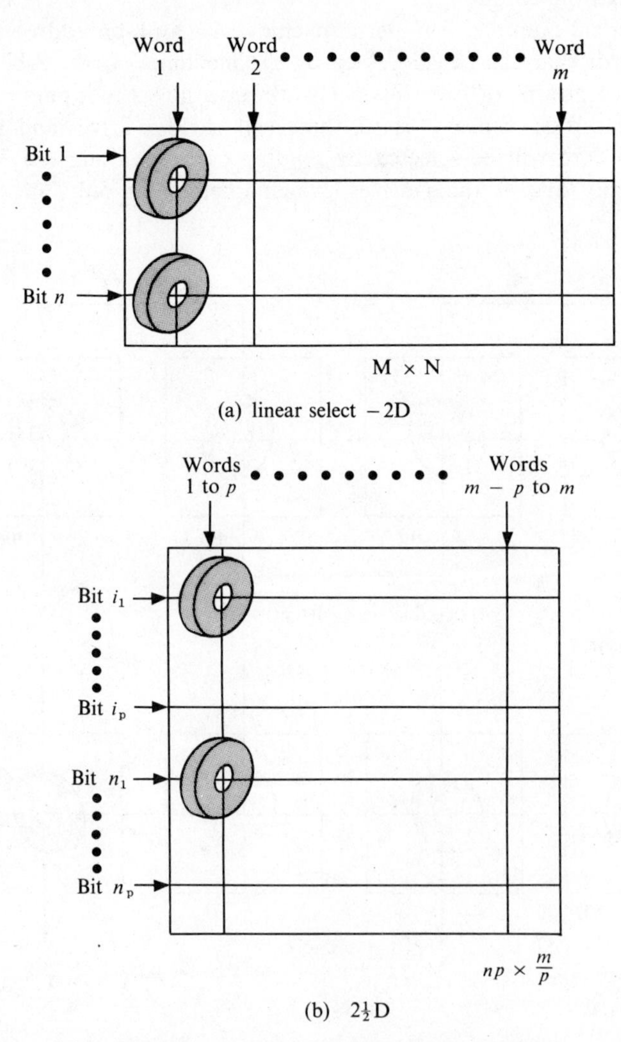

Fig. 12.13. Derivation of $2\frac{1}{2}$D system from linear select (2D) system

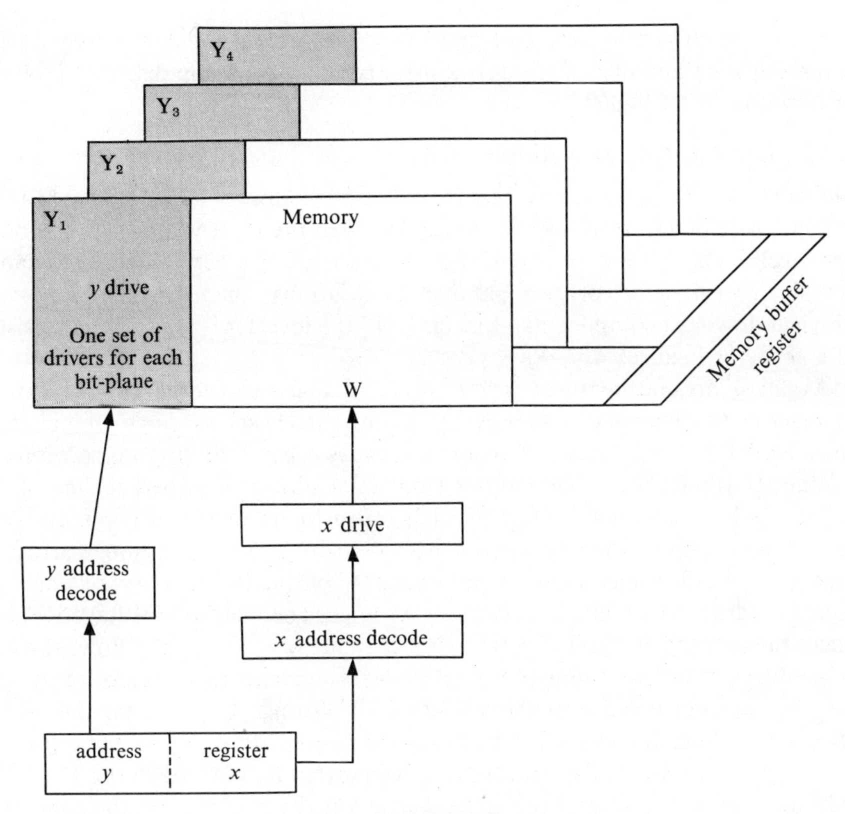

Fig. 12.14. $2\frac{1}{2}$D three-wire memory

For reading, partial current is driven down the appropriate y line, and a coincident partial current is driven down one of the decoded x lines to select a group and a word within that group. This procedure forces all 1 cores in the selected word to 0, and these changes in turn create a voltage on the sense lines in each bit plane. During writing, partial current is driven in the word select lines, and additive current is driven in the lines of the bits selected to be set to 1. Figures 12.12 and 12.13 show how the three organizations compare by deriving $2\frac{1}{2}$D from 2D and 3D systems. Figure 12.14 shows the organization of a memory using a $2\frac{1}{2}$D system.

12.5 SEMICONDUCTOR MEMORIES

Semiconductor memories have supplanted ferromagnetic core memories in some computer systems; in others there may be multiple memories manufactured with different technologies. Sometimes, in fact, the customer has a choice of which memory technology he wants (and is willing to pay for). Selection among competing semiconductor technologies allows the design of relatively expensive high-speed "scratchpad" and buffer memories or lower-cost larger memories which are competitive with ferromagnetic core memories. The basic element of semiconductor memories is the transistor. Various transistor types are used; our discussion follows that introduced in Chapter 11.

The bistable flip-flop as a storage cell

Although there are many devices that may be used as storage elements, the bistable flip-flop, made of two transistor inverters, has been the most widely adopted basic storage cell. The cell is simple to design, has inherently a very high speed, and is generally insensitive to process parameter variations; these properties provide high-yield, low-cost components. In Fig. 12.15 the inverter is made of a transistor and a collector load resistor, R_L.

A gating arrangement gets information in and out of the circuit, writing and reading the memory, respectively. Figure 12.16 shows some of the techniques used for this purpose. Part (a) illustrates a basic flip-flop made with two dual-emitter transistors. One emitter from each transistor is tied to one of the bit lines. The remaining emitter of each transistor is tied to a common word line. Many flip-flops may be interconnected to form a large memory array, in which each flip-flop has a unique address. A particular memory cell may be selected (addressed) for either writing or reading. The word-line voltage is raised to read the content of a cell. The flip-flop current, which normally flows through the word line, transfers to one of the bit lines. A current-sensing amplifier detects the signal current. A cell is similarly selected for writing. Unbalancing the voltage at the two bit lines forces the flip-flop into the desired state. When the cell is not selected, the word-line voltage is low, and cell current flows through the word line. Under this condition, there is no signal current at the bit line, and the content of the cell is not sensed. Similarly, raising or lowering the voltage on the bit line will not affect the state of the flip-flop. The dual emitters perform a local **AND** function.

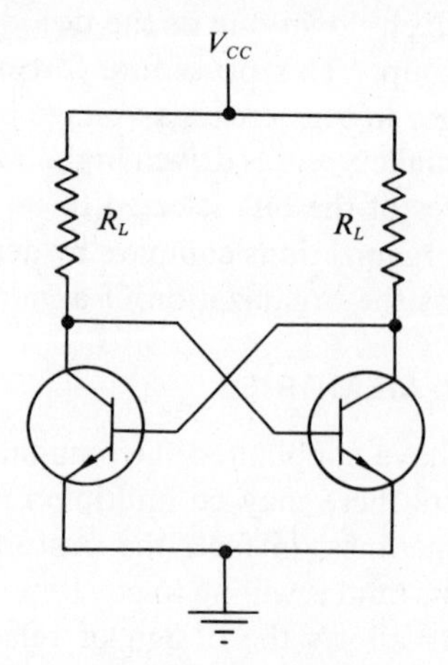

Fig. 12.15. Basic flip-flop memory cell

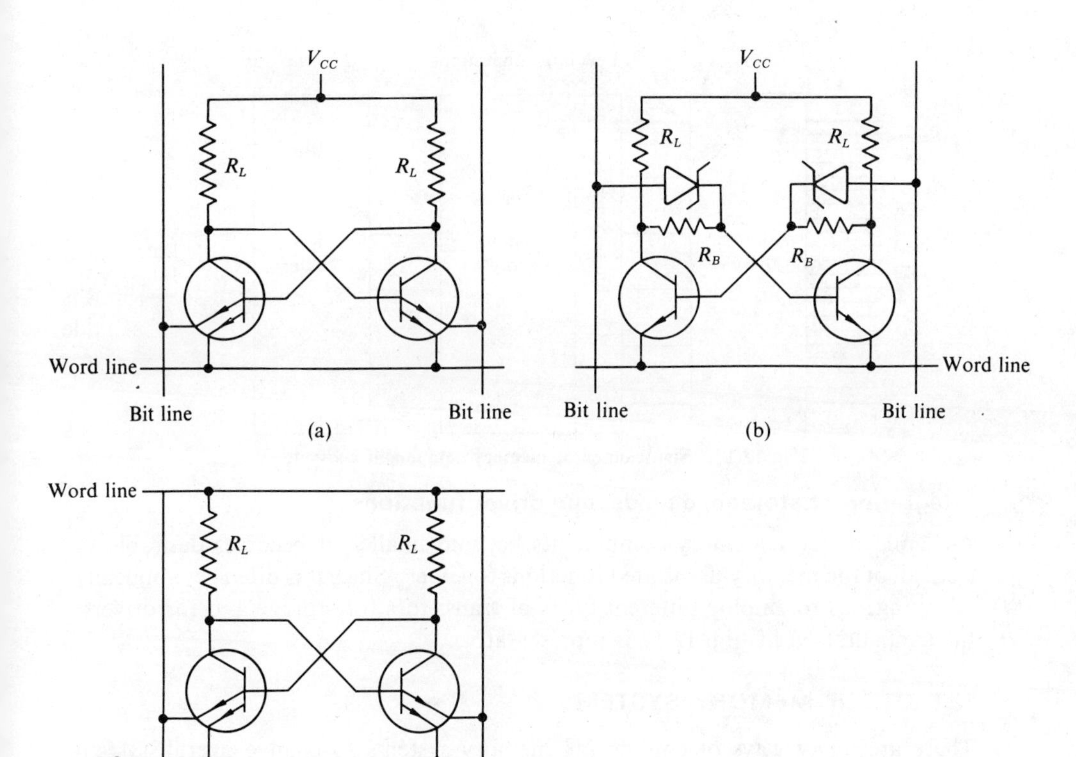

Fig. 12.16. Various flip-flop memory cells: (a) basic flip-flop with two dual-emitter transistors; (b) flip-flop with two added gating diodes; (c) a basic cell for ECL circuits

The circuit shown in Fig. 12.16 (b) operates in a different mode. A cell is selected for reading by lowering the voltage of its word line. Signals may be detected on the bit line through the diode. For writing, a cell is first selected. A large current feeds into one of the bit lines through the diode, which simultaneously turns on the OFF transistor and increases the collector load current of the previously ON transistor, forcing it to turn off quickly.

A third type of memory cell is shown in Fig. 12.16(c). This circuit is popularly used in emitter-coupled logic (ECL) circuits. The memory cell operates as follows: A particular bit is selected by raising its word-line voltage. Writing or reading of the memory cell is very similar to that of the multi-emitter cell of Fig. 12.16(a), except that the voltage across the selected cell is higher than that across the unselected cell; therefore, a large sense current is available from the flip-flop when it is selected. When the flip-flop is not selected, the voltage across its supply terminal is quite low, and low standby power dissipation is thereby maintained.

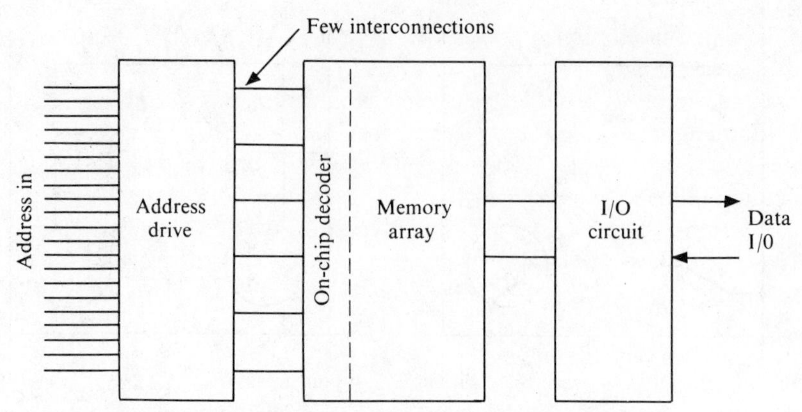

Fig. 12.17. Semiconductor memory component subsystems

Integration of storage, decode, and driver functions

As semiconductor memory components become smaller, it becomes desirable to treat all of the memory-associated functions together. Since it is often economically advantageous to employ different types of transistors for storage and for drivers, the configuration of Fig. 12.17 is representative.

12.6 OTHER MEMORY SYSTEMS

There are many ways of configuring memory systems to change overall system capabilities. Trade-offs among cost, speed, and capacity give rise to modules with different characteristics and applications. Particularly noteworthy are *scratch pad* or *cache* memories, which are small, extremely high-speed memories, used as intermediate buffers for main memory; *read only* memory, used as the control store for microprogrammed control; and *bulk* or *mass core* storage, used as a low-speed, low-cost, high-capacity extension to main memory.

12.7 MOVING MEDIUM MEMORY DEVICES

Primary memory represents a resource of very high quality and utility. In return for an enormous investment in money, the system designer receives several benefits. They are: (1) Rapid access—a microsecond is all that is needed to retrieve a whole word of, say, 32 or 36 bits. (2) Random access—the time to get any word is the same as the time to get any other word. (3) Addressibility—each word has a unique address. Since it is very important for the CPU to have such access to data and instructions, some main memory is a must.

For nearly every computer, however, there are more programs and data than can fit into main memory at one time. Main memory is the most expensive resource in a computer system and at the same time one of the most useful. The problems posed by this situation are partially resolved by the use of other storage devices, in which the three features above have been traded off to some extent in order to decrease cost.

Storage devices that use a moving magnetic medium fall into two basic types: disk or drum and tape. Magnetic surfaces may be plated onto any geometry, but disks and drums are popular because of their large surface area and simplicity of manufacture. Information is recorded on the surface and is laid down in tracks or channels. Some disks or drums have one recording *head* per track; others have fewer heads and have mechanical positioning mechanisms to place them on the desired channel. Disks and drums are characterized by their *access time*, the time required for the head to reach the desired track and for the surface to rotate until the desired information is under the head. In fixed-head systems, the average access time will be the time for one-half revolution.

For increased storage capacity, some disk drives have removable surfaces; or magnetic tape may be used. In both methods, the amount of data that may be stored *off-line* is virtually unlimited, but the access time must include the time an operator takes to find the right tape or disk and mount it on the machine. For tapes, the access time to a particular datum must also include the time spent in winding the tape to the proper position.

Computation is rarely, if ever, done directly to and from tape, disk, or drum. Instead, large blocks of information, constituting sizeable portions of a main memory load, are transferred to and from the moving medium. The CPU will *roll in* a program segment or data bank; operate on it as instructed by the user program, and *roll out* the segment, returning it to the secondary medium when finished.

12.8 FERROMAGNETICS APPLIED TO MOVING MEDIA

Magnetic tapes, disks, and drums are all surfaced with a ferromagnetic square loop material for the recording of information. The supporting substrate on tape is a plastic, such as a polyester; on disks and drums it is usually a metal, such as aluminum. Tape, disk, and drum all work on the principle that a voltage is induced in a stationary conducting coil of wire when a magnetic field is moved past. As noted, the relationship is $E = -d\Phi/dt$. Conversely, a magnetic field may be induced on a medium by moving it past a current-carrying coil. These specially manufactured coils, the *read head* and *write head*, are used respectively for detecting the field on the medium and for storing a localized field on the medium. In many applications a single head is used for both operations.

The unique relationship between the polarity of the coil voltage and the direction of magnetization on the medium makes it possible to create and detect the two stable states in a passive memory device. One of the advantages of the moving medium is that the read cycle is nondestructive. The movement of the medium past the stationary read head will induce a voltage that does not destroy the magnetization.

12.9 INFORMATION STORAGE ON MAGNETIC TAPE

Magnetic tape predates core as a memory device for digital computer systems, but the present state of technology has eliminated it as a primary memory medium. It has been relegated to a status below that of disk and drum in the hierarchy of secondary memory devices. Tape has the advantages of high storage density per unit volume, short- and medium-term archival storage ability, low cost, and presumed interchangeability among computers for information transfer. The primary disadvantage of tape is that information stored on it is only sequentially accessible. To read or write on any particular area of tape, it is necessary to physically position the tape so that the desired information area lies under the heads. Positioning is accomplished by unrolling tape from one reel and rolling it onto another, moving it past the read head, which is constantly reading the tape to determine position.

Tapes have been in use long enough for standards to have been developed. Adherence to standards facilitates transfer of information by making it possible to write a tape on one computer and read it on another. During the evolution of formal standards, there were *de facto* standards, some of which have not yet disappeared, but we shall here restrict discussion to the existing and proposed formal standards.

There are two modes or techniques for recording information on tape: binary and translated. The *binary tape* is written directly from the contents of primary memory. The bit pattern of a word in memory is copied unchanged onto the tape, from which it may be read back into core. Binary tape operations are fast in that only a copying operation is involved; binary tape also provides a very compact organization permitting high storage density.

On a *translated tape* the information is recorded as characters in an externally defined code. The standard character code is ASCII although other codes are in wide use. The internal binary representation in the main memory is translated into ASCII characters, which are written onto the tape. Translation occurs during formatted input/output under program control. Translated tapes can usually be exchanged among computers because of the standardization of the tapes and the ASCII code. However, translated tapes require more processing time than binary tapes because ASCII must be transformed into internal binary and vice versa. Translated tapes also have a lower information packing density than binary tapes. Therefore, for local nonexchange purposes binary tapes are more frequently used than translated tapes.

Magnetic tape organization

The magnetic tape standard deals with half-inch wide tape. The length of the tape is typically about 2400 feet, and it is wound on a reel for easy handling. The tape itself is about 1/500 to 1/1000 of an inch thick, and it consists of a backing of flexible material with a coating of magnetic material (an iron oxide). Information

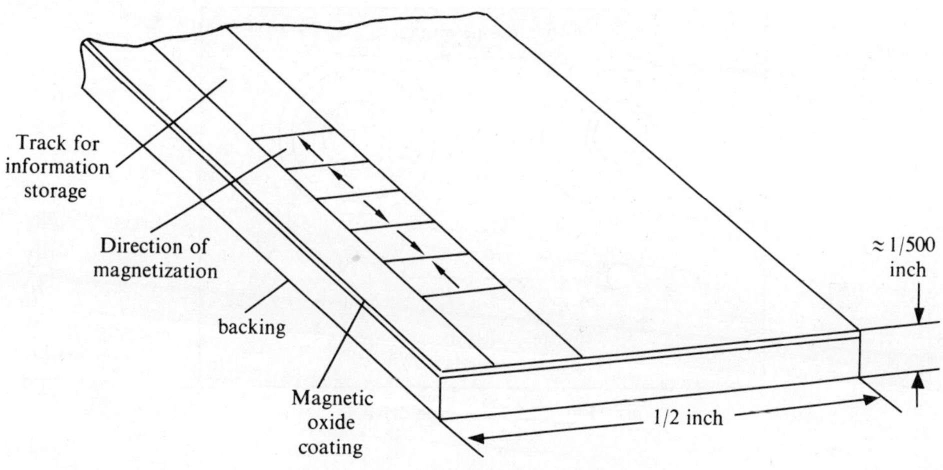

Fig. 12.18. Magnetic tape geometry

is stored on the tape by magnetizing the oxide coating in one direction or the other. A *track* on the tape is a segment of the width of the tape in which the magnetically coded information is stored (see Fig. 12.18). Since a single track is less than 1/16 of an inch wide, several can be placed across the tape.

In accordance with the standard, the surface of the magnetic tape is organized into nine parallel tracks. Each track records the value of a single bit. Thus, across the width at any one place along the tape, there may be values recorded for nine bits, which are referred to as a *frame*. This organization is illustrated in Fig. 12.19. The nine bits per frame are more than adequate for recording one ASCII character. Seven bits of the frame record the ASCII character, one bit is for parity checking, and one is a spare, which may be used as a timing track.

The bits of information are packed very close together along the tape; for example, 200, 556, 800, and 1600 bits to the inch in each track are standard densities. Since the tape moves at speeds of around 120 inches per second, starting or stopping the tape between characters is out of the question. Contiguous characters are grouped into *blocks*. Blocks are separated by a blank tape segment of specified length, known as the *interblock gap* (Fig. 12.19). The interblock gap serves both as

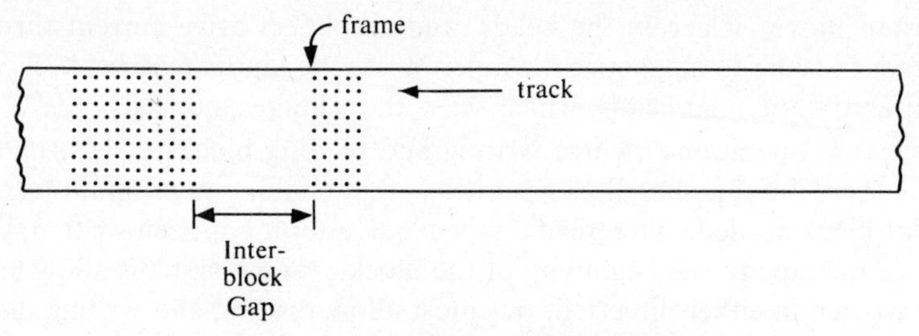

Fig. 12.19. Nine-track magnetic tape

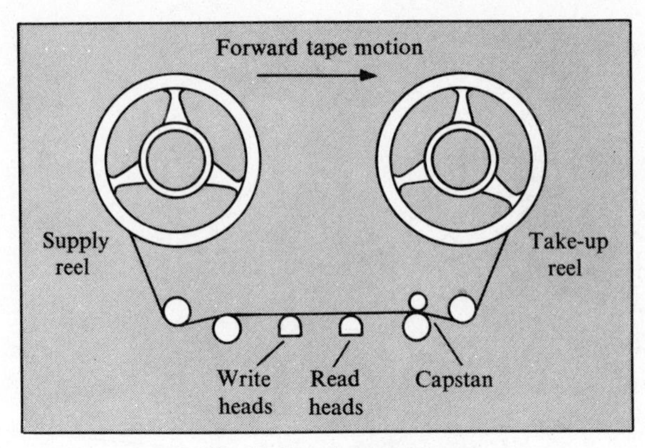

Fig. 12.20. A tape drive

a delimiter and as a coasting area for mechanical starting and stopping of tape motion.

From the commercial data processing jargon comes the term *record.* A *record* is a unit of information, such as one punched card, one line of typing from a typewriter terminal, or one line of printing on a line printer. In many applications one tape block contains one record of information. In others, a record comprises two blocks, of which the first is a header block of fixed size and format containing such information as the record identification number, the number of characters, and the format of the information block that follows.

Just as a group of characters can be organized into a block, a group of blocks can be organized into a *file*, which is a user-defined set of blocks. A special sequence of characters followed by a long gap, which the hardware or software causes to be recorded on the tape, constitutes an *end-of-file mark*. Since end-of-file marks can be sensed by the program, the tape may be positioned by means of them.

Operations with magnetic tape

For reading or writing, a reel of tape is mounted on a *tape drive*, which is a mechanism with two reel drives, *read heads*, *write heads*, and a *capstan* that moves the tape across the heads at a constant speed (see Fig. 12.20). When the tape is written, the capstan moves it across the heads, and amplifiers drive current through the write heads in order to magnetize the tape. When the tape is read back, the capstan moves it across the read heads, which sense the voltages induced.

Basic tape operations include writing and reading blocks of information. In addition, there are tape-handling operations that enable the program to reach the particular block needed. To reread a tape block after it has been written, you must backspace the tape to the beginning of the block. (Some systems allow tape to be read or written in either direction, but most allow reading and writing only in the forward direction.) Since it is sometimes necessary to space forward over one or

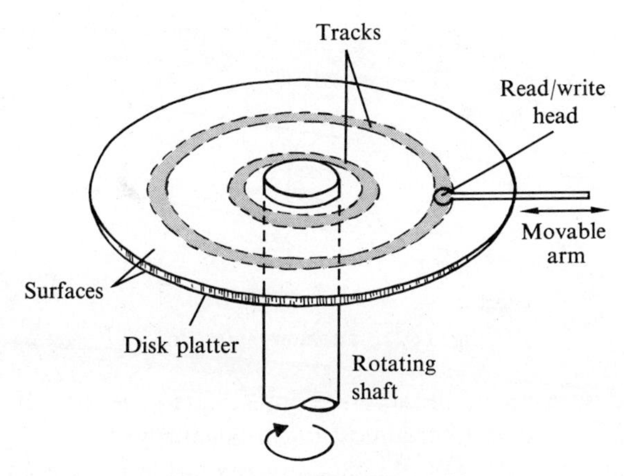

Fig. 12.21. Disk organization

more blocks in order to reach something further down the tape, a forward space operation is desirable.

Magnetic tape is only sequentially accessible; that is, it is more like a scroll than a book. You read a scroll by unwinding it from one end, and if you want to reach a paragraph in the middle, you must roll through all the preceding paragraphs. In general, there are no operations to access an arbitrary block on a tape. The only possible procedure is to skip through all the preceding blocks, one at a time.

12.10 MAGNETIC DISK AND DRUM

A disk is a device resembling a phonograph record and its playback mechanism. Physically, a single *disk platter* is a circular sheet of metal ranging from about 6 to 40 inches in diameter and from 1/16 to 1/2 inch thickness. It is coated on both surfaces with a magnetic material, and its thickness is for mechanical strength only. Information is recorded on the disk surface magnetically, just as information is recorded on magnetic tape. A disk surface contains many tracks arranged in concentric circles (see Fig. 12.21).

The drum was developed before the disk, but it is convenient to think of it as a variant of the disk. Information is stored on the surface of the drum. Drums can achieve much higher rotation speeds than can disk mechanisms. Hence the average access time (a half-revolution) can be reduced. A drum (Fig. 12.22) is like a single cylinder of a disk, but for engineering reasons, higher bit density is possible on a drum than on a disk. Also, as shown, fixed heads may be used to further decrease access time. To achieve high data rates, several tracks may be read in parallel; that is, a large number of words can be read in a single revolution. To keep the blocks a manageable size, the tracks are divided into sectors, each of which contains a block for reading and writing purposes. The computer accesses the drum by specifying a track and a sector address. After one sector has been read or written, the next-

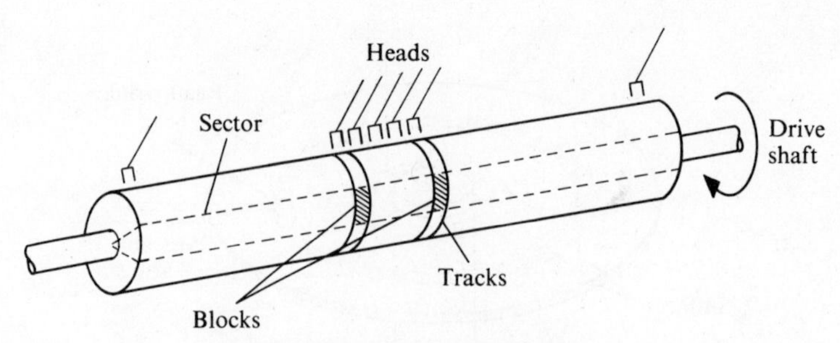

Fig. 12.22. Drum organization

higher-numbered sector can be accessed immediately because it is possible to switch from one head to another almost instantaneously.

While similar in some ways, the mode of operation of disk and drum differs strongly from that of tape. One design factor that requires consideration is the decision whether a separate head is to be provided for each track or whether the head is to be positioned mechanically from one track to another. If the head moves to position itself over more than one track, access time increases.

The most significant feature of disk and drum is near-random access. The access time to any information recorded on either device is relatively short. For fixed-head devices, average access is on the order of 10 milliseconds. For moving heads, times range up to about 200 milliseconds. The addressing of information on drums or disk is more complex than on tape. There may be many parallel tracks which must be located and referenced. Often a *timing track* and one or more *address tracks* are utilized as a reference relative to which all positioning is measured. To free the programmer from the tedious details of drum or track addressing, the system software may provide addressing by file name rather than by location. If so, the system must maintain a translation table or directory to associate file name with drum or disk address.

12.11 MEMORY DESCRIPTION PARAMETERS

One may categorize memory systems according to the following set of parameters.

Access time
Random *or* sequential access
Volatile *or* nonvolatile
Read-write *or* read only
Storage capacity

In this rapidly developing area, it is impossible to assign a cost-per-bit figure, even as a function of the other parameters.

A computer manufacturer will typically offer as part of a system various memory modules characterized by the parameters above. The customer is often left to decide for himself how many modules of each type to include in his configuration.

REFERENCES

Burroughs Corporation, *Digital Computer Principles*, McGraw-Hill, 1969.

D. Eadie, *Introduction to the Basic Computer*, Prentice-Hall, 1968.

J. A. Rajchman, "Integrated Computer Memories," *Computers and Computation*, Freeman, 1971

IEEE Transactions on Electronic Computers, Vol. EC-15, No. 4, August, 1966. Special issue on memories.

Proceedings 1968 Fall Joint Computer Conference, session on memory technology.

USA Standard, *Recorded Magnetic Tape for Information Interchange*, American National Standards Institute, X3.22—1967.

IEEE Journal of Solid-State Circuits, October 1971. Special issue on semiconductor memories and digital circuits.

L. L. Vadasz, H. T. Chua, A. S. Grove, "Semiconductor Random-Access Memories," *IEEE Spectrum*, May 1971.

C. W. Gear, *Computer Organization and Programming*, McGraw-Hill, 1969.

J. Eimbinder, ed., *Semiconductor Memories*, Wiley-Interscience, 1971.

W. B. Riley, ed., *Electronic Computer Memory Technology*, McGraw-Hill, 1972.

QUESTIONS

1. Why are there different memory technologies to choose from?
2. What is meant by primary and secondary memory?
3. What factors enter into memory selection?
4. Why is memory colloquially called "core"?
5. What is the essential property of ferromagnetic material that makes it attractive for use as a computer memory device?
6. Are registers also made from ferromagnetic material? Why?
7. What is the advantage of square-loop material?
8. What parts of the *B-H* loop describe a reversible change? What part describes an irreversible change?
9. Why are we interested in the vector sum of the current through a core?
10. Draw a timing diagram showing the flux density in a core as the magnetizing force varies sinusoidally from $+2H_c$ to $-2H_c$ for several cycles.
11. How can you obtain a $d\Phi/dt$ without changing the flux density?
12. Why is it necessary to have a restore phase in the core memory access cycle?
13. Why must there simultaneously be current in *x*- and *y*-drive lines to select a core using coincident current selection?

14. In the 3D memory organization, what voltage or current appears on the sense-inhibit line when the following conditions apply?
    a) the core plane is selected during reading, and
        1) the bit changes state;
        2) the bit does not change state.
    b) The core plane is selected during writing, and
        1) the bit is to have its state changed;
        2) the bit is not to have its state changed.
15. Compare the cost of memory, cost of associated logic, access speed, and complexity of 3D, $2\frac{1}{2}$D and 2D memory organization.
16. How is a word selected in 3D, $2\frac{1}{2}$D, and 2D memory organization for (a) reading; (b) writing?
17. Explain how the memory buffer register is filled during the read part of the cycle.
18. Explain how the contents of the MBR are employed during the write part of the cycle.
19. What are the advantages and disadvantages of moving medium memory compared with core.
20. Compare moving and stationary heads on disk and drum.
21. Explain the tape-related terms *block*, *record*, *interblock gap*, *file*, *end-of-file mark.*
22. Differentiate between binary and translated tapes.
23. Of what advantage are tape standards?
24. What happens to information stored in semiconductor memory when a power interruption occurs?
25. What criteria should be applied in the selection of memory technology for a system?
26. As noted, core memory read-out is destructive. Is semiconductor memory read-out destructive? Explain why or why not by describing the process and physical device states which occur.
27. One scheme for increasing the operating speed of a computer is to divide the memory into modules, each with its own decoders, drivers, and registers. Addressing is arranged so that consecutive memory references are made to different modules (when possible). The objective might to be decrease the instruction execution time almost to the instruction access time plus the operand access time. Discuss and compare the instruction execution time for a sequence of memory-reference single-address instructions with (a) one, (b) two, (c) three, (d) four, and (e) "sufficiently many" memory modules. How many are "sufficiently many"?

SECTION 4
SOFTWARE DESIGN

13 COMPUTER LANGUAGES

Languages are for communication. When the languages of sender and receiver are dissimilar, communication difficulties develop. The human being and the digital computer are indeed highly dissimilar, and they do experience much difficulty in communication. Many languages have been devised to facilitate communication between man and computer. They are called *artificial languages* to distinguish them from the *natural languages* of written or verbal human communication. The relative inflexibility of the computer imposes requirements on artificial languages which increase the difficulty for humans.

13.1 TRANSLATORS

One may view the field of computer programming languages as a continuum of trade-offs. Languages which can be understood by the computer with a minimum of processing require considerable mental effort for human comprehension. The burden of understanding can be shifted from the human to the computer. When the computer assumes this task, communication becomes less painful for the human computer user and requires additional programs and processing by the computer.

Machine language is very far from the natural language of human communication. Early programmers did learn to communicate with their computers in machine language, but the experience was painful. As we increase our recognition of the problem, as we continually reevaluate the relative value of human versus computer time and effort, and as our competence increases, we provide mechanisms to shift the burden of translation to the computer. Such translators consist of some combination of hardware and software designed to facilitate human communication with the computer.

An overall schematic representation of the translating process is shown in Fig. 13.1(a). The function of the translator alone is shown in Fig. 13.1(b); the input language is called the *source language*, and the output language is called the *object language*. In most of the examples to be considered here, the object language will be some form of machine language.

Figure 13.1(c) shows the translation process in more detail. First, the source language is processed by the translator to produce *relocatable machine language*,

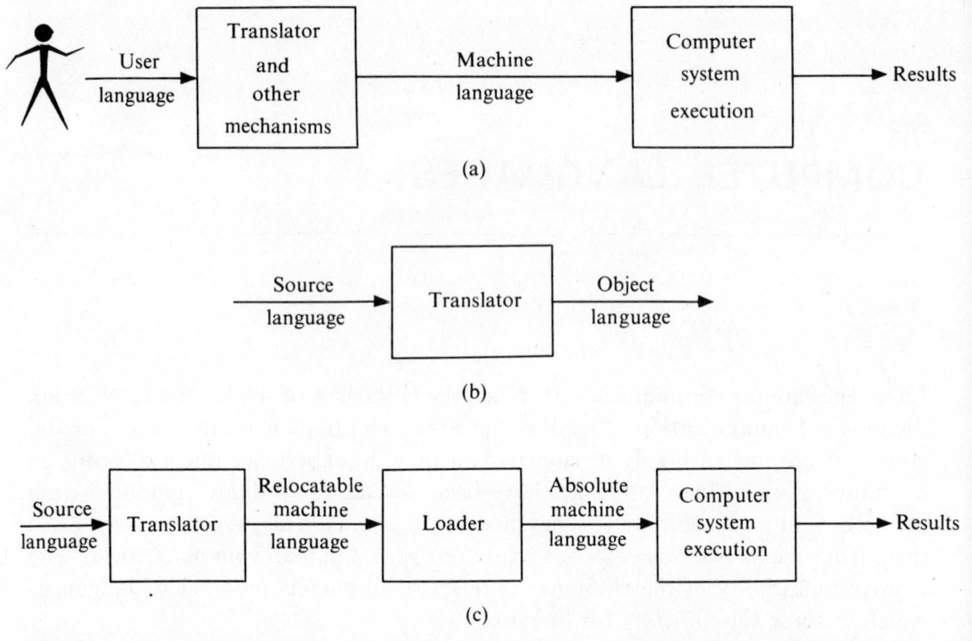

Fig. 13.1. Computer languages: (a) overall computer system function; (b) translation function; (c) more details of translation

which is in turn processed by the *loader* to produce *absolute machine language*. This may then be placed in main memory and *executed* by the computer to produce results. The function of the loader will be further discussed in Chapter 14. The important point here is that the algorithms expressed in the source language are encountered three times by the computer. The time it is processed by the translator is called *translate time*; the time it is processed by the loader is called *load time*; and the time it is executed is called *execute time*, or *run time*.

In considering the processing of an algorithm, we must distinguish which operations take place at translate time, which at load time, and which at run time. There are certain degrees of freedom available for trade-offs. Decisions made at translate time probably reduce the total amount of processing which has to be done to the program. Alternatively, postponing a decision to load or run time may increase the flexibility of the way the computer can handle the algorithm. It may also decrease the resources (in hardware or software) required for the processing.

A major function of translators is to permit the programmer to express himself in a language well suited to the class of operations he wishes to perform. General-purpose or specialized languages can be designed for each new application. A general language should permit the programmer to do whatever he wants in one language, but writing a translator for so flexible a language is very difficult to do well, and mastery of such a flexible language could be very demanding on the programmer. A specialized language may be better suited to the tasks at hand, and it is usually easy to write a translator for such a language. One difficulty of specialized languages is the occasional necessity of using more than one language for a problem, which results in interlanguage communication troubles.

An important benefit of writing in a language removed from machine language is *machine independence*. Once standards have been established (easier said than done), it should be possible to write a program, say in FORTRAN, which will be translated into machine language for any number of different computers. The only requirement is that the source program be written in the *standard language* and that the translator accept this standard language and produce correct machine code from it.

13.2 ASSEMBLY LANGUAGE AND ASSEMBLERS

Source languages have been categorized as to how much they differ from machine language. Languages closest to machine language are called *low-level* languages. The most common low-level language is *assembly language*.

The similarity of assembly language to machine language is demonstrated by the correspondence of the fields of the assembly language to the fields of the machine language, although the label field has no corresponding field in machine language. The introduction of the symbolic label in place of the numerical address is the major contribution of assembly language. In general, one statement in assembly language produces one line of machine code. This one-to-one relationship is a good criterion for defining a low-level language.

Two-pass assembler

We will now consider a simple way to design an assembler. The association of each symbolic label with a memory address occurs during a separate scan, or *pass*, through the source. This *first pass* is dedicated to building a *symbol table*. In building the symbol table, the assembler uses an artificial memory location termed the *location counter*, which associates a sequential integer with each source statement. As each successive source statement is processed, the location counter is incremented. Every time a symbolic label is encountered in the label field of a source statement, the current value of the location counter is assigned to that label, and the resulting symbol-address pair is entered in the symbol table.

Incrementation of the location counter requires that the operation code also be examined during the first pass. Some pseudo-operations, such as RESERVE or ORIGIN will cause the value of the location counter to be changed from the normal sequence.

When the first pass is complete, the statements of the source language are processed again and translated on a one-to-one basis into machine language. Every time a symbolic label is encountered as an operand, the symbol table is searched to find the address associated with the symbol. It is the *second pass* which actually does the translation; the first pass served to gather the information required for the second pass. The operation of the assembler is presented in the flowcharts of Fig. 13.2.

The symbol table is not the only table the assembler has to search. The register designations and mnemonic operation codes are also stored in (separate) tables, which must be searched to find the numeric equivalent of each symbolic representation.

Literals and immediates impose additional requirements on the assembler. Literals may be recognized during the first pass, assigned storage locations, translated from decimal or ASCII to binary, and entered into a literal table, which is searched during the second pass. Recognition of immediates may be postponed until the second pass; inspection of the operation code (or other bits by which an immediate is designated) is necessary to differentiate an immediate from a memory reference.

The presence of arithmetic, as in relative and reflexive addressing, presents additional tasks to be performed during the second pass. When the source language symbolic code is scanned during the second pass, the permitted operators (e.g., +, −, *) must be recognized and their accompanying operands used to produce the proper number in the machine language instruction.

It should be noted that two-pass assemblers do indeed read the source from beginning to end on each pass. The source must therefore be available in its entirety for each pass. Depending on the storage resources of the computer system, the source may be reread from a primary medium, such as punched cards or punched paper tape, or from a secondary storage medium on the system, such as magnetic tape, disk, or drum. The object program need not be retained; as soon as a statement is generated, it may be stored away, ready for loading and execution. The pseudo-operations would most likely be located in the same table as the operations; the assembler directives, however, would more likely be located elsewhere. A table of assembler directives would have to exist, and it would be used to guide the assembler to the proper action being requested by the directive.

One-and-a-half-pass assembler

The two-pass assembler trades nonstorage of object code for retention and multiple scan of source code. An alternate approach is possible and may be advantageous if storage of source code is expensive or unavailable. With a *one-and-half-pass assembler*, all the object code and the tables produced must remain in main memory throughout the assembly process.

The problem in the one-and-a-half-pass assembler is to provide addresses for symbolic labels which have not yet appeared in the label field of an assembly language statement. The occurrence of such a symbolic label as an operand constitutes a *forward reference*.

One scheme is to augment the symbol table with a flag indicating whether the entry is *defined* or *undefined*. A defined entry is one whose symbol has already occurred in the label field, thereby making it possible to associate an address with

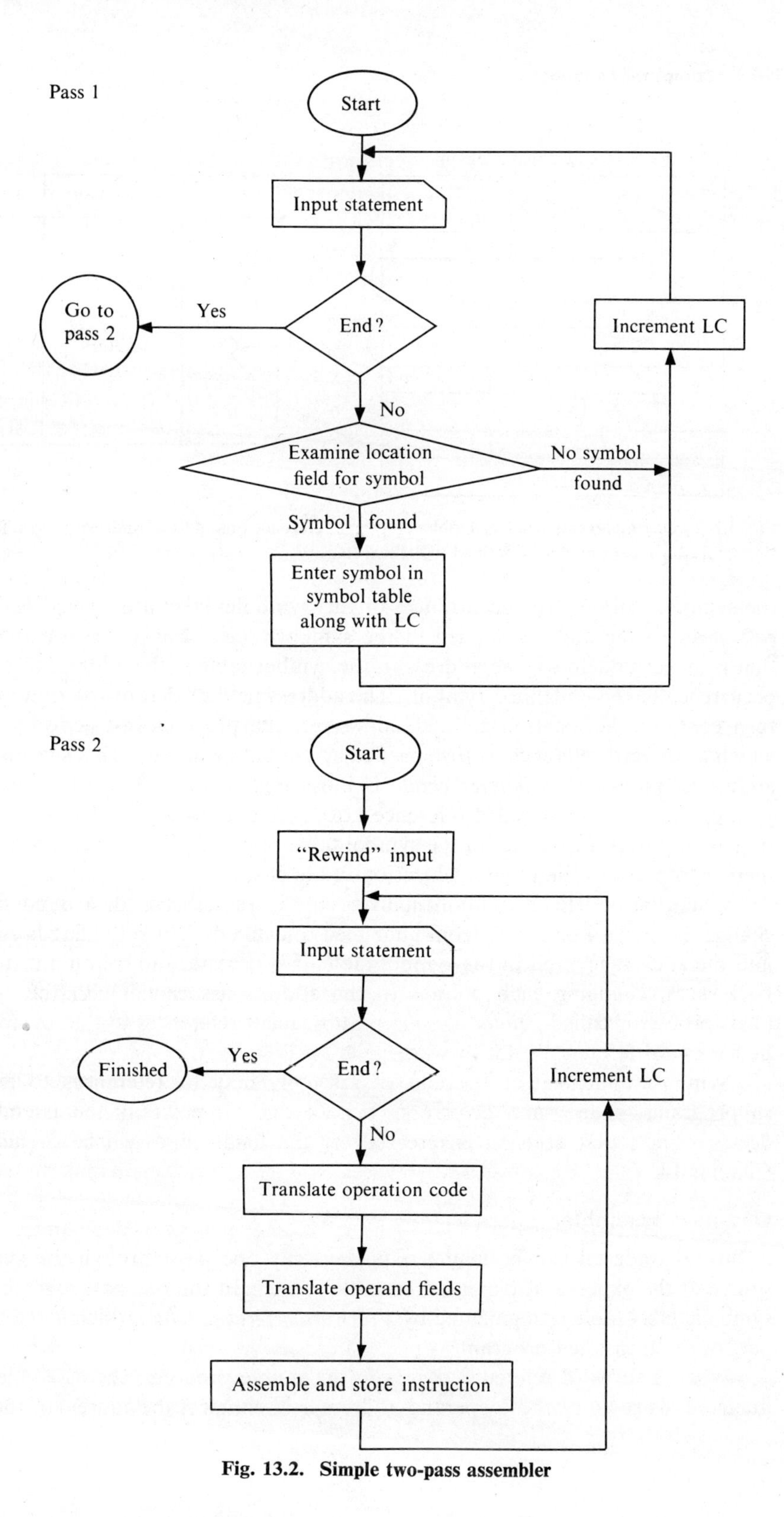

Fig. 13.2. Simple two-pass assembler

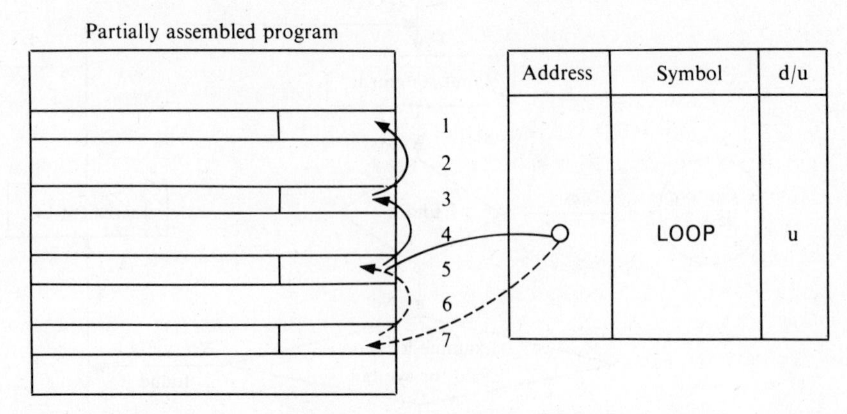

Fig. 13.3. One-and-a-half-pass assembler: forward reference linked list. Address = symbol location or fix-up location; *d/u* = defined/undefined flag.

the symbol. Subsequent occurrences of this symbolic label are termed *backward references* to the address defined in the symbol table. For a forward reference that is as yet undefined, the address in the symbol table is the address of the last occurrence of the undefined symbol. The address field of that instruction word in turn contains the location counter address of the previous instruction word in which a forward reference to that particular symbol occurred. This succession of addresses, known as a *pointer chain*, is illustrated in Fig. 13.3, which gives the pointer chain for the forward reference LOOP. A new forward reference at location 7 will cause the pointer in the symbol table to point to 7 and the address field thereof to point to the previous location in the chain, 5.

When the undefined symbol finally occurs in the label field, it is possible to change the symbol table flag from undefined to defined. When the flag is changed and the address entered in the symbol table, it is necessary to follow the pointers backwards, changing each pointer to the address associated with the symbol. (This process is called *pointer chasing*.) Subsequent references to that symbol will be backward references, not forward.

Symbols undefined at the end of assembly may be references to external subprograms, or they may be errors. It is beyond the power of the assembler to decide which; that decision is reserved for the loader and will be discussed in Chapter 14.

One-pass assembler

A further trade-off can be made, requiring only one pass through the symbolic source at the expense of increased execution time. In the one-pass assembler the symbolic label table is augmented by a *forward reference table*, which is retained as part of the translated program.

When a forward reference occurs in the source program, the assembler produces an instruction whose operand address field contains the address of the entry in the forward reference table associated with that label. The instruction is also set up with the indirect addressing bit turned on. All forward references are executed by indirect addressing *at run time* via the forward reference table. As the labels are encountered in the label field, their addresses become known and are entered into the table. Backward references can be inserted directly in the object instruction during assembly; or, for ease of implementation, all references can be indirect via the table.

When a forward reference table entry is constructed, the table is usually initialized to some value beyond the limit of the address field which serves as a flag. All locations for which this flag remains unchanged at the end of assembly time are external references or errors.

To reduce table size, the assembler may require that the association of symbolic labels with memory location precede the use of the labels as operands except when the operation is a JUMP. In this case, the forward reference table becomes a *jump table.* The jump table technique may be used even if there is no indirect addressing. The operation code field of the jump table is filled with the JUMP op code. Forward-referenced jumps are implemented as two-stage jumps via the jump table.

Macro assemblers

An additional set of features, available in some assemblers but not introduced in Chapter 7, are an important convenience to assembly language programmers. (The reader is referred to manufacturers' manuals for a detailed discussion of these special features, which are quite different for different computers.)

No matter how flexible and powerful the command vocabulary of a programming language is, there will always be situations in which a few additional commands would be very useful. A language translator that incorporates a facility for translating user-defined operations is more complex than one that does not; but the increased flexibility is usually well worth the price. Another justification for these user-defined operations is akin to the rationale used for subroutines, namely that frequently performed operations should be available by simply invoking the identifying name.

For a subroutine, the name *explicitly* refers to a block of code written by the programmer and placed in memory once. By jumping to the subroutine and saving a return path, you can use the same instructions from many places in the program. A user-defined instruction, however, is defined in the assembly language source text, and the assembler creates a new copy of this procedure every time it encounters its name. Such a user-defined op code is called a *macro-instruction*, or just a *macro.* During assembly, each time a user-defined op code appears, it is replaced by one or more lines of machine code. This process is called *expanding the macro.*

Macros may have parameter substitution by association of actual arguments with some formal parameters, known as dummy arguments. As in a subroutine, the formal parameters are part of the definition of the macro, but when the macro is expanded, the actual arguments are substituted for the formal parameters. For example, consider the macro

CALL SUB, ARG1, ARG2

which would be expanded (in our assembly language notation) to

Label field	Operation field	Register designation field	Operand address field	Index register designation	Comments
	STORE PC AND JUMP	XR7	SUB		The CALL becomes this STORE PC AND JUMP on XR7
	NO OPERATION		2		Counted two arguments
	NO OPERATION		ARG1		
	NO OPERATION		ARG2		

A feature called *conditional assembly* makes it possible to change the way a macro is expanded. Conditional assembly allows the programmer to dictate the inclusion or exclusion of designated parts of the macro in each expansion. The parameters which control the process are defined and calculated at *assembly time* and are not the same as those variables or expressions in the program which are translated into machine language. Depending on the results of some conditional test, some instructions from the macro definition may or may not be included in any particular expansion although they are included in others. Conditional assembly allows for selection and modification of the *macro skeleton.* This possibility of a macro definition producing variable assembly language instructions is very much like the programming of tests and branches, but *the testing takes place at assembly time when the macro expansion is substituted for the macro instruction.* The conditions which determine the form of the expanded macro are generally passed along with the actual parameters associated with each macro instruction. The conditional assembly may be dependent on arithmetic operations involving these actual parameters; or the actual parameters may be specification keys, or switches that served to determine the form of the expanded macro.

Macros make an assembly language extensible by allowing the programmer to define new mnemonic operations. These new operations cannot perform operations that are otherwise impossible; they only give a single name for a sequence of assembly language operations. For all intents and purposes, however, the programmer may think of the macro instruction as an extension of the instruction set provided on his computer. It is not uncommon for predefined sets of macros to be available for a given machine. Predefined macros serve to enrich the basic assembly language. For certain applications, a programmer may work almost exclusively with a set of macros; in such a situation the macros in fact define a new language.

From the point of view of the assembler, a macro instruction may correspond to more than one machine instruction. The task of associating addresses with symbolic labels is made more complex by the inclusion of macros. Relative and reflexive addressing spanning a macro instruction are potentially dangerous, especially if there is conditional assembly, since the macro may be expanded into a variable number of machine language instructions. The two-pass assembler must expand each macro it encounters during the first pass, at least to the extent of determining how many storage locations will have to be assigned to it. Various trade-offs are possible to avoid duplication of effort in macro expansion between the two passes.

13.3 COMPILERS

The next higher level of source language complexity exhibits increased structure and produces a greater number of machine language instructions for each source language statement. The translators for such languages are called *compilers.* A compiler language is designed to ease the programmer's task by providing him with a language especially suited to the task he wishes to perform. Just because a compiler language is closer to the application, it is usually further from machine language; consequently, the task of the compiler is much more complex than that of the assembler. Some typical compiler languages that have received wide acceptance are FORTRAN, ALGOL, COBOL, MAD, and PL/1.

Since compilers have a difficult and important job to perform, the art and science of compiler design is an active working area in the computer sciences. The many approaches, techniques, and objectives are quite properly studied as a separate entity. Nevertheless, we shall briefly enumerate a compiler's tasks.

A fundamental task of the compiler is statement classification. The compiler needs to know what the programmer is trying to do before it can even begin translation. Statement recognition and classification are in part a *recursive* process since source language statements may be complex combinations of simple components. Classification may be made easy if the source language was designed with forethought to the accomplishment of unique identification with minimal effort and without ambiguity.

The compiler must recognize the attributes and storage requirements of simple variables and arrays. It must make the necessary association of symbolic name with memory locations. Many compiler languages provide for internal representation of numbers in integer, floating-point, and double-precision forms. The compiler language must provide the ability to specify the internal representation to be employed. These languages usually provide for *compiler directives*, which, like assembler directives, tell the compiler how to do its job. These directives resolve ambiguities or override default operation of the compiler. In many compiler languages, symbolic names serve not only to identify memory locations, but also to declare the internal representation to be used in that location. If a name is declared to identify an integer variable, then the compiler must treat any number stored in that location as an integer.

In general, the compiler must make sure that when a value is stored in the location associated with the identifier, it is of the proper internal representation. If the language permits expressions involving mixed representations, the compiler must convert all the numbers to one representation before arithmetic operations can take place.

One measure of success of a compiler language lies in how much easier it makes the user's programming task. A partial explanation for the proliferation and popularity of various languages is the varying degrees of convenience they offer for specific classes of problems. The task of the compiler is to translate user programs into effective and efficient object code. Two schools of thought represent extremes in the trade-offs between compiler design and language specification. The compiler designer would like to have a language that is easy to translate; the language designer would like to have a language that is flexible and easy to use. At one extreme you have a language so "rich" that it is difficult to compile efficiently. The proliferation of features makes it unlikely that any one program will use most of them, yet the compiler must provide for full use. At the other extreme would be a language so "lean" that it could be used for only one application. Languages exist everywhere along this spectrum, each with its own community of users.

One important function of the compiler is not at all concerned with translating, but rather with the inability to do so. This is the area of diagnostic messages. When the programmer violates the rules of the language, it is essential to give him sufficient information to enable him to recognize and correct his errors. The rules of the source language have been formulated to provide a framework within which the programmer can implement his algorithms. Familiarity with the rules does not always imply compliance; and the beginning programmer may not really know or understand the rules. The challenge to the compiler is to identify the source language error and to print an appropriate, clear, and helpful message that will enable the programmer to find and correct his transgressions. When multiple errors occur in one statement or in closely related statements, it is exceedingly difficult to identify all the errors correctly. Another problem is that errors in key statements, especially declarations, can cause the compiler to produce extraneous error messages, which confuse rather than help the programmer.

Compilers were first implemented as simple translators from compiler language to assembly language, as indicated in Fig. 13.4(a). The compiler output was then assembled in a separate pass. Designers later recognized, however, that most of the work of the assembler is duplicated in the compiler. Compilers are now usually designed to produce machine language directly, as shown in Fig. 13.4(b).

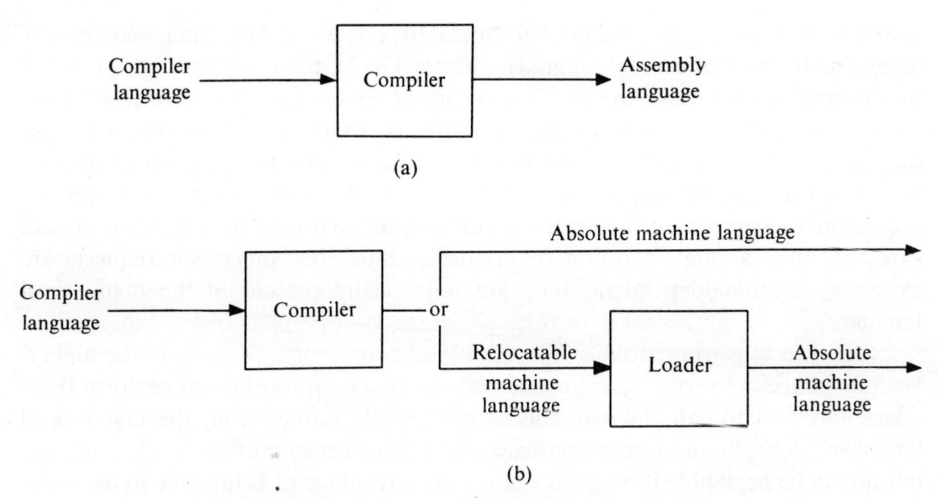

Fig. 13.4. Compiler language construction: (a) early; (b) contemporary

13.4 HIGHER-LEVEL LANGUAGE VERSUS ASSEMBLY LANGUAGE

Now the question to be asked is "When should one use compiler-level language and when should one use assembly language?" The answer has changed over the brief life-span of the computer business. Originally one would have worked exclusively in assembly language or machine language. The compilers that existed were very inefficient and very slow, and they produced very slow running code. However, compiler design has improved, and the claim today is that a good optimizing compiler will produce machine language code as good, or almost as good, as that which could be constructed by the most experienced programmer. If this claim is true, there appears to be no reason for ever writing in assembly language. Writing in a higher-level language usually involves much less work than does writing in assembly language.

Part of the decision depends on the exact kind of operation you are attempting to program the computer to perform. If it is the type of operation that the higher-level language was designed to facilitate, then it certainly should be done in the higher-level language. Languages like FORTRAN exist to make algebraic scientific operations easy. A language like COBOL is designed for business applications, and special-purpose languages exist for many applications. If you are trying to do systems programming, however, you may find your situation a bit more difficult. There is currently no widely accepted higher-level language that can be used for writing a systems program. Efforts have been made to use PL/1 as a systems programming language, but not enough time has elapsed to make an objective evaluation of the success of such efforts.

The very distance of the compiler-level language from the machine language is the root of its disadvantages for use in specific applications. Specific machine instructions that are extremely well suited to a type of operation may not be accessible in the higher-level language. For example, there are routines provided by the supervisor to give the user some sort of service (such as I/O) which he is normally prohibited from initiating. To safeguard the system from the user, and the users from one another, there is a protected set of operations which can be performed only by the supervisor. The user obtains an operation of this type by requesting the supervisor to perform it for him. In so doing, the supervisor makes sure that it is a safe operation to perform. Since the supervisor requests are extremely machine-dependent, they are not usually present in the higher-level language.

One way of getting around the unavailability of certain features in the higher-level language is to write specific assembly language subroutines to perform these operations and to call the assembly language subroutines from the higher-level language. A slight additional overhead is produced because of subroutine linkage, but it may be negligible if weighed against the advantage of being able to use some of the hardware features of the machine. Another approach is to build features into the higher-level language which allow the programmer to get down to assembly language level if he so desires. An *escape convention* will allow the user to insert assembly language code directly into the middle of his higher-level language program; or some kind of interface the compiler provides may enable him to get at these special functions. These are the tools that are necessary in a higher-level language when a programmer is doing particular types of operation, especially those involved in system programming.

Still another way of combating the double problem of possible inefficiency and unavailability of certain features in compiler-generated code is *hand optimization*. After the machine language code has been produced, the human programmer goes over it and makes whatever modifications are necessary in order to improve its performance. If the compiler generates assembly language code in addition to the machine language code, then the job of the programmer is much easier. He can modify the assembly language code and then cause it to be assembled. This certainly constitutes a more efficient use of his time and requires less effort than does the modification of machine language code.

13.5 PROBLEM-ORIENTED LANGUAGES

On a level still further removed from the idiosyncrasies of the computer hardware are the languages designed for use in the solution of specific problems. A *problem oriented language* is designed around an analytic model representing some class of physical system (e.g., electric circuits, trusses). The user has only to specify the topology and parametric values describing the instance he wishes to analyze. The numerical analysis of the physical laws governing the modeled system are built into the language.

13.6 INTERPRETATION

The purpose of using a translator is to let the programmer express what he wants the computer to do in a language that is convenient for his purposes. The translator then interprets his instructions according to pre-defined rules and takes appropriate action to cause the computer to do what has been requested. The technique so far discussed has been to translate the source language into machine language and then to have the machine language instructions executed by the computer. The point of emphasis here is that translation takes place at a time completely unrelated to execution time and that the machine language produced may be saved and executed an unlimited number of times.

If the translation process is implemented so that the translation takes place at execution time, the translator is then known as an *interpreter*. BASIC is a language usually implemented by an interpreter. The process of interpretive execution is illustrated in Fig. 13.5. The characteristic feature of an interpreter is that the source language program is retained and examined every time it is executed. In fact, every statement is examined as it is to be executed; for some loops, therefore, a set of statements may be reexamined quite a few times.

The exact mechanism of interpretation varies among implementations; part of the translation may be into some intermediate language prior to execution, or it may all be accomplished at the time of execution. Some interpreters, known as *direct interpreters*, produce machine language instructions. Other interpreters perform the operations requested by the source language statements; such an interpreter is a program, the execution of which is modified by the source language it reads. This second type is known as an *indirect interpreter*.

Interpreters are smaller and easier to write than conventional translators, but they are much less efficient to run since a large amount of execution time is consumed in the process of interpretation. Interpreters have been used for applications that are experimental, infrequently used, extremely complex, or essentially non-sequential so that direct translation would be quite difficult. Interpreters are especially useful for debugging the source program. Execution can be halted, and *source language changes can be made without destroying the current values of the program variables.* Execution can then be resumed at the point at which it was terminated.

Interpretation and compilation are not mutually exclusive. There is a trade-off between the speed of the compiled code and the relative compactness of the inter-

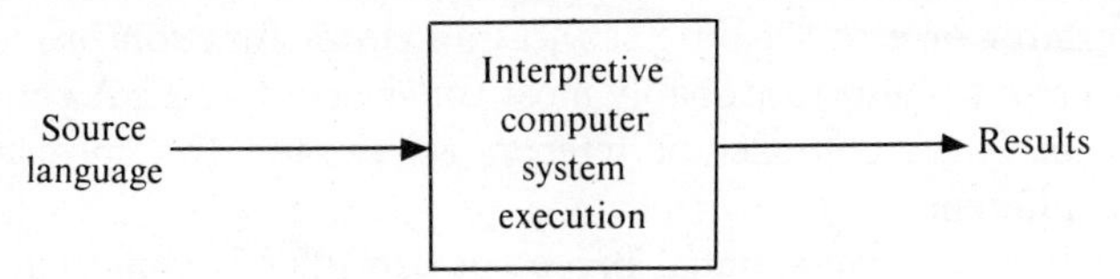

Fig. 13.5. Interpretive execution

preter. Compilers can use interpretation by generating calls to *run-time subroutines* which execute interpretively. (The run-time subroutine, provided by the supervisor and employed during the running of the program, is so named to distinguish it from the *compile-time subroutine* used by the compiler but not retained during execution.) Interpretive subroutines exist either for the convenience of the programmer or because the complexity of the task prohibits compilation. For example, in many languages formatted input/output is accomplished interpretively. Every time I/O occurs, the run-time subroutines scan the editing codes to determine how to format and convert the operands. The programmer can create and modify these editing codes at run time.

Another way in which compilers employ interpretation is to permit the use of symbolic names at run time. If a symbol is retained at run time, it can be employed in I/O to permit transfer merely by stating the variable name. The I/O routines can examine the symbol table to determine not only the storage location but also the internal representation of the numeric value.

13.7 CONVERSATIONAL TRANSLATION

In the traditional *batch mode* of communication with the computer, the user prepares his deck, complete with programs, control cards, and data, and submits it to the computer. The computer reads the cards, acts on the instructions they contained, and produces printed and/or punched output for the user. In the *conversational mode*, by way of contrast, the user types on a communications terminal connected by local wires or telephone cable to the digital computer. The distinguishing feature of conversational mode is the degree of control the user exerts on the services performed by the computer. At almost any point the user can terminate one service and institute another in accordance with his perception of his needs.

Translators designed for conversational mode can and should be more helpful to the user than the batch translator. One way they can be more helpful is to check each statement as it is typed to verify that it obeys the grammatical rules of the language. If it does not, an error message is printed, and the user can retype the offending statement. Some errors cannot be determined on a line-by-line basis. They will be caught only when the user types the translator directive signifying the end of his program unit. Correction of such errors will probably require editing of the typed-in program and retranslation. Editing may be available within the translator or as a separate service; an editor function provided within the conversational translator is probably most convenient to use. A conversational translator may be either compiler or interpreter; the user will probably not be able to tell the difference.

Another feature a conversational processor can offer is *immediate execution*. When a statement is entered for immediate execution, it is immediately interpreted or compiled and executed. Whether interpreted or compiled and executed, the instructions in the statement are carried out then and there, not at some later time. Immediate execution is very useful in program debugging. The user can enter print statements for immediate execution to determine the current values of variables. Execution can be resumed at the next statement, or elsewhere by means of a jump command. Immediate execution requires a run-time symbol table.

13.8 MICROPROGRAMMING

Heretofore we have assumed in the interpretation and execution of the machine language instruction by the operation decoder that a specific set of hardware components was activated to carry out the particular function specified. For instance, if we were doing an ADD operation, signals would be sent to transfer the information from the appropriate memory location to the registers which are part of the arithmetic unit, to perform the arithmetic, and then to transfer the information from these registers back to where it is to be stored. All the operations the machine performs are defined at the time of construction and are implemented as the connections among the various registers and gates that constitute the hardware components of the machine. If at some future time there is a desire to augment or modify the instruction set of the machine, it is necessary to cut into the hardware and to perform physical modifications by the addition or substitution of components. Once the machine has been mass-produced and is out in the field, it is extremely difficult to effect field modification. Doing so would mean sending out a group of highly trained engineers with transistors, resistors, other components, and soldering irons with the intention to make identical physical modifications on every one of the machines. The difficulty is quite evident.

As noted in Chapter 5, an alternative scheme, increasing in popularity, is to provide a special memory unit for this operation decoding. It is very, very fast to read; its generic name is *control store*, and it is sometimes known as a *read-only memory*. Every time the machine language program specifies an operation to be performed, the control store is searched. In the control store under that operation code will be a list of microoperations which are to be performed to implement that machine language instruction. The *microinstruction*, then, is the lowest level of instruction on which the computer can operate; a machine language instruction is implemented by a group of microinstructions.

What, then, are these microinstructions? They specify the register-to-register transfers, the register-to-memory and memory-to-register transfers, and some of the operations of the subsystems of the computer, such as the ALU. The microinstructions that implement a machine language command are collectively described as a *microprogram*. A presently implemented machine instruction can be changed, or a new one can be implemented, by creation or modification of the microprogram which implements that op code.

Microprogramming has introduced a new level of closeness between the programmer and the hardware components of the computer. Before microprogramming, the lowest-level control possible was that afforded by machine language. Microprogramming makes it possible to get to the working registers themselves—certainly a very intimate contact with hardware. Not very many people will want or be able to do microprogramming. The level of expertise of an advanced systems programmer, coupled with hardware design experience, is a reasonable prerequisite for access to microprogramming, which is a very powerful tool when needed.

The microprogram can be changed either by writing into or by physically changing the control store. The control store is often implemented on a set of logic cards; if you want to change a particular microinstruction, all you need do is remove one card and replace it with another. Consider, for instance, the situation in which a manufacturer wishes to change the function of a particular instruction. All that is necessary is that he manufacture a new set of cards and send representatives into the field to remove one old card and replace it with a new one at each installation. Sometimes there is a mechanism for the systems programmer to change a wired microinstruction. He may be able to do it in his own shop, he may have to send a set of specifications off to the manufacturer, or he may have some particular machine that he can instruct to produce the new card that will change the microinstruction. Finally, the computer may have been manufactured with a writable control store. That is, there is a supervisor program that can change the microprogram resident in the control store. It may be very slow, and certainly this facility would be made inaccessible to the casual user to preserve the integrity of the system; but microprograms can be changed. When you change the microcode, you are (typically) changing the architecture of the machine the programmer is seeing and programming for.

In this chapter we have concentrated on the languages used to communicate with the computer. When we introduced the writable control store, we had a new level of complexity in the computer system with which we must communicate. As discussed earlier, computer languages mesh with the resources of the computer system while providing the human the best possible tools to perform the task at hand. Programming a writable control store is quite different from most other programming of a digital computer; therefore a separate language is indicated for this purpose.

Without going into extensive detail of microprogramming for writable control stores, let us examine the properties required in a microprogramming language. It must be capable of describing all the hardware, including registers, buses, memories, and the control store itself. In addition, the language must have explicit timing capabilities, so as to designate which operations may occur in parallel and which must be sequential. It must also be able to control parallel data transfer. Being so close to the hardware, it must produce efficient code. If trade-offs have to be made, they should be in favor of the best utilization of the hardware; sloppy microprogramming will affect every level of software employing that microprogram. Since microprogramming should be attempted only by experienced system programmers, the design of the microprogramming language can take the choice of efficiency at the expense of ease of use.

As may be expected, microprogramming languages are evolving from assembly-type language to compiler-type languages. Since efficiency is very important, it is difficult to produce a compiler that can compete with a micro-assembler, but it will probably be only a matter of time until one is produced. A compiler language would have the advantage of being procedure-oriented, like FORTRAN, ALGOL, or PL/1. In addition, the language itself might be machine-independent; a set of declarations (compiler directives) would then be employed to describe the particular machine being microprogrammed.

The major advantage of microprogramming, as can be inferred from what we have said so far, is its extreme flexibility. Not only can microprograms be changed, but one can introduce into the microinstruction set an extended capability. For instance, if a machine or a family of machines has hardware options that the customer may choose to purchase or not, then different microinstructions will occur, depending on the hardware components that are available. If hardware is available to perform an operation, say, floating-point multiplication, then the microprogram for floating-point multiplication is extremely simple; it enables that particular hardware to function. However, if that hardware does not exist, then a much more complicated microinstruction sequence is required. Such a microprogram has to perform the same operations as the optional hardware, but causes them to be performed by different components. In a way, the microprogram can be thought of as the software for implementing functions not available in the hardware. The term *firmware* has come to be applied to microprograms because they represent some intermediate step between hardware and software. If it is desirable to have one computer simulate another, so that the user may run machine-dependent programs for the machine he does not have, then the microprogram again becomes an extremely valuable tool. One can make microprograms to reproduce some of the hardware function of the target machine not present on the host machine. This process, known as *emulation*, is one of the valuable techniques for assisting programmers in transferring from one machine to another. They simply run their new machine in the mode emulating the previous piece of equipment.

REFERENCES

C. W. Gear, *Computer Organization and Programming*, McGraw-Hill, 1969.

L. L. Constantine, "Integral Hardware/Software Design," *Modern Data Systems*, April 1968 through February 1969.

N. Chapin, "Logical Design to Improve Software-Debugging—A Proposal," *Computers and Automation*, Vol. **15**, No. 2, July 1967.

S. S. Husson, *Microprogramming: Principles and Practices*, Prentice-Hall, 1970.

L. Rakoczi, "The Computer-within-a Computer: A Fourth Generation Concept," *IEEE Computer Group News*, Vol. **3**, No. 2, March 1969, pp. 14–20.

A. B. Tonik, "Development of Executive Routines, Both Hardware and Software," *Proceedings 1967 Spring Joint Computer Conference*.

A. C. Shaw, *Lecture Notes on a Course in Systems Programming*, Federal Technical Information Service, PB 176 762 (1966).

W. Kent, "Assembler-Language Macroprogramming," *Computing Surveys*, Vol. **1**, No. 4, December 1969, pp. 183–196.

D. Gries, *Compiler Construction for Digital Computers*, Wiley, 1971.

D. W. Barron, *Assemblers and Loaders*, American Elsevier, 1969.

J. Tau, ed., *Software Engineering*, Academic Press, 1970.

P. Wegner, *Programming Languages, Information Structures, and Machine Organization*, McGraw-Hill, 1968.

A. J. T. Colin, *Introduction to Operating Systems*, American Elsevier, 1971.

QUESTIONS

1. In Chapter 7, we presented several short program sequences to illustrate various machine instructions. Using the strategies of (a) one-pass assemblers and (b) two-pass assemblers, try to assemble the program segments listed below. Assume in all cases that the assembler directive END appears after the last instruction. As your machine language, use (c) the format in Chapter 7 or (d) the format of the machine you have to program. Generate clear error messages or warning messages whenever necessary.
   Section 7.4
   Section 7.6
   Section 7.7 (conditional jump; both examples of looping).
2. Two complete subroutines, MOD and the set LIST, PUSH, and POP are illustrated in Section 7.8. Repeat Question 1 for these subroutines, explicitly producing all internal tables.
3. Repeat Question 2 for the ARCTANGENT problem of Section 7.14.
4. What advantages does interpretive execution offer when debugging? Are these advantages applicable to both on-line conversational and batch processing?
5. Translators for certain languages are usually implemented as interpreters rather than compilers. Do you know of any interpretive translators? Are there any operations in source languages you know that must be performed interpretively?
6. What FORTRAN operations are performed interpretively? Why?
7. Why are control stores often implemented as read-only memory? What are essential characteristics of a control store?
8. Are there any pre-defined macroinstructions within the assembler and operating system for your computer; what are they and what sequences of code do they cause to be assembled when expanded?
9. Distinguish among the following.
   a) machine language and assembly language
   b) interpretation and compilation
   c) comments (or remarks) and diagnostics
   d) hardware and software.

10. Describe sufficiently to demonstrate your comprehension each of the following terms and groups of terms.

a) Run time	b) Symbol table
c) Forward reference	d) Conditional assembly
e) Conversational translation	f) Microprogram
g) Macro	h) Interpreter
i) Compiler	j) Translator.

11. Under what circumstances should a programmer code his programs in assembler language as opposed to FORTRAN (or a similar higher-level language)?

12. a) Determine the expansion of the macros described below.
    b) If your machine has a macro assembler, write out the complete code required to define the macros.
    c) Assemble your macros and test them out, using driving programs you have specially written for that purpose.
    1) A CALL macro that functions like the FORTRAN subroutine call, i.e., passes control to a subroutine and passes a variable number of parameters in standard linkage form.
    2) Input/Output macros that employ the formatted I/O subroutine provided in the system library. Enumerate any conventions or associated statements and their form.
    3) An instruction that stores a transfer to [the present location + 2] in the location specified by the address field, stores the contents of [the present location + 1] in [1 + the location specified by the address field], and transfers control to [2 + the location specified by the address field]. Describe the use of this instruction.

13. Assume that you have received a microprogrammable computer with no specified machine language. You have decided to implement the set of instructions presented in Fig. 7.6 in assembly language form.
    a) Specify and tabulate the location, size, and values for the op code, register(s) and address field.
    b) Specify the microoperations required to implement some of the instructions specified in (a).

14
THE SUPERVISOR

As we have seen, each level of software—assembler, higher-level language, etc.—serves to isolate the user from the harsh demands of machine code. A modern computer job-shop must provide various services to each user in an orderly, business-like manner. For this purpose, yet another level of software is needed. Like the others, it isolates the user even further from the hardware but compensates with improved services.

Computer time is a valuable resource. On early machines, for example, the FORTRAN compiler had to be read (as a deck of cards), and it was followed by the reading of the source program and then the reading of the data to be processed. Each new user had to load all the subsystems (such as language processors). Very quickly users began to combine jobs in a *stream*, so that one loading of the FORTRAN deck would suffice for all. This procedure was the precursor of the modern supervisor.

A primary function of the supervisor is to schedule and plan the flow of work through the computer. Toward that end, the incoming job stream will be *queued* on intermediate storage (such as a drum). A user's cards are read and saved on drum, but his program may wait there for minutes or even hours while other programs run. Similarly, his output is not printed while his program is running; it is saved in another queue to await the services of a line printer.

The supervisor also measures the resources used by a program (cards read, CPU time occupied, lines printed), and bills the user accordingly. With a supervisor in charge, the CPU is better utilized because it never has to wait for a user to load his job or for the line printer to finish a line. Such an organization requires very sophisticated combinations of hardware and software.

Clearly a user cannot, in this context, issue a machine language command to, say, read one card. This would subvert the entire scheme, which requires that his input card deck actually be located on drum. The input/output commands (and some others) are therefore *protected* from the user, and he must request I/O via the supervisor. Somewhere in the hardware is a *protect bit*, which allows only *privileged* programs, such as the supervisor, to have access to these functions.

14.1 SUPERVISOR COMMAND LANGUAGE

A computer user first becomes aware of the supervisor in the process of telling it what to do. Depending on one's viewpoint, the user orders, instructs, or requests the supervisor to perform certain operations in a certain specified sequence to accomplish his objectives. We will identify the user's statements specifying the operations to be performed as *commands*. For ease of identification, commands often have a special character in the first column (or perhaps a sequence of special characters in the first few columns). Commonly used characters include $, *, /, and @. This convention places a slight restriction on the user in limiting the characters he can put in column 1, but it trades this restriction for unambiguous recourse to the supervisor.

The set of supervisor commands constitutes a language, which we shall call the *supervisor command language*. Although supervisor commands have been around for a long time, recognition of the fact that they constitute a language is a relatively recent development. This recognition should lead to more systematic command sets for future machines. It may even be possible someday to standardize system command languages in much the same way that FORTRAN and COBOL are standardized.

Supervisor command languages are generally interpreted (see Chapter 13). There is a component of the supervisor, the command interpreter, whose function is to decipher the commands and take appropriate action. All input from the user is scanned for supervisor commands before being passed on to a user program. When no user program is active, a supervisor command is used to initiate one. Figure 14.1 is a schematic representation of the control and information paths through the supervisor.

What does the supervisor do for the user? What are the types of commands to which it responds? The answers depend on each particular supervisor, but there is a large area of commonality.

14.2 SUPERVISOR SERVICES

We can divide the functions of the supervisor into five general categories.

1. Direct user requests for service
2. Requests for service by subsystems (such as FORTRAN)
3. Requests for service by user's program at run time
4. Requests for service by subsystems at their run time (such as FORTRAN I/O)
5. Housekeeping and overhead.

Direct user requests for services

Direct user requests for service are usually accomplished by writing entirely in the supervisor command language. For example, a user might wish to have the contents of the first file on a magnetic tape read and listed on the line printer. Since

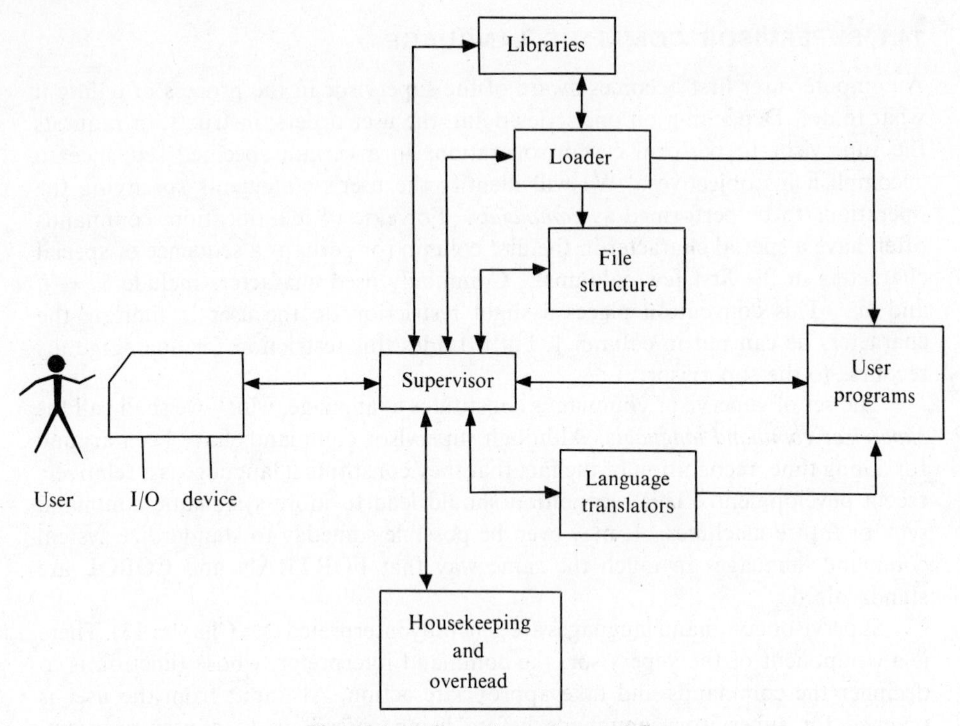

Fig. 14.1. Information and control flow

the data formats, data codes, and data transfer rates on these two disparate devices are different, writing a user program to obtain such output would be difficult indeed. In the context of a supervisor, however, the task is very simple. The supervisor will have machine language *modules* whose purpose is to communicate with each input/output device and convert the data stream to or from a standard form. Thus in principle it is simple to make the tape drive's output data stream be the line printer's input data stream, and the job is done. The supervisor may also take charge of allocating resources. In the cited job it would then defer execution until the tape drive and the printer were both available at the same time. Actually, the example of reading and printing the contents of a tape is over-simplified. While the supervisor does in effect perform the indicated actions, the most common way of doing so is to copy the tape information onto some intermediate storage medium—usually a disk file—then copy it onto the printer output file, and finally, copy to the printer proper. It is also not necessary to wait until both the tape drive and the line printer are simultaneously idle. Most important, however, the above intricacies require no special action on the part of a user. He simply writes a straightforward request in the supervisor command language and returns later for his printout.

Subsystem service requests

Under the category of user requests for subsystem services, the user issues system commands which cause a language translator or other subsystem (such as the linking loader) to be run. The user passes parameters to the subsystem via the supervisor. In a FORTRAN job, the user must specify the location of the source and object files, and he may in addition request such special services as a memory map, assembly listing, and compile-time statistics. Each subsystem will have its own defined set of parameters, which are passed to it by the supervisor.

Programmed requests for supervisor service

The supervisor command language provides the means for direct communication between the user and the supervisor. However, there are also many circumstances when a user program will require a supervisor service while running and must therefore communicate with the supervisor. To differentiate these run-time supervisor communications from the supervisor commands, we will call them *supervisor requests*. They have also been referred to as *extracodes*; the rationale for the latter name is that supervisor requests extend the set of machine functions available from assembly language programs.

The set of services available through supervisor requests is distinct from that available through supervisor commands, but their intersection is nonempty. Typical run-time supervisor services include input/output, storage management, and error handling. The actual implementation of supervisor requests may employ any of the subroutine linkages presented in Chapter 7, or a special op code may be provided just for supervisor requests. In systems where certain operations are reserved for the supervisor (input/output for example), the supervisor request is usually the only way for a user program to perform these operations.

Implicit supervisor services

Frequently a systems or applications programmer will invoke the supervisor without realizing it. If he is writing a FORTRAN program that requires input/output, the object code produced by the FORTRAN compiler will actually use the supervisor to provide the I/O, but the programmer need not be aware of this fact.

Housekeeping and overhead

The supervisor has the responsibility of scheduling the various tasks in its input *queue* to take best advantage of the machine resources. Programs may be classified in two ways. If a program's processing speed is limited by input/output, it is referred to as *I/O-bound*. If its speed depends on how fast the CPU can perform the required arithmetic, it is referred to as *compute-bound*. A job that requires a great deal of I/O to or from slow devices, such as tapes, may be interleaved efficiently with short compute-bound jobs, such as translation of small programs.

Requests for high priority (at high cost) from the user must be taken into account. Scheduling is done according to a *scheduling algorithm*, which is a subject of great interest among users and manufacturers, since it can greatly influence the efficiency of a machine.

The supervisor must also take responsibility for the cost-accounting involved—identifying users, acquiring and accumulating data on their use of the machine, and even rendering invoices.

The user must be able to communicate with the *operator*, the man or woman who runs the machine. Operator intervention is necessary when a tape or disk is mounted or when some special service is required. Often the operator must also be informed of what the supervisor is doing, and there may be a single supervisor command for both functions.

14.3 SUPERVISOR ORGANIZATION

The supervisor is not just one great big program. It is a set of routines, each of which has a particular function. When that function is to be performed, control is passed from a master program to the appropriate subsystem. Upon completion, control is returned to the master. Not all of the supervisor needs to be in main memory at one time. For example, there is no need to retain a compiler when the object code it produced is running. Some sort of file system is employed for supervisor subsystems. The whole process of bringing a subsystem into core and passing control among subsystems is designed to be transparent to the user.

File systems

If the system design includes a mechanism for retaining machine-readable sets of user data on bulk storage, a new area of supervisor services is opened up. The user data may include programs (source, relocatable, and absolute); data on which programs will operate; and even sets of supervisor commands. The subsystem responsible for managing all this material is known as the *file system*. A file system differs from one in which cards or tapes are the media for bulk storage in that the supervisor must maintain a file directory for each user. Users may then refer to a data set by name, and the supervisor takes the responsibility of locating and accessing the data.

The user must be able to create and alter files, both with supervisor commands and under program control. One of the trends in computer use has been away from program and data storage on paper tape and punched cards and toward the use of file systems. File systems are very effectively combined with remote-access typewriter-like terminals to produce *conversational*, *time-sharing* services. The philosophy of design of conversational supervisors and subsystems differs considerably from that of batch-processing systems. Among the challenges is to maintain system efficiency while satisfying many users each wanting immediate service for their program.

14.4 THE LINKING LOADER

The output of most language translators does not consist of completed machine-executable commands. The requirements of modern supervisor-controlled systems dictate that main memory be allocated to more than one program at a time. For this reason, user object programs are stored in *relocatable* form. Such programs, with suitable action by yet another supervisor subsystem, may be located anywhere in memory. Once the actual memory addresses have been placed in the binary instructions, the program will run only at those addresses. Programs which have been through this process are called *absolute*. Frequently more than one relocatable program unit must be combined to form one absolute program. This would be true, for example, if a user invoked a library subroutine.

The supervisor subsystem that combines and translates relocatable programs into absolute programs is called the *linking loader*. Alternative names in current use include *linkage editor*, *linkage loader*, *collector*, *librarian*, and *loader*. The function of the loader is to collect a set of relocatable machine language programs and transform them into one absolute machine language program.

Global references

During translation, each program fragment is separately translated from its source code. In general, when a program unit is translated, the translator is (almost) unaware of the existence of other program units. At the end of translation if some symbolic labels remain which have not been referenced, the translator becomes aware of the possible existence of other program units. These remaining symbols are called *undefined labels*. One source of undefined labels is errors. A spelling or typing error will transform an internal symbolic label reference into an undefined label. There is no guarantee, however, that subtle spelling differences are accidental rather than intentional. Therefore, the entire search procedure described below must be followed; if the undefined label is still not found, then it is assumed to be an error, and (possibly) corrective action can be taken.

Besides errors, there are other reasons for undefined labels, notably *external references* and *global references*. These are labels which are defined in another program unit. The problem is to find out where.

Creating global labels

In higher-level languages, global labels are generally associated with subroutines or external subprograms of the function type. The global label is declared by means of such commands as SUBROUTINE, FUNCTION, and ENTRY. There is usually a predefined subroutine linkage convention employed to access such external subprograms (see Chapter 7).

In assembly language programs it is usually possible to designate any symbolic label as global. Some assemblers provide an assembler directive for this purpose; others suffix a specified character to labels which are to be globally defined. Commonly, global labels in assembly language programs are employed as subroutine *entry points*. These labels may be reached by employing the same subroutine standard linkage used by higher-level language compilers, or by making nonstandard use of the various subroutine instructions in the machine's vocabulary. Of course, there must be agreement between called and calling program as to the convention to be used for passing arguments and return location.

Another use of global labels is for data. It is sometimes useful to have arrays of data globally referenced to avoid renaming them or passing their addresses as subroutine arguments. Global labels are also employed to identify locations in which the supervisor will deposit data. A global label will be understood by each relocatable element to refer to the same data or instruction set.

Common blocks

Another way of globally referencing data is by means of a special translator directive, the COMMON directive, which associates a global label with a block of contiguous memory used for data storage. Within each program unit, local symbolic labels may be assigned to the same locations which constitute the common block. Such local labels may represent only one data storage location in the common block, or they may be array references. The size of the common block is the sum of the locations referred to by the local labels.

Common blocks are clearly identified as data; globally defined labels are not. It is generally impossible to execute the data stored in common as if they were instructions although there are no formal restrictions on transferring control to any globally defined label. The use of common also reduces the number of global labels. Only the names of the common blocks are global; the identifiers within the block are local to the subprogram in which they occur. It is possible to use the same labels in different program units to refer to the same locations in a common block, but it is certainly not necessary. Since the labels are local, any other label can just as well be used. The local labels need not be cross-referenced by the loader. In using common, you can also create different boundaries and definitions of scalars and arrays in separate subprograms; this can be either an error or a sophisticated programming technique, depending on whether it was done accidentally or intentionally.

Satisfying external references

The job of the loader is to satisfy all the external references by matching them with global labels. This process involves making changes in each relocatable machine language program. One characteristic of the relocatable program is that it does not know the whereabouts of the external references. It can only make their names available to the loader. When the loader does its job, the global labels are located. Then it is possible to change the address fields so that they do indeed have the correct address for each external reference. External references are the ultimate form of forward reference; thus the various techniques for resolving forward references in assembly language programs apply here also. Since the linking loader is a generally available supervisor service, the relocatable format must be standardized so that the loader can work on all relocatable programs. As mentioned in Chapter 13, the pointer chain and the jump table are common techniques employed for satisfying external references.

When all linking is completed, all remaining undefined labels are errors. If incorporated into the design of the loader, corrective action can be taken. One simple test consists of examining the remaining labels for spelling errors. However, it is not unusual to allow a set of programs with remaining undefined labels to go into execution. The first attempt to reference an undefined label produces a run-time error, of course. Allowances must be made for linking subprograms at any of the following four times during the process from program statement to job execution:

1. *Translation* time, when symbolic references are established and relocation indicators identified.
2. *Linkage* time, when interprogram references for different segments are resolved.
3. *Load* time, when program instructions and data are placed in absolute memory locations.
4. *Execution* time, when actual machine instructions are performed.

In earlier systems, completion of a linking loader's work occurred primarily at load time. In more recent systems, linking loader functions may be required at any of the four times noted above.

Memory allocation and relocation

The loading part of the job is to take the set of linked relocatable programs and to store them in main memory. All external references have been resolved, all relative addressing is made absolute, common blocks are assigned storage, and a starting address is specified. After being loaded, the program is ready for execution.

The major task in transforming a set of relocatable machine language programs into one executable absolute program is the adjustment of all addresses to the actual memory location where the program unit will be loaded. This process is known as *relocation*. We have already discussed the mechanism of setting external reference addresses at load time. But often internal references must also be adjusted. The most common mode of translating a program from source to relocatable form is to assume that the first statement of the relocatable program will be loaded into location 0 of the memory and that the program will be successively laid down in an increasing address sequence. The task of the translator is simplified because it need not consider the actual memory locations into which the absolute program will be loaded. Indeed, if the translator had to be aware of the location in memory, it might be necessary to retranslate the program unit every time it was employed in a new program. The task of fitting a relocatable program into a specified block of memory is postponed until load time.

When a relocatable program unit is loaded, it generally will not begin at memory location 0. All instructions which reference memory locations are therefore potentially addressing the wrong location. We say "potentially" because some of the addresses may be correct. Addresses that do not change when a program unit is relocated are known as *absolute addresses.* When registers have memory addresses, these addresses are absolute. Similarly, address references into the supervisor are usually absolute. Immediate operands may also be absolute. Addresses that change when a program unit is relocated, known as *relocatable addresses*, point to locations within the program unit. Typically, the addresses involved in such instructions as JUMP and STORE are relocatable. Before run time, relocatable addresses must be *offset* by the address of the first statement. That is to say, the address of the first statement (the actual *origin*) must be added to the relocatable address (formed with origin = 0) to form the absolute address.

An interesting alternative possibility is the use of *relative addresses.* In relative addressing, the effective address is calculated by adding the contents of the operand address field, called the *displacement*, to the contents of a special register, known as the *base address.* Typically, the register selected is the program counter, and thus we have run-time relative addressing similar to that used in assembly language programming. The process of adding the contents of a register to the contents of an instruction address field, forming an effective address, is also similar to index register address modification. The hardware must support this address calculation at run time.

It is not *a priori* obvious which addresses are relocatable and which are absolute. It is therefore necessary to provide relocation information as part of the translation into relocatable machine language. As a minimum, there must be one bit for each potentially relocatable address. The bit is used as a flag indicating whether or not the field contents are to be displaced by an offset or a base address. More than one bit per field may be necessary, depending on the complexity of the services provided. Some decision has to be made concerning the whereabouts of this *relocation information.* It cannot be stored in the same word as the instruction because the instruction, by definition, occupies an entire word. If the relocatable machine language program is to be retained in main memory, then space is at a premium. It is then common to collect the relocation information for several instructions into one word and to locate relocation words at a fixed frequency throughout the relocatable machine language program. Alternatively, all the relocation information can be collected in a table which is prefixed to the machine language part of the relocatable program. If the relocatable machine language program is to be stored on some external medium, space is relatively inexpensive, but access time becomes important. To minimize time, the relocation information is commonly placed adjacent to the instruction it describes. Space may be inefficiently used, but loading can proceed rapidly because all the necessary information is together.

Design considerations

The following questions, whose answers strongly affect the design of a linking loader, originate from the hardware properties of a given computer system.

1. How can the available main storage of the computer be used most efficiently?
2. How does the method of addressing memory affect linking-loader functions?

Space considerations

The first consideration, that is, the efficient use of available space, is a primary one in designing a linking-loader program. Three possibilities exist as to the availability of space.

1. A sufficient memory capacity is provided, so that the space consideration is not a controlling design factor.
2. The available memory capacity is limited, so that the allocation of space is constrained.
3. Memory capacity is a variable including sizes in the first and second categories above.†

The first is of course the ideal situation. The second and third possibilities, however, introduce significant design choices. For example, the second raises the question whether the loader should be capable of handling input programs of all sizes or of a limited size only. The variable case asks whether or not one design should be used for the full range of possible memory sizes.

In any but the ideal situation, a basic design decision must be made as to how the available space should be distributed between the storage requirements of the loader program and its input stream. It is not evident, for instance, that the loader itself must be fully resident in memory. Similarly, input to the loader consists of several types of data, each of which has different characteristics affecting memory allocation. In general, linking-loader input may be divided into four classes.

1. *Loader instructions*, which are instructions to the loader to specify what functions are to be performed.
2. *Program text*, which comprises the user's program instructions and constant data.
3. *Linkage indicators*, which define references among separately translated subprograms.
4. *Relocation indicators*, which identify the different relocation properties of each element in the program text.

---

† A software designer is often confronted with problems of this type. His program, in this case a loader, must be suitable for a whole family of machines which are program-compatible (have the same hardware instruction set) but come in very different configurations. Some small versions would therefore not have enough room in main memory for the more space-inefficient loader schemes although a typical large configuration would have plenty. The designer may provide more than one loader in order to supply all users with an appropriate system.

Because they do not require much space, loader instructions are normally condensed and stored in memory for use throughout processing. Linkage and relocation indicators are also more efficiently used if they are kept in memory. Since the basic work of a loader is relocating program text and linking interprogram references, these indicators are the items used with highest frequency, and they should be the last to be relegated to auxiliary storage.

Program text, like the loader program itself, may, but need not, be kept in memory. This choice is identical to the text-in-core versus processor-in-core decision seen in the design of a language translator. An effective solution, of course, is dependent on the answers to questions raised earlier. For example, if the space allowance is severely limited but program size is correspondingly restricted, then a text-in-core design would be efficient since all input programs could be left in memory for immediate execution, if desired. On the other hand, if the space allowance is limited but program size is usually large, then a processor-in-core emphasis is perhaps more appropriate since leaving program text in memory will often be impossible anyway.

Addressing considerations

Three basic schemes of addressing memory are in common use, and the differences among them result in somewhat different loader designs. The first, called the *full n-bit addressing* method, is the simplest concept, maintaining an address field of n bits. The range of 0 to 2^n provides access to all available memory locations.

The second method, known as the *base-plus-displacement addressing* (or *base register addressing*) technique, provides for addressing a memory of large size without requiring a complete n-bit address field in each instruction with an address reference. A general-purpose register or an index register is set aside and loaded with a base address value. Memory locations are then determined by adding this quantity to a displacement of less than n bits.

The third method, known as *sector addressing*, also utilizes an address field with fewer than n bits. An effective address is assembled, with the upper bits of the program counter providing the sector and the address field providing displacement within the sector. Addressing across sector boundaries is by indirect addressing.

Systems which use the full n-bit addressing method typically require, first, the identification of address relocation types during the translation process and, second, the inspection of all address quantities during loader processing. The following relocation types are usually distinguished during the translation process.

1. *Program relocatable*, which indicates the normal relocation property; i.e., the address reference is to be modified by adding to it the difference between the absolute program origin at load time value and the origin specified at translation time.

2. *Absolute*, which means that the value specified in the address field is a fixed quantity and should not be altered for relocation purposes.

3. *Negative relocatable*, which is primarily used when large data blocks, common to all subprogram units, are stored at one end of memory, and individually translated subprograms need to place certain data areas at locations either preceding or within these common blocks.

At some point during loading, an evaluation of the relocation property of each full *n*-bit address quantity is made, and the appropriate action is performed. In other words, a program relocation factor is calculated and added to program relocatable address fields; absolute terms are not changed; and so on.

One of the advantages of base register addressing is the relative ease with which sections of code not exceeding the maximum displacement amount may be relocated. This is possible because address symbols are defined as the sum of the contents of a specified base register and a displacement value. In order to relocate a program by 1000 locations, it is necessary only to add 1000 to the quantity in the base register. However, because distinct sections of code are individually relocatable, some means of identifying those addresses used as base values must be provided, so that (1) they can be appropriately relocated at the time memory positions are fixed, and (2) inter-section referencing will remain valid despite the relocation process. This difficulty is often solved by the use of an *address constant*, which by definition contains what is known about the location of a given symbol at each step in the translate, link, load, and execute cycles; naturally, its most useful value occurs at execution time, when it contains the absolute location of the given symbol after the instruction or datum it references has been placed in memory.

Using the base-plus-displacement technique, then, we can accumulate address constants during the translation process rather than specify relocation types for each address reference; and during loading, we need to insert absolute address values for only the required address constants. We do not need to inspect all address fields.

A program written in a single sector for a sector-addressed machine will run *without address modification* in any other sector so long as it does not reference addresses outside its sector. This is possible because all addresses are understood by the hardware as being relative to the first address of the sector. The loader must therefore locate and modify only those locations used as indirect addresses for out-of-sector references.

14.5 LIBRARIES

A companion to the linking loader is the library, which is a set of programs. Although it is certainly proper to speak of a library of symbolic source language programs, we are here concerned with the library of relocatable machine language program units, known as the *relocatable library*.

A library exists to be used to satisfy external references. The most common example is the supervisor library. Whether one or many, the supervisor library, often called the *system library*, contains the subroutines most often called. A major source of subroutines for the system library consists of the run-time subroutines used by compiler languages. In general, subroutines are not explicitly called in the user's program. Rather, the compiler generates subroutine calls for those operations which must be taken care of at run time. As noted in Chapter 13, formatted I/O is a typical example. Also found in the system library are the subroutines available in a particular compiler language. FORTRAN, ALGOL, etc., list a set of subroutines which the programmer may call.

There may be multiple system libraries if such partitioning adds to the operating efficiency of the computer system. For example, there may be separate FORTRAN and ALGOL libraries. Furthermore, subroutines that are not often used but are nevertheless to remain available may be kept in separate libraries on different secondary storage media in the interests of saving both space and money.

Library searching

The purpose of having a library is to satisfy external references. The linking loader first loads all program units explicitly requested, and then it searches the library for implicitly named programs. Given multiple libraries, there must be some means for specifying which library is to be searched. Perhaps the compiler will insert such an instruction. It is often necessary to search a library at least twice. Some subroutines in the library may call other subroutines in the library. It may not be possible to arrange the library so that these references are satisfied in one pass.

User-created libraries

The library concept is naturally extended to users. A user may have one or more sets of subroutines that are used in his applications. It is much more convenient to be able to identify a set of subroutines as a library than to specifically load the subroutines employed in a particular application program. The creation of separate libraries allows the user to reduce library search time if he limits each library to a certain class of applications.

Another use of libraries is program-sharing. The user may make his library available to others in whatever ways are possible under the supervisor. Such *common* or *public libraries* serve to reduce program duplication as well as to increase information distribution.

The existence of multiple-user and system libraries raises another question related to library searching: "In which order should libraries be searched?" The most common design decision calls for searching libraries in whatever order the programmer specifies, but leaving the system library until last to allow for the possibility that the user may want to provide his own subroutines in place of some in the system library. Because the user's library is searched first, his own subroutines will be loaded in preference to those in the system library. Each successive library is scanned only for remaining unsatisfied external references; therefore, the system subroutines will be loaded only if no others have been found. Thus the user can establish priorities among subroutines of the same name by specifying the order of library searches.

14.6 MULTIPROGRAMMING

One of the major innovations in contemporary systems is the concept of *multiprogramming*, which may be defined as the technique of having more than one program in memory capable of being executed at any given time. This means that if one program stops executing for some reason, it is possible to immediately begin executing another program. There are many reasons why a program may stop executing. The most common is that it is performing an input or output operation. Because input/output occurs at a rate much slower than the CPU is capable of sustaining, a program must cease operation while I/O is occurring. If more than one program is in memory, when the first becomes blocked by I/O operations, the second uses the idle CPU. Indeed, if the second program becomes blocked and there is a third available, it can commence operation. The number of programs that it is possible to multiprogram is a function of the memory capacity and the allowed size of such worker programs. It is also recognized, however, that there is an upper limit on the number of programs that can be successfully multiprogrammed because of their competition for computer system resources, especially high-speed secondary storage. Contemporary studies establish such an upper limit near 10; a common limit is 5 or 6.

Relocation registers

One of the complications of having more than one worker program in memory at a given time is that they cannot all be located at the same place. This obvious physical impossibility does present an additional problem, which the hardware or software must solve. Let us first look at the single program technique to define the problem more exactly.

With only one program loaded in memory at a time, the location for the beginning of the program was known, and therefore every location in the program could be handled relative to that beginning. When the system is going to multiprogram, however, the compiler, assembler, or loader cannot possibly know where the program will be at execution time. In fact, it is quite possible that the program will be loaded at some different place in memory at each different execution time.

Like most other problems of a computer system, this one can be solved with either hardware or software. It is possible to solve the address transformation problem either when the program is loaded into core or during execution. If address transformation takes effect when the program is loaded into core, then all addresses that refer to locations within the program must be displaced by the distance from the beginning of memory to the beginning of the program. That is to say, a constant must be added to all addresses that reference locations within the program. However, it is not possible to add this constant to the address field of every machine word. Some of the machine words may contain data rather than program instructions. Even if they do contain instructions, however, the address field may be used as an immediate operand or an address referencing some absolute location in the executive system rather than the location relative to the program. Also to be considered is the possibility that more than one field may have to be relocated. There may be special instructions that contain two or more fields, each of which references a location relative to the program. For each possible field of relocation, it is necessary to associate a flag indicating whether loading time relocation is or is not to take place. This relocation information is in the form of extra bits, which are set or not, according to whether relocation is to occur. Since, in general, an instruction fills an entire machine word, it will be necessary to find some extra place to put these bits. One common technique is to allot every nth word to contain relocation information for the following block of n words (n is a function of the number of bits of relocation information required and the word size). The relocation scheme described above is similar to the one used by the loader, discussed in a previous section.

The alternative to loading-time relocation is execution-time relocation, which requires the existence of a memory location, preferably a high-speed register, for each user. Such a register contains the address of the first memory location being used by that user. Thus, whenever the user makes a memory reference, the contents of this register, the *base register*, are added to the memory address specified to form the effective address. In order to maintain the speed of the machine, the whole mechanism is usually implemented in hardware. When the program is loaded into core, the base register is loaded with the address of the beginning of the area into which the user program goes. This technique is very similar to index register modification.

The supervisor (protected) state

Multiprogrammed computer systems have a variety of facilities that are either only partially available to the programmer or not under his command at all. Such *programmer-inaccessible* features may take the form of special instructions which can be implemented only when the computer is in a special *supervisor state.* They may also take the form of special-purpose hardware, the function of which is to "keep an eye on" the activities of the programmer (actually on his product, the program under execution). Most of the special hardware facilities to be described in the rest of this chapter were originally developed to make multiprogramming possible with less effort on the part of the supervisor itself.

The reason such special facilities exist is twofold: (1) to protect the supervisor from tampering by the user and (2) to protect the users from interference from one another. There are many things a programmer can inadvertently do that can "spoil" his program and cause him to consume much more of the computer system resources than he has originally intended. Since the resources cost money (either directly or indirectly), it is important to provide means to minimize the effect of misbehaving programs.

The kinds of mistakes the hardware protection mechanisms can actually protect against fall into several general categories.

Addressing faults. The user makes a reference to a memory address that is illegal because it was specified as not belonging to the user, or does not exist.

Instruction faults. The individual user program attempts to execute a nonexisting, illegal, or protected instruction. (A protected instruction is one reserved for use only when the supervisor has control of the computer.)

Result faults. The result of the execution of some instruction is a number outside the permitted range, that is, too long (or too short) or in some way not permitted for that particular computer.

Execution-time-limit faults. The program has taken more than its allotted time for some reason. Typically, an execution-time-limit fault occurs because the program is in an infinite loop. The programmer is usually allowed to specify the time limit.

Machine faults. The computer itself has detected a mistake from which it cannot recover.

A machine fault is a relatively rare occurrence, but it needs some explanation. Many modern computer systems have provisions for detecting actual hardware mistakes. This capability, which is called *automatic error checking*, may involve checking results generated in the arithmetic unit, keeping tabs on the validity of information retrieved from the computer memory system, or monitoring data transferred to the computer from external media. (The details of these checking procedures are beyond the scope of this book.) Almost invariably, automatic error checking is not under the control of the programmer, since the correction of a machine fault, if one does occur, is usually outside his capability.

Attributes defining processor state

Most of the types of faults listed above—addressing faults, instruction faults, result faults, execution-time-limit faults, and machine faults—are the results of a user program's failing in some way. Because in a multiprogramming system there may be many users sharing the computer system (either in the space or in the time domain), and because it it important that no user interfere with any other or with the supervisor, computer systems have special facilities to monitor the occurrence of exceptional events within any one user's program. However, this monitoring facility requires that the limits describing what *is* appropriate for one user's program be known to the system. Only when the information about each user is at the ready disposal of the supervisor can it actually detect the occurrence of an exceptional event.

The computer system monitors such events with a set of *program state registers*. This set of special-purpose registers contains sufficient information about the program currently in execution to ensure the detection of any mistake that program (or the hardware) makes. In principle, the set of program state registers describes the capabilities and resources of the machine to which any particular user is permitted access. This information is usually, but not necessarily, a subset of the full capabilities of the machine.

For example, when the supervisor itself is in control, the program state registers may be set so that the supervisor can look at any portion of the memory space, perform any protected instructions which the ordinary user programs cannot execute, etc. In some computer systems, when the supervisor has control, it can also invoke special diagnostic features which are used to help the computer engineers locate the source of a recurrent hardware fault.

The most important reason for the existence of program state registers is to make it relatively easy to switch among various programs, i.e., to multiprogram various users as their individual demands on the computer system change. This capability of performing a *program switch* is the main reason that multiprogramming is so successful a concept.

The information kept in the program state registers includes:

Program execution address (contents of the program counter)

Program memory limits permitted

Contents of arithmetic and index registers

State of input/output operations

Other miscellaneous information

The important point about this list is that the information it specifies is sufficient to describe the "state" of a particular program, whether in execution or temporarily suspended. In a multiprogramming system, where more than one user program is active—that is, in the system and a possible candidate for execution—there is still only one set of values of the program state registers actually available to the computer. The values of the program state registers for all the other users are stored away in some pre-selected place in memory. Usually, the locations reserved for storing the state registers of dormant programs are kept separate; only the supervisor has access to that information.

These concepts are probably best made understandable by giving some examples. The next two sections illustrate two extremes of the use of program state registers. These examples are still another instance of trade-offs between hardware and software, but in this case there is not only the decision about how a function is to be performed, but the decision whether it is to be performed at all.

The small program status word

In the first example of the use of program state registers, the state of an individual program in execution is described by the contents of one register. Figure 14.2

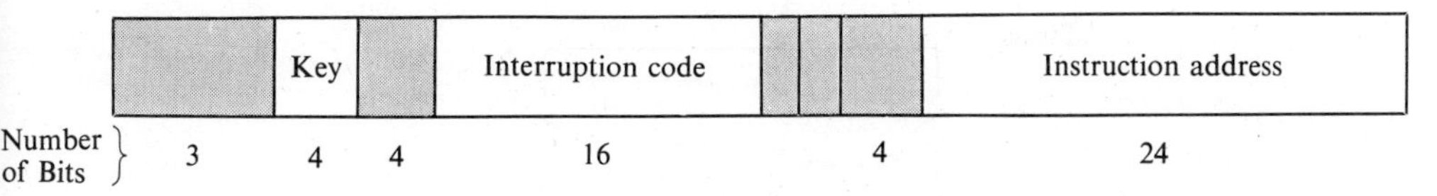

Fig. 14.2. The small PSW

shows the organization of the fields within this register, which is called the *program status word* (PSW). The PSW fields that are important follow.

Key. This field indicates the *protection key* for the current program or for a suspended one; since the key has four bits, 16 different keys are possible. The key with all zero bits is reserved for use by the supervisor. Thus, at most, 15 user programs can be in the system at one time. The memory is organized into blocks, each of which is 4096 eight-bit bytes long. Associated with each block of memory is a four-bit register containing the *storage protection key* for that block. If the key in the PSW does not match the one associated with the block of storage in which a requested address resides, the executing program is interrupted (see below).

Interruption Code. The executing program can be interrupted—that is, the hardware can make it give up control—for a wide variety of reasons. This field of the PSW contains a combination of bits enabling each possible type of interruption.

Instruction address. This field is essentially equivalent to the program counter. It contains the address in memory from which the last instruction executed was taken.

Although there are many ways in which a change from one PSW to another PSW can happen, let us assume that only five basic classes exist.

External interrupts. Something (or somebody) requests that the currently executing program be interrupted.

Supervisor requests. The executing program specifically requests that it be suspended for some reason by use of an extracode.

Program interrupts. Something goes wrong with the executing program which makes further execution senseless. Program interrupts fall into three classes.

1. The program makes an arithmetic mistake, producing an illegal number.
2. The program attempts to execute an illegal instruction—(one that does not exist or is reserved for the supervisor only).
3. The program attempts to reference a memory address for which the protection key in the PSW does not match the protection key for the block of memory containing the given address.

Machine check interrupts. The computer hardware detects an error that in some way renders the result of a computation invalid.

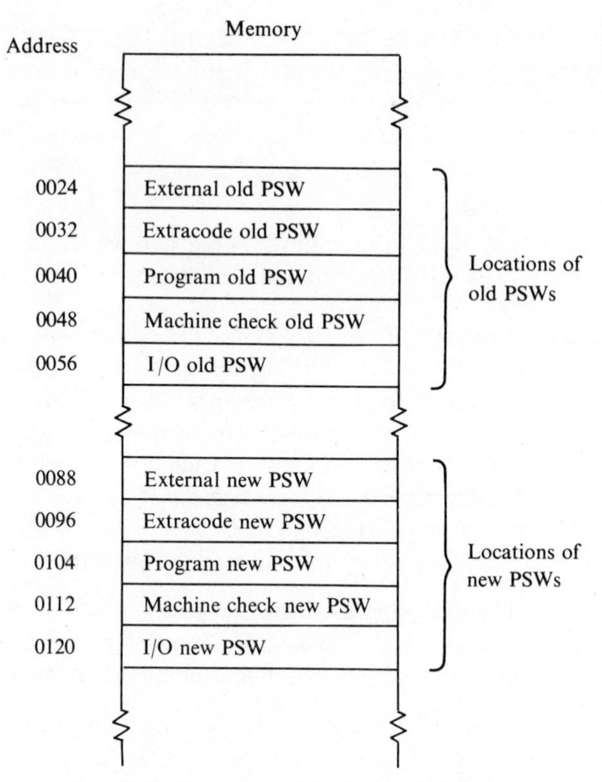

Fig. 14.3. Pre-assigned locations of PSW

I/O interrupts. One of the myriad of input/output devices connected to the computer requires service.†

When any of the foregoing events occurs, the hardware executes an EXCHANGE PSW instruction; that is, it exchanges the current PSW for a new one selected from a location in memory according to the class of the interrupt. At the same time, the values of the current PSW—with the interruption code inserted for the convenience of the supervisor program—are stored in a similarly predetermined location.

Figure 14.3 shows a portion of the memory which is byte-addressable. Each word is 64 bits long; each byte has 8 bits. At the top are five words reserved for holding the *old* program status word whenever an interruption occurs; at the bottom are five words that hold the *new* program status word to be used in response to any of the five classes of interrupt.

† I/O operations generally have precedence over central processor computations because holding up a slow I/O device is more costly than momentarily interrupting the activities of the central processor.

To make the operation of the PSW exchange clear, let us take a specific example. Suppose that the user program needs the services of some program embedded within the supervisor; it can force the supervisor to take control by executing the extracode mentioned before. When it does that, the PSW which represents the current state of the user program is placed in the memory location labeled "Extracode old PSW." The contents of the PSW stored in the position labeled "Extracode new PSW" are then placed in the registers which actually control the machine. The new PSW has a program counter value associated with it (see Fig. 14.2), and from that address the first instruction associated with the new program is taken. Since this program switch was initiated by an extracode, it is the supervisor program which begins executing.

Now the supervisor completes whatever operation the original user program wanted to have done. Since all the information about the interrupted program was put in the "Extracode old PSW" location (as indicated in Fig. 14.3), the supervisor program then executes a protected instruction (one reserved for use only by the supervisor) called LOAD PSW. The address portion of this instruction points to the "Extracode old PSW." As a result, *that* PSW is put in the program status registers. The computer then resumes execution of the user program at the point where it left off when the interruption was requested.

The same sort of *PSW swap* occurs for each of the five possible classes of interrupt. The values stored in the new portion of the memory shown in Fig. 14.3 usually remain constant, since they point to particular places in the supervisor which are designed to react appropriately to the class of the interrupt. One important point is that the "Machine check new PSW" points to a special-purpose section of the supervisor from which return to the user program usually does not occur, because a machine check indicates a serious problem with the computer hardware itself.

It is important to note that the contents of the accessible registers are not part of either the program status word when it is active or the other program status words awaiting use. It is the responsibility of the supervisor either (1) to save the contents of the operating registers or (2) to be especially careful not to disturb their contents. If these registers are disturbed, it is unlikely that the interrupted program, once resumed, can continue to operate coherently.

The extended program status word

In this second example we introduce a larger PSW, which can contain more information, and an alternative method for memory allocation and protection. In many ways, the extended PSW approach to keeping control of a user's executing program is much simpler than the mechanisms employed in the preceding example. As you might expect, the overall capabilities one has with a simpler system are fewer than those one has with a more complex and more adaptable system.

The extended PSW describes the state of a program completely, whereas the small PSW describes it only partially. The information in the extended PSW falls into three general categories.

1. The contents of all of the machine's programmer-accessible registers.
2. The base address and size of the region of the computer memory assigned to the user.
3. The address of the current instruction relative to the base address for that user region.

When a user program is executing, all the instructions and all the memory accessed have addresses between the value contained in the base register and the sum of the assigned size and that value. Since all program address references are relative to a fixed beginning place in the memory, this is called *relative addressing*.

Figure 14.4 shows the arrangement used when the extended PSW is stored at some address, N, in memory. A total of 16 words is required to store the values of all the registers, the base address, and the assigned size.

When the program represented by this set of values is actually in execution, of course, all of the values shown in Fig. 14.4 reside in actual registers. However, the base address, field length, and program counter registers are not accessible to the programmer. Transfer of control from one user program to another occurs through the EXCHANGE PSW instruction. As in the preceding example, this is an instruction that can be executed only under very special conditions—namely, under the direct control of the supervisor. The effect of an EXCHANGE PSW instruction is to store all the values in the extended PSW at the address specified in the operand address field and, at the same time, to take new values for the entire set of control and operating registers.

For an executing program there are only three types of errors which can cause an EXCHANGE PSW to be executed.

1. An instruction issues a request for an address outside the range defined by the base address and field length registers.
2. An instruction execution produces a result out of range, (for example, an exponent that cannot fit into the floating-point format used on the machine).
3. An indefinite operation is attempted, such as trying to divide by zero.

When any of these errors occurs, the EXCHANGE PSW instruction switches control from the user program to the supervisor. There the appropriate actions—perhaps including terminating the job completely—are taken.

In addition to effecting PSW exchanges because of errors, the supervisor can also respond to the demands of input/output operations and user program-generated extracodes.

Another matter of importance concerns the fact that there is no actual hardware requirement that two user programs span different regions of memory. It is

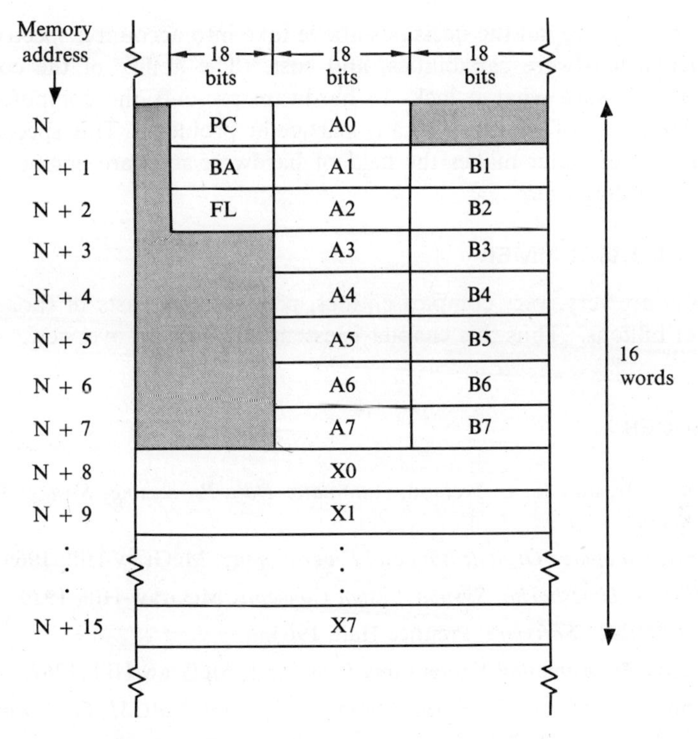

PC = Program counter
BA = Base address
FL = Field length
Ai = Address registers, $i = 0,7$
Bi = Index registers, $i = 1,7$
Xi = Accumulator registers, $i = 0,7$

Fig. 14.4. Extended PSW

the responsibility of the supervisor to see that the base address and the size registers (see Fig. 14.4) define sections of the memory that do not overlap. These registers can be set to any value; therefore the memory system can actually be used more efficiently than in the preceding example since only that amount of memory space required by each user need be actually assigned to him. By comparison, the preceding example allocated memory in 4096-byte blocks, the span of one of the individual protection keys.

Design alternatives

The last two examples have shown possible approaches to the design of multiprogramming machines. Each system made has been designed to trade off hardware and software in the support of multiprogramming. There are many solutions extant in addition to the two given here.

In a larger sense, all the decisions above take into account a spectrum of user requirements, hardware capabilities, and cost. The ability of the computer to perform in software what it lacks in hardware presents the computer architect with a wide range of solutions to any particular problem. This special property of the digital computer makes the field of hardware/software engineering a particularly rewarding one.

14.7 FINAL DISCLAIMER

Supervisors are very, very complex entities, produced by casts of thousands with budgets of billions. Thus this chapter is essentially only an overview.

REFERENCES

F. P. Brooks, Jr. and K. E. Iverson, *Automatic Data Processing, System 360 Edition*, Wiley, 1969.

C. W. Gear, *Computer Organization and Programming*, McGraw-Hill, 1969.

R. W. Watson, *Timesharing System Design Concepts*, McGraw-Hill, 1970.

I. Flores, *Computer Software*, Prentice-Hall, 1965.

S. Rosen, ed., *Programming Systems and Languages*, McGraw-Hill, 1967.

P. J. Denning, "Third Generation Computer Systems," *ACM Computing Surveys*, Vol. **3**, No. 4, December 1971, pp. 175–216.

D. W. Barron, *Assemblers and Loaders*, American Elsevier, 1969.

L. Presser and J. R. White, "Linkers and Loaders," *ACM Computing Surveys*, Vol. 4, No. 3, September 1972, pp. 149–167.

QUESTIONS

1. How could you, as a user, gain access to privileged instructions? If you did, could you then prevent the supervisor from charging you for computer time? How?
2. Why is it necessary to have a special card character to define supervisor commands?
3. Why must a supervisor command precede all user commands in a large batch system?
4. Give a *specific* example of each of the following types of supervisor service.
   a) Direct user request
   b) Request from subsystem
   c) Request by user at run time
   d) Request by subsystem at its run time
   e) Housekeeping and overhead.
5. When rewinding a tape, would you use an extracode or would you issue the machine language command directly? Why?

6. Specify the kind of protection your user files need if you are to have the responsibility for their integrity. Specify the same for system files.

7. For each of the following assembly language commands, indicate whether the operand address field is changed when the program is relocated.

   ADD TO REGISTER
   STORE REGISTER
   LOAD REGISTER, IMMEDIATE
   STORE PC AND JUMP
   SUPERVISOR CALL

8. Compare a batch-processing job stream with an individual user hands-on operation. Identify as many steps as possible, and weigh the efficiency of one operating technique against the other.

9. Many contemporary computer systems offer a service called *time-sharing* or *conversational processing*, wherein the user operates a typewriter-like terminal in direct communication with the computer. In some ways this mode of operation is like the hands-on operation of a single user in earlier systems. Explore in detail the similarities. What tasks of a batch-processing supervisor carry over? What new tasks must be performed by a conversational supervisor?

10. What is the format of the supervisor command language(s) you have used? What property (rule) exists to permit easy identification of supervisor commands?

11. List all the supervisor commands you know in tabular form. Include in the table a brief description of the action invoked by each command. If you know sets of commands for more than one computer, append additional columns to your table, entering similar commands on the same line. What criteria should you use for deciding that commands are similar? Classify each command according to the categories given in Section 14.2 or a set of categories of your own devising.

12. The programs constituting the supervisor require resources. Much effort is required to write and debug them. When loaded, they consume significant portions of main memory. If only part of the supervisor programs are kept in main memory, resources must be expended in transferring them from secondary memory when they are needed. Why is the supervisor worth it? What ensures that it "pays its own way"? What prevents the supervisor from taking over the full machine?

13. List the supervisor requests you have used in your programming. If there is more than one way of issuing a supervisor request on your computer(s), subdivide your list by the way of issuing the commands and compare the various ways.

14. From the system documentation provided by the manufacturer (and perhaps users' groups or the organization that operates the computer), devise a classification of supervisor requests by function. Give several examples of each function.

15. If the higher-level language translators of your computer system permit it, obtain assembly language listings of the results of translation. Tabulate how many *implicit* supervisor services are provided as a result of such translation. If you can gain access to listings of the run-time subroutines, continue your tabulation for

each subroutine; otherwise, consider the run-time subroutines as supervisor services themselves.

16. Answer the following questions about the file systems you have on your computer system(s).
    a) How do you name the smallest unit in the file system?
    b) May more than one unit have the same name? If so, how is confusion eliminated?
    c) Are there any parts of the naming process understood implicitly unless otherwise specified? If so, what are they?
    d) Are names separated into subdivisions? If so, what are the subdivisions called, what purpose do they serve, and how are they used? Are there different default rules for certain subdivisions?
    e) Is there a name and common use for sets of the units? If so, answer questions (a) through (d) for these sets.
    f) Are there any entities in the file system which are exceptions to the organization implied by your answer to the preceding questions? If so, describe them fully and explain why you consider them exceptions.
    g) How do you create a user-defined library?
17. Do all language translators produce output which must be further processed by the linking loader? If not, why not?
18. Distinguish between relocatable and absolute object programs. What names are employed for these by the manufacturer of your computer system(s).
19. In each of the compiler-level languages you know, how do you create external references? Can you create an external reference by mistake? Answer these two questions for assembly language.
20. How do you go about creating global labels in assembly and compiler languages? Are there any differences in their functions?
21. What is the function of the linking loader with respect to COMMON blocks?
22. How are external references and global labels identified to the loader? What does it do with (for) them?
23. Why must the loader change addresses which are not external addresses? How does it know when and when not to change the contents of an address field?
24. What does the loader on your system(s) do when it discovers that the results of its labors will be too large to fit in available main memory? Are there special instructions for the loader to cover this situation?
25. Explain how "base address" relative addressing reduces the task of the linking loader.
26. Is there a special command language for the linking loader on your computer(s)? If so, what is the form taken by the commands that invoke the loader? List the ones you know and explain their function.
27. Would you ever want to specify the same library to the loader more than once? Why?

APPENDIX A
NUMBER SYSTEMS

A.1 POSITIONAL NUMBER SYSTEMS

Positional notation using radix r (alternatively called base b) is defined by the rule

$$N = (\cdots a_3 a_2 a_1 a_0 . a_{-1} a_{-2} \cdots)_r$$
$$= \cdots + a_3 r^3 + a_2 r^2 + a_1 r^1 + a_0 + a_{-1} r^{-1} + a_{-2} r^{-2} + \cdots. \quad (1)$$

For example, $(530.3)_6 = 5 \cdot 6^2 + 3 \cdot 6^1 + 0 + 3 \cdot 6^{-1} = 198\frac{1}{2}{}_{10}$. Our conventional decimal number system is, of course, the special case when r is ten, and each a represents one of the decimal digits 0, 1, 2, 3, 4, 5, 6, 7, 8, 9; in decimal notation the subscript r in (1) may be omitted.

The simplest number systems are obtained when we take r to be an integer greater than 1 and when we require the a's to be integers in the range $0 \leq a_k < r$. This gives us the standard binary ($r = 2$), ternary ($r = 3$), quaternary ($r = 4$), quinary ($r = 5$),... number systems.

The dot between a_0 and a_{-1} in (1) is called the *radix point*. (When $r = 10$, it is also called the decimal point. When $r = 2$, it is sometimes called the binary point, etc.) In some countries, it is conventional to use a comma instead of a dot to denote the radix point.

The a's in (1) are called the *digits* of the representation. A digit a_k for large k is often said to be "more significant" than the digits a_k for small k; accordingly, the leftmost or *leading* digit is referred to as the *most significant digit*, and the rightmost or *trailing* digit is referred to as the *least significant digit*. In the standard binary system, the binary digits are often called *bits*. In the standard hexadecimal system (radix 16) the hexadecimal digits 0 through 15 are usually denoted by

0, 1, 2, 3, 4, 5, 6, 7, 8, 9, A, B, C, D, E, F.

A.2 NUMBER BASE CONVERSION

Four number systems are most frequently used in digital computers: binary (base 2), decimal (base 10), octal (base 8), and hexadecimal (base 16). Occasions

Decimal	Hexa-decimal	Octal	Binary	Decimal	Hexa-decimal	Octal	Binary
1	1	1	1	51	33	63	110011
2	2	2	10	52	34	64	110100
3	3	3	11	53	35	65	110101
4	4	4	100	54	36	66	110110
5	5	5	101	55	37	67	110111
6	6	6	110	56	38	70	111000
7	7	7	111	57	39	71	111001
8	8	10	1000	58	3A	72	111010
9	9	11	1001	59	3B	73	111011
10	A	12	1010	60	3C	74	111100
11	B	13	1011	61	3D	75	111101
12	C	14	1100	62	3E	76	111110
13	D	15	1101	63	3F	77	111111
14	E	16	1110	64	40	100	1000000
15	F	17	1111	65	41	101	1000001
16	10	20	10000	66	42	102	1000010
17	11	21	10001	67	43	103	1000011
18	12	22	10010	68	44	104	1000100
19	13	23	10011	69	45	105	1000101
20	14	24	10100	70	46	106	1000110
21	15	25	10101	71	47	107	1000111
22	16	26	10110	72	48	110	1001000
23	17	27	10111	73	49	111	1001001
24	18	30	11000	74	4A	112	1001010
25	19	31	11001	75	4B	113	1001011
26	1A	32	11010	76	4C	114	1001100
27	1B	33	11011	77	4D	115	1001101
28	1C	34	11100	78	4E	116	1001110
29	1D	35	11101	79	4F	117	1001111
30	1E	36	11110	80	50	120	1010000
31	1F	37	11111	81	51	121	1010001
32	20	40	100000	82	52	122	1010010
33	21	41	100001	83	53	123	1010011
34	22	42	100010	84	54	124	1010100
35	23	43	100011	85	55	125	1010101
36	24	44	100100	86	56	126	1010110
37	25	45	100101	87	57	127	1010111
38	26	46	100110	88	58	130	1011000
39	27	47	100111	89	59	131	1011001
40	28	50	101000	90	5A	132	1011010
41	29	51	101001	91	5B	133	1011011
42	2A	52	101010	92	5C	134	1011100
43	2B	53	101011	93	5D	135	1011101
44	2C	54	101100	94	5E	136	1011110
45	2D	55	101101	95	5F	137	1011111
46	2E	56	101110	96	60	140	1100000
47	2F	57	101111	97	61	141	1100001
48	30	60	110000	98	62	142	1100010
49	31	61	110001	99	63	143	1100011
50	32	62	110010	100	64	144	1100100

Figure A.1. First 100 Integers in Four Bases

arise when a number expressed in one system must be converted to another. Figure A.1 shows the relationships among the four systems for the first 100 integers.

Binary to octal and binary to hexadecimal (hex) conversion

Since $8 = 2^3$, a binary number is easily written in the octal system by dividing it into groups of three digits, starting from the right-hand side. Then each group is transformed to an octal digit separately. Referring to Fig. A.1, for example, the binary number 11010111 is transformed as follows:

$$(11010111)_2 = (011)(010)(111) = (327)_8.$$

Note that in groups-of-three-bits it is sometimes necessary to introduce zero(s) in the most significant position(s). Obviously this does not change the value of the number.

Because of that relationship between the two systems, the octal system acquires a particular importance in computer arithmetic. It suggests an efficient "shorthand" way of writing binary numbers. In conversion in the opposite direction, each octal digit is replaced by an equivalent three-digit binary number. For hexadecimal to binary conversion, each hex digit is represented as a *four*-bit binary group ($2^4 = 16$).

$$(110010110001)_2 = (1100)(1011)(0001)_2 = \mathrm{CB1}_{16}$$

Decimal to binary integer conversion

For decimal to binary conversion of an integral number, the *repeated-division-by-2 method* is used. Let N be the decimal integer. When it is converted into an n-digit binary number of digits b_i, it can be written

$$N = b_{n-1}2^{n-1} + \cdots + b_1 2^1 + b_0 2^0. \tag{2}$$

When both sides of the equation above are divided by 2, we have the quotient Q_1, which is an integer, and a remainder, b_0, as follows:

$$\frac{N}{2} = Q_1 + b_0 2^{-1} \qquad \text{where } Q_1 = b_{n-1}2^{n-2} + \cdots + b_1 2^0. \tag{3}$$

The remainder b_0 thus obtained is the desired least significant digit, because the given number N and the radix 2 are both integers; the quotients and remainders on both sides of (3) can be equated, respectively. Next, we divide Q_1 by 2 and thus obtain Q_2 and remainder b_1,

$$\frac{Q_1}{2} = Q_2 + b_1 2^{-1} \qquad \text{where } Q_2 = b_{n-1}2^{n-3} + \cdots + b_2 2^0.$$

The remainder b_1 thus obtained is the desired next least significant digit. This process continues until Q_i is equal to 0.

Examples

26_{10}		18_{10}		32_{10}	
Quotients	Remainders	Quotients	Remainders	Quotients	Remainders
26		18		32	
13	0 (LSD)	9	0 (LSD)	16	0 (LSD)
6	1	4	1	8	0
3	0	2	0	4	0
1	1	1	0	2	0
0	1 (MSD)	0	1 (MSD)	1	0
				0	1 (MSD)
$26 = 11010_2$		$18 = 10010_2$		$32 = 100000_2$	

Decimal to binary fraction conversion

For decimal to binary conversion of a fractional decimal number, the *repeated-multiplication-by-2* method is used. Let N be the decimal fraction. When it is converted into a binary number, it can be written

$$N = b_{-1}2^{-1} + b_{-2}2^{-2} + \cdots + b_{-m}2^{-m}. \tag{4}$$

When both sides of (4) are multiplied by 2, we have a product which consists of an integer part, b_{-1}, and a fractional part, F_1, as follows:

$$2N = b_{-1} + F_1 \qquad \text{where } F_1 = b_{-2}2^{-1} + \cdots + b_{-m}2^{-m+1}. \tag{5}$$

The integral digit b_{-1} thus obtained is the desired most significant digit, because the integral and fractional parts on both sides of (5) can be equated, respectively. Next we multiply F_1 by 2 and obtain

$$2F_1 = b_{-2} + F_2 \qquad \text{where } F_2 = b_{-3}2^{-1} + \cdots + b_{-m}2^{-m+2}. \tag{6}$$

The integral bit b_{-2} thus obtained is the desired next most significant digit. This process continues until F_m becomes zero.

Examples

0.8125_{10}		0.7893_{10}		0.218_{10}	
Product	Integer part	Product	Integer part	Product	Integer part
0.8125		0.7893		0.218	
1.6250	1	1.5786	1	0.436	0
1.2500	1	1.1572	1	0.872	0
0.5000	0	0.3144	0	1.744	1
1.0000	1	0.6288	0	1.488	1
		1.2576	1	0.976	0
		0.5152	0	1.952	1
		1.0304	1	1.904	1
				1.808	1
				1.616	1
				1.232	1
				0.464	0
$0.8125 = 0.1101_2$		$0.7893 = 0.1100101+_2$		$0.218 = 0.00110111110+_2$	

Binary to decimal integer conversion

For binary to decimal conversion of an integral binary number, the repeated-multiplication-by-2 method may be used. Let N be the binary integer with six digits,

$$N = b_5 2^5 + b_4 2^4 + b_3 2^3 + b_2 2^2 + b_1 2^1 + b_0 2^0. \tag{7}$$

To convert this binary number into its equivalent decimal number, (7) is rewritten

$$N = (\{[(b_5 2 + b_4)2 + b_3]2 + b_2\}2 + b_1)2 + b_0. \tag{8}$$

The most significant digit is doubled if the next digit is a 0, or doubled and a 1 added if the next digit is a 1. This result is doubled or doubled and a 1 added, depending on the value (1 or 0) of the next digit in the descending order of digits in the binary number. This process is continued until the operation indicated by the least significant digit has been performed.

Examples

110101_2

Binary digit	Result
(MSD) 1	
1 = (1 × 2) + 1	= 3
0 = (3 × 2) + 0	= 6
1 = (6 × 2) + 1	= 13
0 = (13 × 2) + 0	= 26
(LSD) 1 = (26 × 2) + 1	= 53

$110101_2 = 53$

101010_2

Binary digit	Result
(MSD) 1	
0 = (1 × 2) + 0	= 2
1 = (2 × 2) + 1	= 5
0 = (5 × 2) + 0	= 10
1 = (10 × 2) + 1	= 21
(LSD) 0 = (21 × 2) + 0	= 42

$101010_2 = 42$

110001_2

Binary digit	Result
(MSD) 1	
1 = (1 × 2) + 1	= 3
0 = (3 × 2) + 0	= 6
0 = (6 × 2) + 0	= 12
0 = (12 × 2) + 0	= 24
(LSD) 1 = (24 × 2) + 1	= 49

$110001_2 = 49$

Binary to decimal fraction conversion

The repeated-division-by-2 method may be used to convert binary fractions to their decimal equivalents. Let N be the binary fractional number with four fractional digits,

$$N = b_{-1}2^{-1} + b_{-2}2^{-2} + b_{-3}2^{-2} + b_{-4}2^{-4}. \tag{9}$$

To convert this binary number into its equivalent decimal number, (9) is written

$$N = 2^{-1}\{b_{-1} + 2^{-1}[b_{-2} + 2^{-1}(b_3 + b_4 2^{-1})]\} \tag{10}$$

The least significant binary digit is divided by 2 if the next significant digit is 0, or divided by 2 and 1 added if the next significant digit is 1. This first result is divided by 2, with the next-order digit determining whether or not the 1 is added. This process of dividing the result by 2, with the ascending-order digits controlling the operation, continues until the 0 digit to the left of the binary point effects the final result.

Examples

0.0101_2

Binary digit	Result
(LSD) 1	
0	$= (1 \div 2) + 0 = 0.5$
1	$= (0.5 \div 2) + 1 = 1.25$
0	$= (1.25 \div 2) + 0 = 0.625$
0.	$= (0.625 \div 2) + 0 = 0.3125$
	$0.0101_2 = 0.3125$

0.1011_2

Binary digit	Result
(LSD) 1	
1	$= (1 \div 2) + 1 = 1.5$
0	$= (1.5 \div 2) + 0 = 0.75$
1	$= (0.75 \div 2) + 1 = 1.375$
0.	$= (1.375 \div 2) + 0 = 0.6875$
	$0.1011_2 = 0.6875$

0.1100_2

Binary digit	Result
(LSD) 0	
0	$= (0 \div 2) + 0 = 0$
1	$= (0 \div 2) + 1 = 1.0$
1	$= (1 \div 2) + 1 = 1.5$
0.	$= (1.5 \div 2) + 0 = 0.75$
	$0.1100_2 = 0.75$

Mixed-number conversion

The procedures for conversion discussed above may be used to convert mixed numbers from decimal to binary and from binary to decimal. When a decimal mixed number is converted to its binary equivalent, the integer and fractional parts are converted separately by using the methods above, and the separate results are then combined. For example, for the decimal mixed number 528.27, the integer part (528) and the fractional part (0.27) are converted to binary separately. Combining the results, we obtain 1000010000 + 0.010001 as the binary equivalent of 528.27.

In conversions from binary to decimal, the appropriate procedures may be selected for the conversion of the integer and fractional portions of the number.

Example

528.27_{10}

Division-by-2 method		Multiplication-by-2 method	
Integer part		Fractional part	
Quotients	Remainders	Product	Integer part
528		0.27	
264	0	0.54	0
132	0	1.08	1
66	0	0.16	0
33	0	0.32	0
16	1	0.64	0
8	0	1.28	1
4	0		
2	0		
1	0		
0	1		

$528 = 1000010000_2$ $\qquad 0.27 = 0.010001+_2$

$528.27 = 1000010000.010001_2$

Binary to decimal conversion: an alternative

A binary mixed number may be written in the form of (1) as a combination of (7) and (9):

$$N = \cdots + b_5 2^5 + b_4 2^4 + b_3 2^3 + b_2 2^2 + b_1 2^1 + b_0 + b_{-1}2^{-1} + b_{-2}2^{-2} + b_{-3}2^{-3} + \cdots \quad (11)$$
$$= \sum b_n 2^n.$$

One may directly apply (11) to convert a number from binary to decimal. The powers of 2 are multiplied by their coefficients, b_n, and the results are summed. Since the coefficients are either 0 or 1, the multiplication reduces to a decision whether or not to add the corresponding power of 2.

Examples

110101.0101_2

Binary digit				Result
(LSD) 1	$= (1 \times 2^{-4})$	$= (1 \times \frac{1}{16})$	=	0.0625
0	$= (0 \times 2^{-3})$	$= (0 \times \frac{1}{8})$	=	0.0
1	$= (1 \times 2^{-2})$	$= (1 \times \frac{1}{4})$	=	0.25
0	$= (0 \times 2^{-1})$	$= (0 \times \frac{1}{2})$	=	0.0
.				
1	$= (1 \times 2^{0})$	$= (1 \times 1)$	=	1.
0	$= (0 \times 2^{1})$	$= (0 \times 2)$	=	0.
1	$= (1 \times 2^{2})$	$= (1 \times 4)$	=	4.
0	$= (0 \times 2^{3})$	$= (0 \times 8)$	=	0.
1	$= (1 \times 2^{4})$	$= (1 \times 16)$	=	16.
(MSD) 1	$= (1 \times 2^{5})$	$= (1 \times 32)$	=	32.

$110101.0101_2 = 53.3125$

101010.11$_2$

Binary digit		Result
(MSD) 1 = (1×2^5)	$= (1 \times 32)$	= 32.0
0 = (0×2^4)	$= (0 \times 16)$	= 0.0
1 = (1×2^3)	$= (1 \times 8)$	= 8.
0 = (0×2^2)	$= (0 \times 4)$	= 0.
1 = (1×2^1)	$= (1 \times 2)$	= 2.
0 = (0×2^0)	$= (0 \times 1)$	= 0.
1 = (1×2^{-1})	$= (1 \times \frac{1}{2})$	= 0.5
(LSD) 1 = (1×2^{-2})	$= (1 \times \frac{1}{4})$	= 0.25
101010.11 = 42.75		

Conversion to decimal from any radix

The selection of a method for conversion of a number expressed in any radix to the same number expressed in decimal is based on human convenience. *We choose a method in which all arithmetic is done in decimal.* It is possible to do arithmetic in any base, of course, but we assume that manual operation will be most convenient in the decimal system. When the conversion is by computer, that assumption is no longer valid.

We convert from any radix r to decimal by direct expansion of (1),

$$\begin{aligned} N &= (\cdots a_3a_2a_1a_0 \,.\, a_{-1}a_{-2}a_{-3}\cdots)_r \\ &= \cdots + a_3r^3 + a_2r^2 + a_1r^1 + a_0 + a_{-1}r^{-1} + a_{-2}r^{-2} + a_{-3}r^{-3} + \cdots \qquad (12) \\ &= \sum a_nr^n. \end{aligned}$$

We use the technique of binary to decimal conversion: The powers of the radix are converted to their decimal form and then multiplied by the coefficients representing the number in that radix. The resultant products are summed (in decimal) to yield the decimal equivalent of the radix r number.

Conversion from decimal to any radix

When converting a number from its decimal representation to any other radix, we again choose to do our arithmetic in decimal. We must handle the integer and fractional parts separately.

The integer part is converted by repeated division by the radix. Following (2) and (3), we have

$$N = a_nr^n + a_{n-1}r^{n-1} + \cdots + a_1r^1 + a_0,$$

$$\frac{N}{r} = Q_1 + a_0r^{-1} \qquad \text{where } Q_1 = a_nr^{n-1} + a_{n-1}r^{n-2} + \cdots + a_1r^0. \qquad (13)$$

As before, the remainder a_1 is the least significant digit. The partial quotient, Q_1, is then divided by the radix to obtain the next digit, and the process is repeated until Q_n is zero.

Similarly, the fractional part of the decimal number is converted by repeated multiplication of the radix. Following (4) and (5), we have

$$N = a_{-1}r^{-1} + a_{-2}r^{-2} + \cdots + a_{-m}r^{-m},$$
$$rN = a_{-1} + f_1 \qquad \text{where } F_1 = a_{-2}r^{-1} + \cdots + a_{-m}r^{-m+1}. \tag{14}$$

The integer digit a_{-1} is the most significant digit of the fractional part. The partial fraction F_1 is then multiplied by r to obtain the next digit, and the process is repeated until F_m is zero.

Finally, the integer and fractional parts are combined to form the representation in the new radix.

APPENDIX B
LOGIC DIAGRAM CONVENTIONS

Many schemes have been developed to represent logic circuits on paper. Two standards have been devised: MIL-STD-806C and ANSI Y32.14. Neither has achieved its goal in that it is not used widely enough to be understood by most workers in the field.

The conventions used in this book, which are derived from MIL-STD-806C, are the following.

A

represents a signal named A, positive logic ($V_1 > V_0$), where 1 and 0 are both DC levels, as defined in Fig. 10.1.

A

represents a signal named A, negative logic ($V_0 > V_1$), where 1 and 0 are both DC levels, as defined in Fig. 10.1.

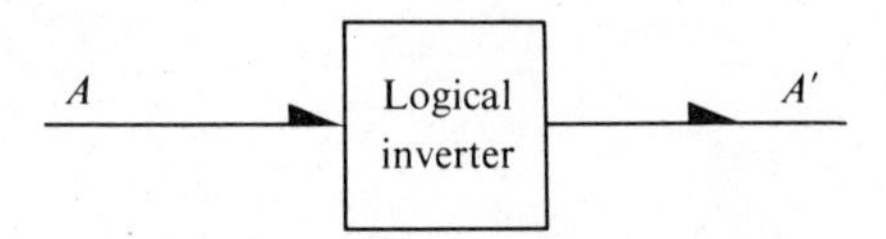

Note that the complement of A is called A' but is still positive logic ($V_1 > V_0$) because the *logical* sense of A is inverted. It is then called A'.

A logical inverter is actually an electrical inverter, and so it is equally valid to write

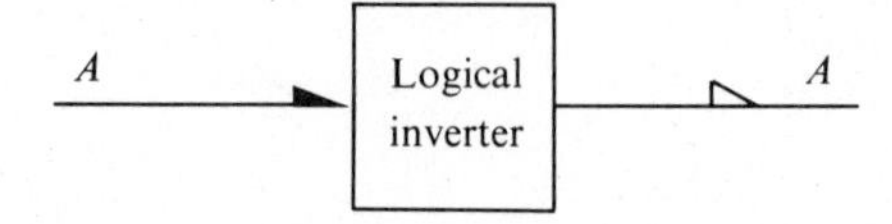

Which symbol is used for the output of an inverter depends only on how the signal will be used later in the diagram. For two-level **NAND-NOR** logic, level inversion is shown

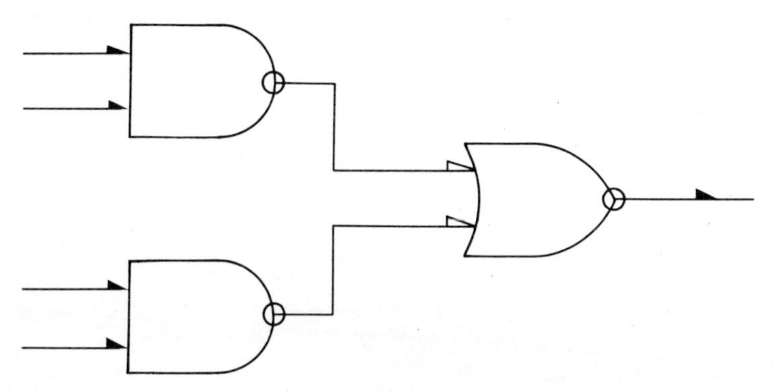

All three of the gates above are implemented with the same circuit. It is a **NAND** when the inputs are —▶, and **NOR** when the inputs are —▷.

The remaining conventions for logic signals are the following.

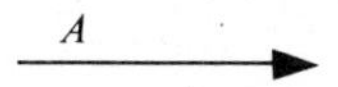

represents a signal named A, positive logic ($V_1 > V_0$), where logic 1 consists of a transition from V_0 to V_1.

A ———▷

represents a signal named A, negative logic ($V_0 > V_1$), where logic 1 consists of a transition from V_0 to V_1.

A flip-flop using positive-logic J and K inputs, but triggering on the trailing edge of the clock, would be shown as

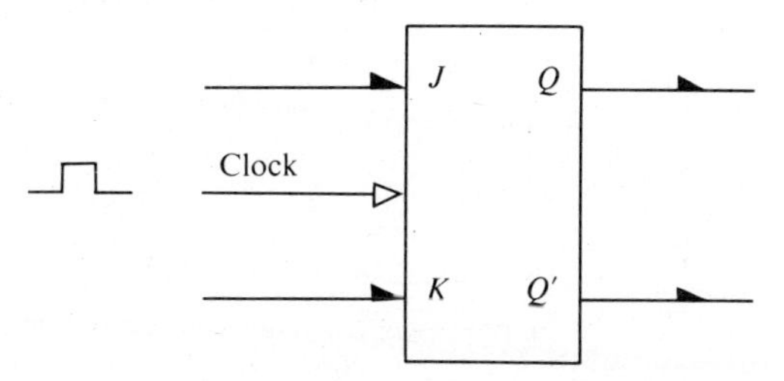

since the action takes place on a transition from a more positive voltage to a less positive one.

It is also possible and occasionally necessary to show

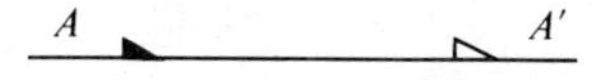

and

A ▶———▷ A'

which are logically consistent but difficult to grasp.

The symbols for gates are

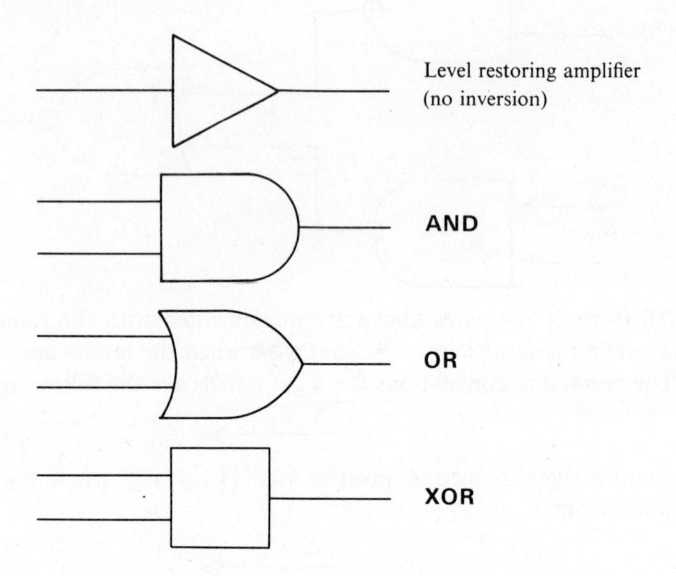

Internal inversion on a stage is shown with a circle on the output.

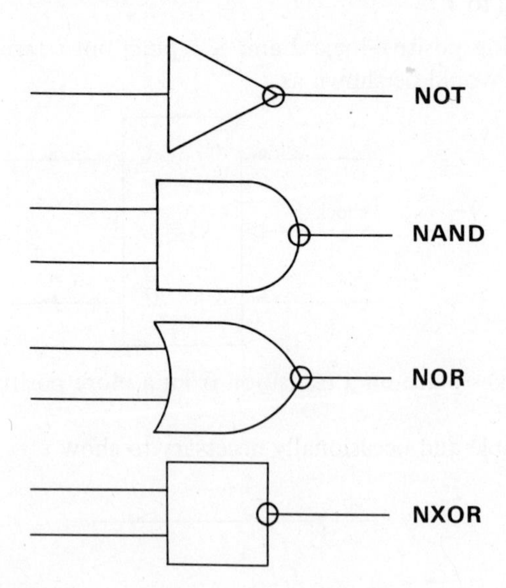

If a collector resistor has been left out to allow wired **OR**s, an $\times$ is placed on the output.

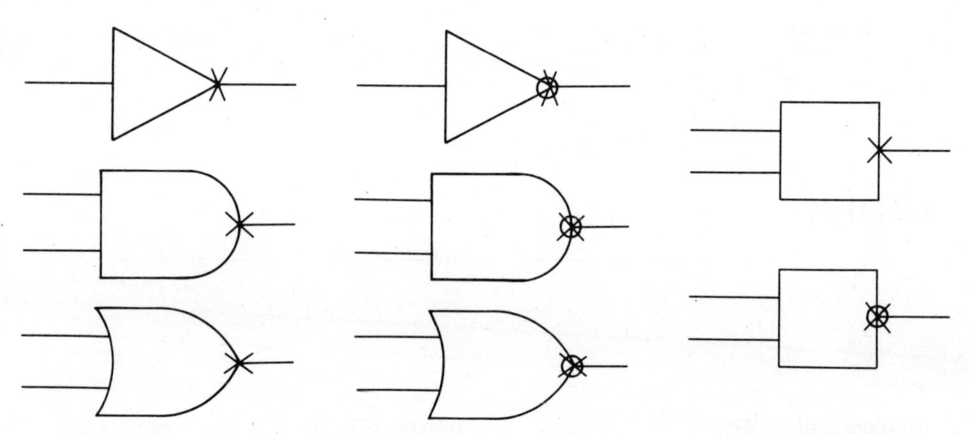

Again, whether arrowheads are open or filled depends more on the intended use of a signal than on the convention used for other signals in the same circuit.

INDEX

GENERALIZED ASSEMBLY LANGUAGE INSTRUCTIONS (Continued)

Operation code	Description of operation
JUMP IF REGISTER ZERO	IF (REGISTER) = 0, ADDR → (PC) †
JUMP IF REGISTER NONZERO	IF (REGISTER) ≠ 0, ADDR → (PC) †
JUMP IF REGISTER POSITIVE	IF [sign bit] = 0, ADDR → (PC) †
JUMP IF REGISTER NEGATIVE	IF [sign bit] = 1, ADDR → (PC) †
DECREMENT REGISTER JUMP IF NONNEGATIVE	(REGISTER) − 1 → (REGISTER) IF [sign bit] = 0, ADDR → (PC) †
INCREMENT INDEX REGISTER JUMP IF LESS THAN REGISTER	(XR) + 1 → (XR) IF (XR) < (REGISTER), addr → (PC) †
STORE PC AND JUMP	(PC) + 1 → (REGISTER), ADDR → (PC)
STORE RETURN AND JUMP	(PC) + 1 → $(\text{ADDR})_{\text{addr}}$, ADDR + 1 → (PC)
PUSH RETURN AND JUMP	(PC) + 1 → STACK, (ADDR) → (PC) (stack pushed)
POP RETURN AND JUMP	(stack popped), then STACK → (PC)
PUSH PARAMETER	(ADDR) → (STACK), (stack then pushed down)
POP PARAMETER TO REGISTER	(stack popped up), then (STACK) → (REGISTER)
Input/output operations	
TRANSMIT DATA TO DEVICE	Initiate data transfer from main memory to device specified in addr
RECEIVE DATA FROM DEVICE	Initiate data transfer from device specified in addr to main memory

Operation code	Description of operation
SENSE STATUS OF DEVICE	Load status word of device specified in addr into register specified
OUTPUT COMMAND TO DEVICE	Send command word in register specified to device specified in addr
Shift instructions	
SHIFT REGISTER {LEFT / RIGHT} {SINGLE / DOUBLE} {ARITHMETIC / LOGICAL} {CYCLIC / (null)}	Shift (REGISTER) addr places as specified
Pseudo-operations	
BINARY	Exactly reproduce the binary number in the operand as a machine word
OCTAL	Transform the octal number in the operand into a machine word
DECIMAL	Transform the decimal number in the operand into a machine word
HEXADECIMAL	Transform the hexadecimal number in the operand into a machine word
LOAD ZERO	0 → (REGISTER)
NO OPERATION	(PC) + 1 → (PC)
Assembler directives	
END	Marks end of program unit
RESERVE	(Location counter) + addr → (Location counter)

† Otherwise (PC) + 1 → (PC) in all cases